Polymer Composites for Electrical and Electronic Engineering Application

Polymer Composites for Electrical and Electronic Engineering Application

Editor

Shaojian He

MDPI • Basel • Beijing • Wuhan • Barcelona • Belgrade • Manchester • Tokyo • Cluj • Tianjin

Editor
Shaojian He
School of New Energy
North China Electric Power
University
Beijing
China

Editorial Office
MDPI
St. Alban-Anlage 66
4052 Basel, Switzerland

This is a reprint of articles from the Special Issue published online in the open access journal *Polymers* (ISSN 2073-4360) (available at: www.mdpi.com/journal/polymers/special_issues/Polym_Compos_Electr_Electron_Eng_Appl).

For citation purposes, cite each article independently as indicated on the article page online and as indicated below:

LastName, A.A.; LastName, B.B.; LastName, C.C. Article Title. *Journal Name* **Year**, *Volume Number*, Page Range.

ISBN 978-3-0365-4068-9 (Hbk)
ISBN 978-3-0365-4067-2 (PDF)

© 2022 by the authors. Articles in this book are Open Access and distributed under the Creative Commons Attribution (CC BY) license, which allows users to download, copy and build upon published articles, as long as the author and publisher are properly credited, which ensures maximum dissemination and a wider impact of our publications.
The book as a whole is distributed by MDPI under the terms and conditions of the Creative Commons license CC BY-NC-ND.

Contents

About the Editor . . . . **vii**

Songming Li, Hao Huang, Sibao Wu, Jiafu Wang, Haijun Lu and Liying Xing
Study on Microwave Absorption Performance Enhancement of Metamaterial/Honeycomb Sandwich Composites in the Low Frequency Band
Reprinted from: *Polymers* **2022**, *14*, 1424, doi:10.3390/polym14071424 . . . . **1**

Hailan Kang, Sen Luo, Hongyang Du, Lishuo Han, Donghan Li and Long Li et al.
Bio-Based *Eucommia ulmoides* Gum Composites with High Electromagnetic Interference Shielding Performance
Reprinted from: *Polymers* **2022**, *14*, 970, doi:10.3390/polym14050970 . . . . **15**

Xiaobo Meng, Liming Wang, Hongwei Mei, Bin Cao and Xingming Bian
Streamer Propagation along the Insulator with the Different Curved Profiles of the Shed
Reprinted from: *Polymers* **2022**, *14*, 897, doi:10.3390/polym14050897 . . . . **29**

Qi Li, Rui Liu, Li Li, Xiaofan Song, Yifan Wang and Xingliang Jiang
The Dynamic Behaviour of Multi-Phase Flow on a Polymeric Surface with Various Hydrophobicity and Electric Field Strength
Reprinted from: *Polymers* **2022**, *14*, 750, doi:10.3390/polym14040750 . . . . **45**

Meng Song, Xiulin Yue, Chaokang Chang, Fengyi Cao, Guomin Yu and Xiujuan Wang
Investigation of the Compatibility and Damping Performance of Graphene Oxide Grafted Antioxidant/Nitrile-Butadiene Rubber Composite: Insights from Experiment and Molecular Simulation
Reprinted from: *Polymers* **2022**, *14*, 736, doi:10.3390/polym14040736 . . . . **55**

Yinjie Dong, Zhaoyang Wang, Shouchao Huo, Jun Lin and Shaojian He
Improved Dielectric Breakdown Strength of Polyimide by Incorporating Polydopamine-Coated Graphitic Carbon Nitride
Reprinted from: *Polymers* **2022**, *14*, 385, doi:10.3390/polym14030385 . . . . **71**

Yunqi Xing, Yuanyuan Chen, Jiakai Chi, Jingquan Zheng, Wenbo Zhu and Xiaoxue Wang
Molecular Dynamics Simulation of Cracking Process of Bisphenol F Epoxy Resin under High-Energy Particle Impact
Reprinted from: *Polymers* **2021**, *13*, 4339, doi:10.3390/polym13244339 . . . . **81**

Hongchuan Dong, Yunfan Liu, Yanming Cao, Juzhen Wu, Sida Zhang and Xinlong Zhang et al.
Terahertz-Based Method for Accurate Characterization of Early Water Absorption Properties of Epoxy Resins and Rapid Detection of Water Absorption
Reprinted from: *Polymers* **2021**, *13*, 4250, doi:10.3390/polym13234250 . . . . **93**

Yingjie Jiang, Yujia Li, Haibo Yang, Nanying Ning, Ming Tian and Liqun Zhang
Deep Insight into the Influences of the Intrinsic Properties of Dielectric Elastomer on the Energy-Harvesting Performance of the Dielectric Elastomer Generator
Reprinted from: *Polymers* **2021**, *13*, 4202, doi:10.3390/polym13234202 . . . . **109**

Xiaobo Meng, Liming Wang, Hongwei Mei and Chuyan Zhang
Influence of Silicone Rubber Coating on the Characteristics of Surface Streamer Discharge
Reprinted from: *Polymers* **2021**, *13*, 3784, doi:10.3390/polym13213784 . . . . **125**

Hanwen Ren, Qingmin Li, Yasuhiro Tanaka, Hiroaki Miyake, Haoyu Gao and Zhongdong Wang
Frequency and Temperature-Dependent Space Charge Characteristics of a Solid Polymer under Unipolar Electrical Stresses of Different Waveforms
Reprinted from: *Polymers* **2021**, *13*, 3401, doi:10.3390/polym13193401 . . . . **139**

Xupeng Song, Xiaofeng Xue, Wen Qi, Jin Zhang, Yang Zhou and Wei Yang et al.
Research on the Compound Optimization Method of the Electrical and Thermal Properties of SiC/EP Composite Insulating Material
Reprinted from: *Polymers* **2021**, *13*, 3369, doi:10.3390/polym13193369 . . . . **153**

Qiu-Wanyu Qing, Cheng-Mei Wei, Qi-Han Li, Rui Liu, Zong-Xi Zhang and Jun-Wen Ren
Bioinspired Dielectric Film with Superior Mechanical Properties and Ultrahigh Electric Breakdown Strength Made from Aramid Nanofibers and Alumina Nanoplates
Reprinted from: *Polymers* **2021**, *13*, 3093, doi:10.3390/polym13183093 . . . . **175**

About the Editor

Shaojian He

Dr. Shaojian He was born in 1984 and received his PhD in Materials Science and Engineering from Beijing University of Chemical Technology, China in 2010. He is currently an associate professor at the School of New Energy at North China Electric Power University. He was a visiting scholar at State University of New York, University at Buffalo, USA, during March 2019 – August 2020. His current research is focused on high-performance rubber nanocomposites and high voltage insulation material. He has undertaken several research projects including two research funds from the National Natural Science Foundation of China (NSFC). He is a recipient of Alan Glanvill Award by the Institute of Materials, Minerals and Mining (IOM3) in 2012. He has published more than 60 papers in academic journals and applied more than 20 Chinese invention patents. He is also a member of the Rubber & Plastic Green Manufacturing Professional Committee of the Chemical Industry and Engineering Society of China (CIESC), a guest editor of the SCI journal *Polymers*, an academic dissertation evaluation expert of Academic Degree Center of Ministry of Education, a communication evaluation expert of the National Natural Science Foundation of China, and an evaluation expert of the Beijing Science and Technology Commission.

Article

Study on Microwave Absorption Performance Enhancement of Metamaterial/Honeycomb Sandwich Composites in the Low Frequency Band

Songming Li [1,2], Hao Huang [1], Sibao Wu [1], Jiafu Wang [3], Haijun Lu [1,2] and Liying Xing [1,2,*]

1. Composite Technology Center, AVIC Beijing Aeronautical Manufacturing Technology Research Institute, Beijing 101300, China; lisongming860206@126.com (S.L.); thu_huanghao@163.com (H.H.); wusibao0507@163.com (S.W.); haijunlu_hjl@163.com (H.L.)
2. National Key Laboratory of Advanced Composites, AECC Beijing Institute of Aeronautical Materials, Beijing 100095, China
3. Department of Basic Sciences, Air Force Engineering University, Xi'an 710051, China; wangjiafu1981@126.com
* Correspondence: vcd4321@sina.com

Abstract: With the rapid development of electronic technology and modern radar detection system, there is increasingly urgent demand for microwave absorbing composites working efficiently in the low frequency range (e.g., 1–2 GHz). In this work, a type of metamaterial/honeycomb sandwich composite (MHSC) was proposed and fabricated, which exhibited a light weight structure and excellent wave-absorbing performance in the low frequency band. The relationship between the wave-absorbing properties and the design parameters of the composite, such as the thickness of the wave-transmitting skin, the thickness and dielectric properties of the wave-absorbing honeycomb, was systematically investigated. The electromagnetic coupling interference between the honeycomb absorber and metamaterial resonator proved to be a crucial factor that affects synergistic wave-absorbing performance in the low-frequency band. Under the rational design, the incorporation of subwavelength-sized phase-gradient metamaterial units in the composite can significantly improve low-frequency wave-absorbing performance for greater than 5 dB (an increment larger than 100%); and the obtained MHSC exhibits averaged reflectivity (R_a) less than -10 dB in the low frequency band of 1–2 GHz as well as outstanding performance ($R_a < -14.6$ dB) over an extremely wide frequency range (1–18 GHz). The MHSC reported in this study could be a promising candidate for the key material in high-performance radar stealth and other related applications.

Keywords: honeycomb sandwich composites; metamaterial; radar stealth; microwave absorbing material; low frequency

1. Introduction

In recent years, microwave-absorbing materials (MAMs) have attracted much attention in view of their practical applications in electromagnetic shielding, camouflage, radar stealth and other advanced technologies for both commercial and military purposes [1–8]. Compared with other types of MAMs (e.g., wave-absorbing coating), structural wave-absorbing composites show comprehensive superiorities, including high mechanical strength, heat resistance and broadband absorption ability [9]. They exhibit dual functions for load-bearing and wave-absorbing in an integral structure and are widely used in radar stealth equipment [10]. Nowadays, with the rapid evolution of electronic technology, modern radar systems can effectively detect a target within a broad frequency band (e.g., 1–18 GHz). In particular, low frequency early warning radar (1–2 GHz) has become the most predominant threat for conventional military planes and even stealth aircraft [11–14]. As a result, there is an increasingly urgent demand for structural wave-absorbing composites having light weight structure, broadband absorption ability, and high performance low-frequency radar stealth function at the same time [14]. This presents a great challenge in the design and fabrication of such material systems within the bounds of current knowledge.

Honeycomb sandwich composites (HSCs) are widely used in aeronautical structures and automobiles due to their advantages, such as being light weight, having high stiffness-to-mass ratio, good heat-insulation performance and so on [15]. By coating or filling hexagonal honeycomb cores with lossy agents, an HSC can become an efficient MAM [16]. The wave-absorbing performance of an HSC is affected by a series of design parameters, such as thickness, core size and orientation of the honeycomb [17–19], coating thickness and dielectric properties of the lossy agents [20–22], and even the fabrication/molding process [23]. Vast work and progress have been made by using varied lossy agents, including dielectric materials (e.g., carbon black, graphene and multi-wall carbon nanotube) and magnetic materials (e.g., carbonyl iron, ferrite and nickel metal) to fabricate wave-absorbing HSCs with low reflectivity and high absorptivity [17,21,24–31]. Nevertheless, early studies mainly focused on enhancing the wave-absorbing properties of HSCs in a relatively high frequency band within 2–18 GHz (mostly 8–18 GHz). By contrast, there is still a lack of effective methodologies to realize satisfactory radar stealth performance (averaged reflectivity less than -10 dB) in the low frequency band of 1–2 GHz for conventional HSC. Although increasing the thickness of honeycomb, using a high content of lossy agents, or employing magnetic materials with high complex permeability may be helpful in improving low-frequency wave-absorbing performance [14,32], this will inevitably result in drastic weight increase for the total structure, which is infeasible for practical applications. Consequently, it is necessary to introduce new technical approaches and mechanisms to fundamentally solve this problem.

Metamaterials are kinds of artificial materials which consist of subwavelength structural units and possess many peculiar electromagnetic properties unattainable with natural materials [33–35]. By arranging resonant units in a periodic, quasi-periodic or disordered/coded manner, the design of metamaterials can exhibit artificially customized permittivity and permeability following predetermined spatial distribution, resulting in full control of the amplitude, phase, polarization, propagation and dispersion of the electromagnetic wave [36–38]. Compared with the traditional microwave absorber, metamaterial not only provides new perspectives on the wave-absorbing mechanism but also offers more practical approaches to realize high performance radar stealth with a light-weight structure [39–41]. Among metamaterials, phase-gradient metasurface (PGM), which is based on generalized laws of reflection and refraction (Snell's Law), is a promising candidate to solve the bottleneck problem of traditional wave-absorbing material in the low frequency region [42–45]. Through the abnormal reflection properties of PGM, the main lobe of a reflected radar wave can be deviated in a non-threatening direction to reduce radar cross section (RCS); meanwhile, due to abnormal refraction, the effective thickness of the absorber below PGM is significantly increased to enhance the energy dissipation of the electromagnetic wave [46,47]. Although it is expected that the incorporation of metamaterial can possibly improve absorption efficiencies, or widen the absorption band of conventional MAMs, at the current stage, there is still lack of study concerning the fabrication and wave-absorbing performance of honeycomb sandwich composites loaded with metamaterials [48,49]. Huang et al. [48] proposed a type of wave-absorbing honeycomb containing "H-shaped" metamaterial, where the inner wall of wave-transmitting honeycomb cores was first posted with metal units and then impregnated with wave-absorbing resin. The reflectivity of the metamaterial-loaded honeycomb is less than -5 dB in the frequency band of 2~18 GHz. However, electromagnetic interaction and coupling interference between the metamaterial units and honeycomb absorber, especially in the low-frequency microwave region, is still ambiguous. It is of vital significance and urgent demand to investigate the synergistic/conflict effect of the two absorbing mechanisms and to optimize design parameters to the composites, in order to achieve satisfactory low-frequency and broadband wave-absorbing performance within a light-weight structure.

In this work, we reported on a type of metamaterial/honeycomb sandwich composite (MHSC) which exhibited excellent microwave absorption performance in the low frequency band of 1–2 GHz. The metamaterial units designed, following the phase-gradient principle, were incorporated into MHSC via a pattern-transfer method and molded together with the composite. The influences of the design parameters (including the thickness of skin, and thickness and dielectric properties of honeycomb) on the low-frequency wave-absorbing properties of the composites were systematically investigated, and the related mechanisms were elucidated. It was demonstrated that coupling interference between the honeycomb absorber and metamaterial resonator was the most crucial factor that affected their synergistic wave-absorbing performance in the low-frequency band. On the basis of the above understanding, an optimized radar stealth performance with averaged reflectivity (R_a) less than −10 dB in the band of 1–2 GHz was successfully obtained for MHSC, while there was no significant weight increase compared with the original HSC. On this occasion, the MHSC also exhibited excellent radar stealth performance ($R_a < -14.6$ dB) over an extremely wide frequency range of 1–18 GHz, which shows great opportunities for real applications.

2. Materials and Methods

2.1. Materials

Quartz fiber/bismaleimides wave-transmitting prepregs (QW280/5429), wave-absorbing honeycomb (SKuF), wave-transmitting Nomex honeycomb (NH-1-2.7-48) and carbon fiber/bismaleimides prepregs (ZT7H/5429) were purchased from AVIC Composite Corporation Ltd., Beijing, China. The poly(ether-ether-ketone) (PEEK) film with the thickness of 10–15 μm was fabricated by solution casting, following the literature [50]. Epoxy glue film (J-116) was purchased from the Institute of Petrochemistry, Heilongjiang Academy of Sciences, Harbin, China. All the reagents and materials were purchased from commercial sources and used as received without further purification.

2.2. Fabrication of MHSC

The fabrication process of the metamaterial/honeycomb sandwich composites (MHSCs) is schematically shown in Figure 1. First, the PEEK film was plated with a thin copper layer by magnetron sputtering, and the periodic metal units were then fabricated by wet-etching, according to the designed patterns, to obtain the metamaterial layer. The wave-transmitting skin and the reflection skin were prepared by the autoclave vacuum bag molding method. To fabricate the wave-transmitting skin, several layers of the QW280/5429 prepregs were stacked together with the metamaterial layer, which were cured at 180 °C for 3 h, followed by post-curing at 200 °C for 5 h. During the curing process, a vacuum and a pressure of 0.7 MPa was applied. A photograph of the fabricated wave-transmitting skin containing metamaterial is shown in Figure S1 in the Supplementary Materials, where the dark patterns are the periodic metal units. The reflection skin was prepared by curing ZT7H/5429 prepregs using the same procedure. Finally, the wave-transmitting skin, containing metamaterial units, the wave-absorbing honeycomb and the reflection skin were bonded together by J-116 glue film at a temperature of 180 °C under a pressure of 0.3 MPa for 2 h [51], by which means the MHSCs were successfully fabricated. To investigate the effect of metamaterial on wave-absorbing properties, conventional HSC was also fabricated, which uses the same procedure as above, except that the wave-transmitting prepregs were cured alone without adding the metamaterial layer.

2.3. Characterization

The dielectric property (i.e., the complex relative permittivity, $\varepsilon_r = \varepsilon_r' + i\varepsilon_r''$) of the wave-absorbing honeycomb was measured by the free space method in a frequency range of 1–2 GHz. Two spot-focusing horn lens antennae (transmitting and receiving antennae) were placed face-to-face at a distance of twice the focal length. A honeycomb sample with an area of 600 mm × 600 mm was mounted on the focal plane of both the antennae. The S-parameters (Scattering parameters) of the sample in free space was measured by a vector network analyzer (Keysight PNA-L). The complex relative permittivity was calculated from the input reflection scattering parameter (S_{11}) and the forward transmission scattering parameter (S_{21}), by using the Nicholson-Ross algorithm [52]. The reflectivity (R, unit: dB) of the MHSC and HSC was measured by the NRL-arc method in a frequency range of 1–2 GHz. The transmitting antenna and receiving antenna, connected to a vector network analyzer (Keysight PNA-L), were mounted on top of the sample at normal incidence. Wave-absorbing pyramids were placed beneath the sample to suppress the reflection of the background. The value of R was calculated using the equation $R = 10\log(P/P_m)$, where P and P_m are respectively the reflection power of the composite with an area of 600 mm × 600 mm and the reflection power of the metal sheet with the same dimensions. All the tests were conducted at a temperature of 25 °C.

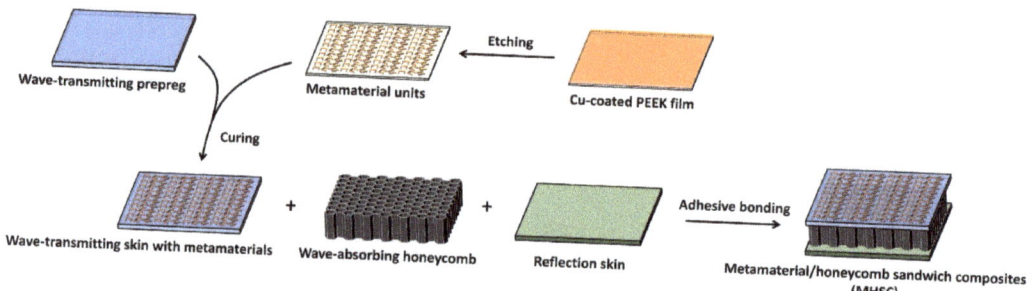

Figure 1. Schematic illustration about the fabrication of the metamaterial/honeycomb sandwich composites (MHSCs).

3. Results and Discussion

3.1. Wave-Absorbing Properties of HSC

Honeycomb sandwich composite (HSC) is generally composed of a wave-transmitting skin, a wave-absorbing honeycomb and a reflection skin. The wave-absorbing mechanism of HSC is schematically illustrated in Figure S2 in the Supplementary Materials. When the electromagnetic wave is incident onto the HSC, it will first interact with the wave-transmitting skin. The material of this layer often exhibits low dielectric constant and low loss tangent (tan $\delta = \varepsilon_r'' / \varepsilon_r'$) [53,54], in order to realize good impedance matching with the free space and to decrease reflection loss (increase transmittance) on the surface of the HSC. With good performance of the wave-transmitting skin, the electromagnetic wave energy can enter into the HSC as much as possible. As the electromagnetic wave enters into the wave-absorbing honeycomb, it will be scattered in the periodic hexagonal structure and attenuated by the wave-absorbing resin adhering to the honeycomb surface (Figure S2). To obtain rapid and sufficient attenuation of the electromagnetic wave, the honeycomb should have relatively large dielectric loss and sufficient thickness. Meanwhile, impedance matching between the skin layer and the honeycomb is also important to suppress reflection on the interface and realize the maximum absorption ratio of the radar waves. The reflection skin consists of carbon fiber reinforced bismaleimides resin, which is the bottom layer of the HSC. Due to the high conductivity of carbon fiber, this layer can be regarded as a reflection layer and shields electromagnetic signals on the back side of the HSC. According to the above analysis, it is realized that the rational design of the wave-transmitting skin and the wave-absorbing honeycomb are very important for wave-absorbing performance of the HSC. The effects of thickness of wave-transmitting skin, dielectric properties of wave-absorbing honeycomb and thickness of wave-absorbing honeycomb on wave absorbing properties of the HSC are presented as follows.

3.1.1. Effect of the Thickness of the Wave-Transmitting Skin

For wave-absorbing HSC, the wave-transmitting skin is the first layer of material that interacts with the incident electromagnetic wave. The impedance matching between the skin and free space determines direct reflection on the surface and affects the integral wave-absorbing performance of HSC. Herein, the quartz fiber ($\varepsilon_r' = 3.7$, $\tan \delta = 0.0005$) and bismaleimides ($\varepsilon_r' = 3.3$, $\tan \delta = 0.01$) with low dielectric constant and low loss tangent were respectively selected as the reinforcing fiber and resin matrix of the wave-transmitting skin. Under the same composition of material, the thickness of the skin layer is the main factor in influencing transmittance of the electromagnetic wave and wave-absorbing performance. Wave-transmitting skins with different thicknesses were fabricated by adjusting the number of plies of the stacked prepregs in the curing step. Figure 2 shows the reflectivity of the HSC with different thicknesses (0.5–2.5 mm) of the wave-transmitting skin in the low frequency range (1–2 GHz), where the same type and thickness of honeycomb is used. The results illustrate that the wave-absorbing performance of HSC is slightly enhanced by increasing the thickness of the wave-transmitting skin. This phenomenon is ascribed to the long wavelength of the electromagnetic wave in this frequency band, which can sufficiently penetrate through the skin layer with negligible dielectric loss. However, increasing the thickness of the wave-transmitting skin can also weaken wave-absorbing properties in the high frequency range (e.g., 12–18 GHz) [13], and increase the total weight of the HSC structure, which works against the demand for wide-band radar stealth using light-weight material. To balance wave-absorbing performance, weight and mechanical property of the HSC, a moderate thickness (1 mm) of wave-transmitting skin is selected and used in the following sections.

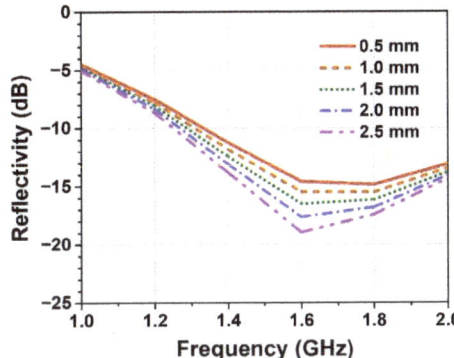

Figure 2. Reflectivity of the honeycomb sandwich composites (HSCs) with different thicknesses of the wave-transmitting skin.

3.1.2. Effect of Dielectric Properties of Wave-Absorbing Honeycomb

The honeycomb structure is the core functional layer of the wave-absorbing HSC, whose dielectric property and thickness show significant influence on radar stealth performance of the composite. Wave-absorbing honeycombs were fabricated by repeatedly immersing the Nomex honeycomb in a dispersion of conductive carbon black mixed with bismaleimides resin and acetone for several cycles to reach a predetermined weight increment, and finally cured at an elevated temperature in the oven. The Nomex honeycomb is a kind of wave-transmitting material with small dielectric constant ($\varepsilon_r' = 1.1$) and low loss tangent ($\tan \delta = 0.003$). Conductive carbon black can attenuate the radar wave through electronic polarization, interfacial polarization and other effects, during which electromagnetic energy is converted into conduction current and finally dissipated into heat. The effective dielectric constant and loss tangent of honeycomb can be promoted by increasing the loading amount of wave-absorbing agent. Figure 3a,b respectively show the real part and imaginary part of the complex relative permittivity of the eight wave-absorbing honeycombs (1#~8#) used for this study. In the frequency range from 1 GHz to 2 GHz, ε_r' of the honeycombs show the order of 1# < 2# < 3# < 4# ≈ 6# < 5# < 7# < 8# and ε_r'' show the order of 1# < 2# < 3# < 4# < 5# < 6# < 7# < 8#. The reflectivity of the HSC with honeycombs 1#~8# was measured in the frequency range of 1–2 GHz, and the results are shown in Figure 3c. The thickness of wave-absorbing honeycomb is fixed at 30 mm. The results demonstrate that, by increasing the dielectric properties (ε_r' and ε_r'') of honeycomb, the wave-absorbing performance of HSC gradually increases at first, while rapidly decreasing when complex permittivity exceeds a certain level. The effect of the dielectric property of honeycomb on wave-absorbing performance originated from the following mechanisms. When the real part and imaginary part of permittivity are both small (e.g., 1#~5#), the impedance difference between the wave-transmitting skin and the honeycomb is relatively low, and the electromagnetic wave can effectively pass through their interface and enter into the honeycomb. In this condition, the increasing of ε_r'' enhances the attenuation rate and absorption ratio of the radar wave in the honeycomb structure, which contributes to decrease in the reflectivity of HSC. However, when the dielectric property of honeycomb is sufficiently large (e.g., 7# and 8#), impedance mismatching between the skin and honeycomb leads to considerable reflection of the incident radar wave on the interface, which decreases energy transmission efficiency and weakens wave-absorbing performance of the HSC. Limited by the above reasons, even for honeycomb 6#, which has the best performance among the overall samples, reflectivity in 1–1.3 GHz is still larger than −10 dB, which is unsatisfactory in satisfying the demand for high-performance radar-stealth in the low frequency band. Moreover, wave-absorbing honeycomb with high dielectric properties loads a large amount of the wave-absorbing agent and shows a large density (Table 1), which significantly increases the total weight of the composite structure. Consequently, increasing the dielectric properties of honeycomb is of limited effect in improving low-frequency wave-absorbing performance of the composites in line with the demand for a light weight structure.

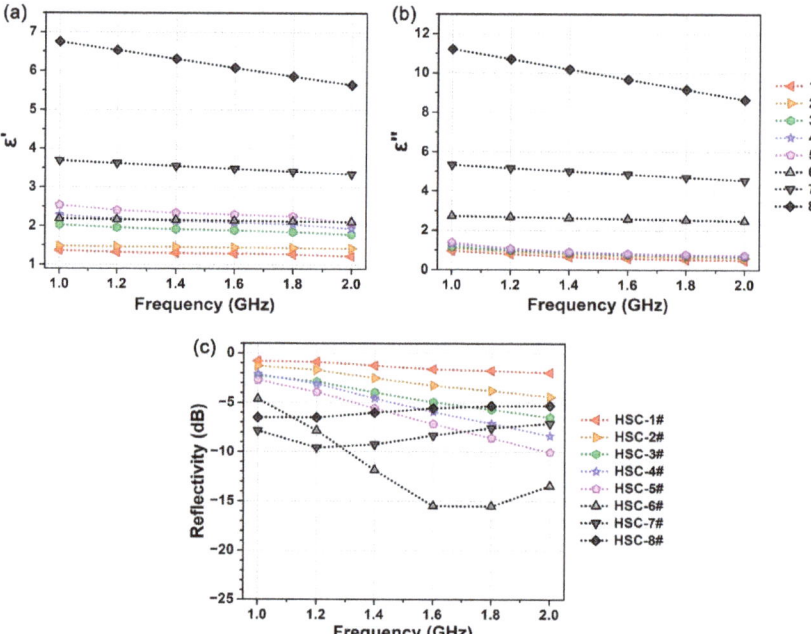

Figure 3. (a) Real part and (b) imaginary part of the complex relative permittivity ($\varepsilon_r' + i\varepsilon_r''$) of different wave-absorbing honeycombs (1#~8#). (c) Reflectivity of the honeycomb sandwich composites fabricated from the wave-absorbing honeycomb 1#~8#.

Table 1. Density of the wave-absorbing honeycomb.

Wave-Absorbing Honeycomb	1#	2#	3#	4#	5#	6#	7#	8#
Density (g/cm^3)	0.06	0.07	0.08	0.09	0.1	0.15	0.2	0.3
Surface density (kg/m^2) [1]	1.8	2.1	2.4	2.7	3	4.5	6	9

[1] The surface density was calculated for the honeycomb with the thickness of 30 mm.

3.1.3. Effect of the Thickness of Wave-Absorbing Honeycomb

As the low-frequency radar wave has a long wavelength and large penetration depth, the thickness of wave-absorbing honeycomb also affects the wave-absorbing property of HSC to a large extent. The influence of honeycomb thickness was investigated using honeycomb 6#, which exhibits optimal dielectric property (Figure 3c) in the low frequency range compared with the others. Figure 4 shows the reflectivity of HSC with the thickness of honeycomb (6#) ranging from 10 mm to 50 mm. It demonstrates that the honeycombs with moderate thickness of 30 mm and 40 mm possess the best wave-absorbing performances among the samples, where the averaged reflectivity of the corresponding HSC is respectively −11.5 dB and −11.2 dB in the frequency band of 1–2 GHz. For the fixed dielectric property, the honeycombs with lower thickness (e.g., 10 mm and 20 mm) could not sufficiently attenuate the incident radar wave in the transmission length. In this case, increasing honeycomb thickness is beneficial to enhance wave-absorbing performance. However, when honeycomb thickness reaches a much larger level (e.g., 50 mm), the resonant absorption peak of honeycomb moves to a lower frequency region ($f < 1$ GHz), which weakens its wave-absorption ability in the band of interest (1–2 GHz). By balancing these two opposite factors and considering the demand for a light-weight wave-absorbing composite, the honeycomb with moderate thickness of 30 mm was chosen to fabricate the wave-absorbing MHSC in the following sections.

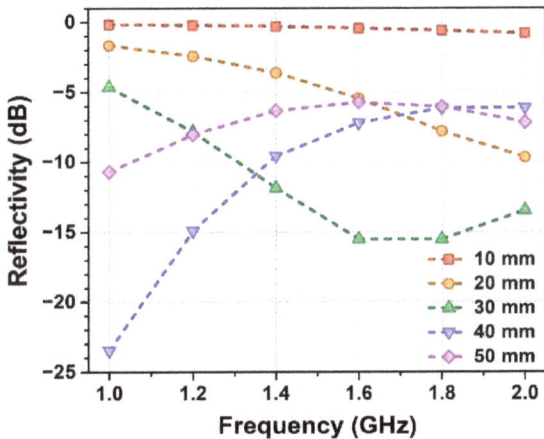

Figure 4. Reflectivity of HSC-6# with different thicknesses of the wave-absorbing honeycomb.

3.2. Wave-Absorbing Properties of MHSC

According to the above discussion, increasing the dielectric property (Figure 3) and increasing the thickness (Figure 4) of honeycomb both have limited effect in improving the wave-absorbing property of HSC in the low-frequency band. To solve this problem, we designed a type of MHSC that contains metamaterial structural units with unique electromagnetic response in the frequency range of 1–2 GHz. MHSC was fabricated by incorporating a PEEK layer loading the periodic metal arrays into the conventional HSC structure, which is schematically illustrated in Figure 1. The PEEK film acts as the supporting substrate of the periodic metal units and guarantees position accuracy in the pattern-transfer process. Moreover, during the curing step, the PEEK film can be dissolved in the resin matrix and fused into wave-transmitting skin, which prevents mechanical property loss brought about by the redundant film layer.

As shown in Figure 5, the metamaterial layer consists of open square metal rings with subwavelength sizes, which were designed on the basis of the phase gradient principle [45]. These anisotropic metal units can effectively resonate in the electromagnetic field under the specific wavelength, inducing a phase shift to the incident radar wave and producing a phase gradient on the meta-surface. In consequence, when the radar wave is incident on to the metamaterial units, both the reflected wave and the transmitted wave will be deflected with an abnormal reflection/refraction angle. The mechanism is schematically shown in Figure 5. On one hand, abnormal reflection can make the reflected wave deviate from the incident direction, which changes the direction and shape of the reflected lobe and reduces the RCS (Radar Cross Section) of the target. On the other hand, the abnormal transmission can increase the propagation distance (i.e., the effective thickness) of the radar wave in the honeycomb, which can significantly enhance electromagnetic loss and improve the wave-absorbing property of the composites in the low-frequency band.

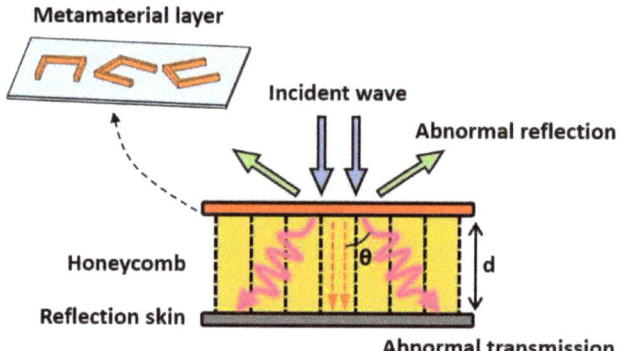

Figure 5. Schematic illustration of the wave-absorbing mechanism of metamaterial/honeycomb sandwich composites (MHSCs).

3.2.1. Properties of MHSC with Wave-Transmitting Honeycomb

Firstly, the effect of metamaterial units on the radar stealth performance of composites was investigated using the wave-transparent Nomex honeycomb with different thicknesses ranging from 10 mm to 50 mm. The thickness of the wave-transmitting skin of MHSC was fixed at 1 mm. As shown in Figure 6, increasing the thickness of Nomex honeycomb can significantly decrease the reflectivity of HSCM in the frequency range of 1–2 GHz. This phenomenon is ascribed to the abnormal transmission of the electromagnetic wave on the metamaterial surface. As demonstrated in Figure 5, the effective thickness of the honeycomb ($d/\cos(\theta)$, propagation distance of the refracted wave) shows positive correlation with its actual thickness d. Consequently, for a certain electromagnetic response ($\cos(\theta)$) of metamaterial units, a relatively large thickness (d) of honeycomb is conducive to fully utilize the radar stealth function of the metamaterials in MHSC. According to Figure 6, when the thickness of Nomex honeycomb is larger than 30 mm, the composite with metamaterial shows relatively small reflectivity (less than −5 dB) in the band. However, the wave-transparent Nomex honeycomb without loading absorbing agents has low dielectric loss and cannot effectively attenuate the electromagnetic wave in the high frequency range (e.g., 8–18 GHz). Consequently, in the next section, wave-absorbing honeycombs with different dielectric constants and dielectric loss were complexed with metamaterial structure to fabricate MHSC, in order to realize more efficient wave-absorbing property in the low frequency range and balanced performance in a wide-frequency range.

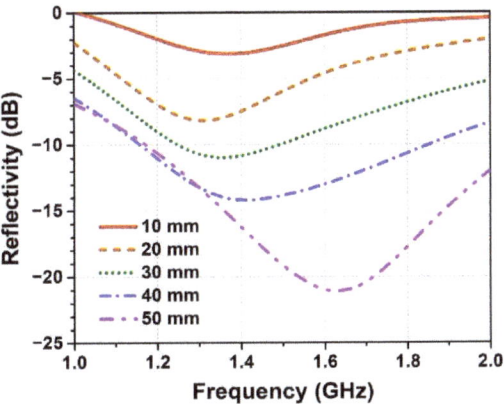

Figure 6. Reflectivity of the MHSC with the different thicknesses of the wave-transmitting Nomex honeycomb.

3.2.2. Properties of MHSC with Wave-Absorbing Honeycomb

Figure 7a–e show the reflectivity curves of five different MHSCs (colored solid lines) in the frequency band of 1–2 GHz, which were compared with corresponding HSCs without metamaterials (black dash lines). The HSCs and MHSCs were fabricated from wave-absorbing honeycombs (1#~5#) with different dielectric properties (see Figure 3a,b) and with a fixed thickness of 30 mm. The results illustrate that, for the five honeycomb samples, incorporating the metamaterial layer in the composite can decrease reflectivity in the low frequency band to a large extent. In fact, the metamaterial designed in this paper is a kind of subwavelength anisotropic structure. Due to the distinct phase-frequency response of the resonant units with different rotational orientation, the metamaterial can generate a gradual phase distribution (i.e., phase gradient) for the reflected and transmitted waves [44–46]. This unique electromagnetic effect equivalently changes the shape and direction of the reflected lobe and increases the effective thickness of the honeycomb absorber, which significantly enhances the low-frequency radar stealth performance of the composites. However, as revealed in Figure 7f, the wave-absorbing performance of MHSC is still correlated with the dielectric property of the wave-absorbing honeycomb in a complex manner. To quantitatively evaluate this issue, the averaged reflectivity (R_a) of the HSC and MHSC in the frequency range of 1–2 GHz is extracted from the curves and listed in Table 2. HSC without metamaterial increasing the dielectric property of the honeycomb (1#~5#) within a certain range can significantly enhance radar-stealth performance, where R_a is decreased from −1.35 dB to −6.29 dB from 1# to 5#. This effect is attributed to the more intensified wave absorption ability of the honeycomb with the larger dielectric loss. By contrast, when incorporated with metamaterials, it is the MHSC-4# with the moderate dielectric property of honeycomb that exhibits the best wave-absorbing performance (R_a = −10.59 dB) among the five MHSC samples. As shown in the second row of Table 2, for honeycombs 1#, 2#, 3#, 4# and 5# with increased dielectric property, averaged reflectivity increases first and then decreases. In other words, the wave-absorbing performance of the metamaterial units is strongly restricted by the dielectric properties of the honeycomb structure. The value of ($R_{a,HSC} - R_{a,MHSC}$) is the reflection reduction of MHSC compared with HSC, which represents the enhancing effect of the radar stealth performance contributed by the metamaterial layer in the composite. As shown in the last row of Table 2, this value is negatively correlated with the dielectric property of the wave-absorbing honeycomb. Specifically, for the wave-transmitting Nomex honeycomb with the lowest dielectric constant and lowest dielectric loss, the metamaterial decreases the averaged reflectivity for 8.14 dB in the low frequency band, while the value is only 4.09 dB for the wave-absorbing honeycomb 5# with the largest dielectric property among the samples. The mechanism is that, although the honeycomb with larger dielectric loss exhibits better wave-absorbing property, it will also show more significant interference on the electromagnetic response of the metamaterial layer. To achieve optimal radar-stealth performance in the low frequency band, the metamaterial units should be complexed with the wave-absorbing honeycomb that exhibits well-matched dielectric properties.

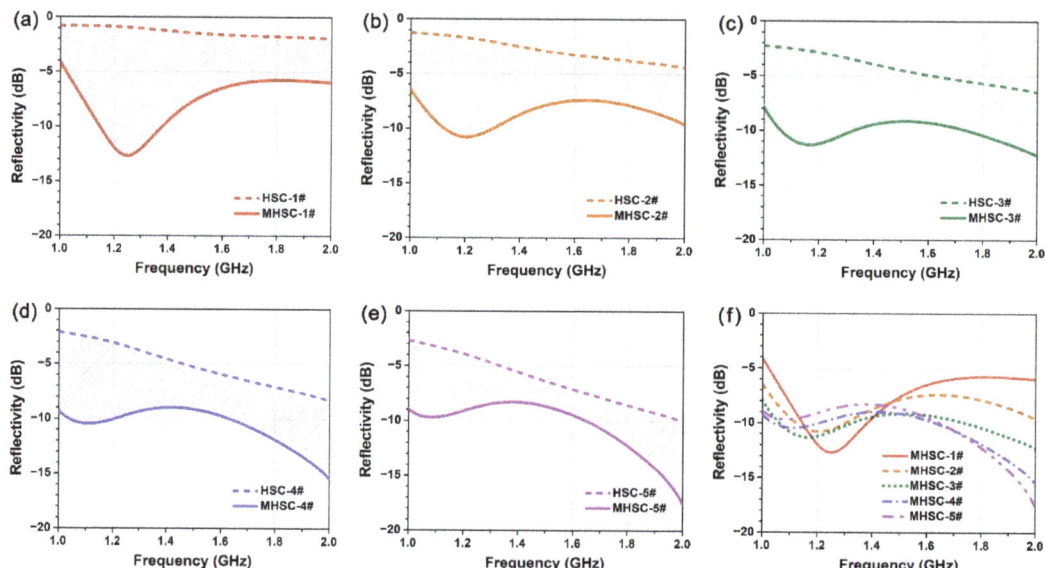

Figure 7. (a–e) Reflectivity of HSC (dash line) and MHSC (solid line) in the low frequency range (1–2 GHz). The wave-absorbing honeycomb is respectively (**a**) 1#, (**b**) 2#, (**c**) 3#, (**d**) 4# and (**e**) 5#, whose dielectric properties show an ascending order of 1# < 2# < 3# < 4# < 5#. The shadow area in each figure represents the reduction of reflectivity by inserting the metamaterial units in the honeycomb sandwich composite. (**f**) Comparison of the wave-absorbing performance of MHSC fabricated from the honeycomb with different dielectric properties.

Table 2. Averaged reflectivity of HSC and MHSC with different honeycombs in the frequency band of 1–2 GHz.

Honeycomb	Nomex	1#	2#	3#	4#	5#
$R_{a,HSC}$	0	−1.35	−2.80	−4.35	−5.16	−6.29
$R_{a,MHSC}$	−8.14	−7.85	−8.64	−10.18	−10.59	−10.38
$R_{a,HSC} - R_{a,MHSC}$	8.14	6.50	5.84	5.83	5.43	4.09

According to the above investigation, the fabricated honeycomb sandwich composite with metamaterial (e.g., MHSC-4#) can realize excellent low-frequency wave-absorbing performance with averaged reflectivity as low as −10.59 dB in the band of 1–2 GHz. Compared with HSC-4# with the same thickness of wave-transmitting skin (1 mm) and wave-absorbing honeycomb (30 mm), incorporating the phase-gradient metamaterial into the composite can significantly enhance the wave-absorbing performance for 5.43 dB, with an increment larger than 100%. Meanwhile, the thin metamaterial layer with a surface density of only 0.19 kg/m^2 (much lower than that of the wave-absorbing honeycombs, see Table 1) shows no obvious weight-increase to the total structure. Moreover, under dielectric match between the honeycomb and metamaterial, the MHSC also possesses a more satisfactory and enhanced wide-band wave-absorbing property than the corresponding HSC. The reflectivity of MHSC-4# and HSC-4# in the frequency range of 1–18 GHz was also characterized, and the curves are shown in Figure 8. According to the results, averaged reflectivity in the frequency band of 1–4 GHz is −11.18 dB for MHSC-4# but only −7.51 dB for HSC-4#. The −10 dB bandwidth in the frequency band of 1–8 GHz is 5.4 GHz for MHSC-4# but only 3.16 GHz for HSC-4#. Moreover, over an extremely wide frequency range of 1–18 GHz, the averaged reflectivity of the HSCM-4# is lower than −14.8 dB, with the lowest reflectivity of −33.4 dB at the frequency of 9.5 GHz. These results indicate that, when matched with a suitable honeycomb, incorporation of the phase-gradient metamaterial layer can significantly enhance the wave-absorbing properties of the composites in the low frequency band while effectively maintaining original wave-absorbing performance in the high frequency range. It is strongly believed that the MHSCs reported in this study show promising potential to be applied as high-performance radar stealth composite materials.

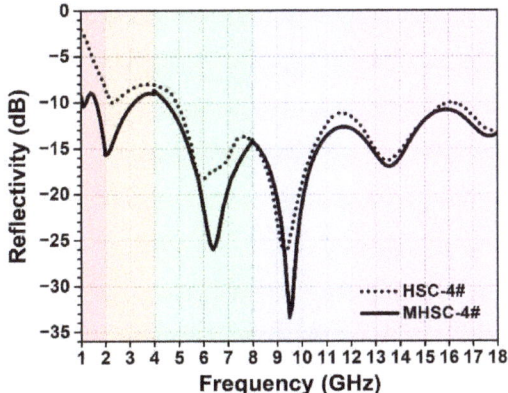

Figure 8. Wide-band wave-absorbing performance of the MHSC (solid line) and HSC (dot line) in the frequency range from 1 GHz to 18 GHz. Both the composites used the wave-absorbing honeycomb 4# with a thickness of 30 mm. From left to right, the five regions with the colors of red, yellow, green, blue and purple respectively represent frequency bands of 1–2 GHz, 2–4 GHz, 4–8 GHz, 8–12 GHz and 12–18 GHz.

4. Conclusions

In summary, we have fabricated a new type of metamaterial/honeycomb sandwich composite (MHSC) and systematically investigated wave-absorbing performance in the low frequency band of 1–2 GHz. It is proved that, simply increasing dielectric properties and heights of the wave-absorbing honeycomb both show a limited effect in improving low-frequency radar stealth performance of the conventional HSC, which is ascribed to the long wavelength of the electromagnetic wave in the band. This challenging problem is successfully solved by incorporating a phase-gradient metamaterial layer in the composite via a pattern-transfer method and optimizing the designing parameters of the fabricated MHSC. When the height of the wave-absorbing honeycomb and the height of the wave-transmitting skin are respectively 30 mm and 1 mm, the fabricated MHSC can possess averaged reflectivity less than −10 dB in the frequency band of 1–2 GHz, whose wave-absorbing performance is promoted for larger than 5 dB compared with the corresponding HSC. It is discovered that enhancement of radar stealth performance in the low frequency band contributed by the metamaterial layer shows negative correlation with dielectric properties of the wave-absorbing honeycomb, which is caused by coupling interference between the two response structures. The optimal low-frequency wave-absorbing property of MHSC is achieved by elaborately balancing the dielectric loss of the honeycomb and the electromagnetic response of the metamaterial units. Under rational design, the MHSC loaded with a thin and light-weight metamaterial layer not only possesses more excellent low-frequency (1–2 GHz) wave-absorbing properties compared with the conventional HSC but also shows more satisfactory radar stealth performance over a wide frequency range of 1–18 GHz. This study both proposes an effective methodology to enhance the low-frequency wave-absorbing performance of conventional HSCs and provides deep insight into electromagnetic coupling between the honeycomb absorber and metamaterials. It is highly expected that the discoveries and design methods of this work can be extended to fabricate light-weight and wide-band MAMs based on metamaterials and other composite structures, such as laminates, foams, fabrics, aerogels, and so on.

Supplementary Materials: The following supporting information can be downloaded at: https://www.mdpi.com/article/10.3390/polym14071424/s1, Figure S1: Photograph of the fabricated wave-transmitting skin containing metamaterial units; Figure S2: Schematic illustration about the structure and wave-absorbing mechanism of the honeycomb sandwich composites (HSC).

Author Contributions: Conceptualization, S.L. and L.X.; methodology, S.L. and H.L.; formal analysis, J.W.; investigation, S.L., H.H. and S.W.; resources, H.L. and L.X.; writing—original draft preparation, S.L. and H.H.; writing—review and editing, L.X.; visualization, H.H.; supervision, L.X.; project administration, L.X. All authors have read and agreed to the published version of the manuscript.

Funding: This research received no external funding.

Institutional Review Board Statement: Not applicable.

Informed Consent Statement: Not applicable.

Data Availability Statement: Not applicable.

Conflicts of Interest: The authors declare no conflict of interest.

References

1. Qin, F.; Brosseau, C. A review and analysis of microwave absorption in polymer composites filled with carbonaceous particles. *J. Appl. Phys.* **2012**, *111*, 061301. [CrossRef]
2. Cao, M.; Han, C.; Wang, X.; Zhang, M.; Zhang, Y.; Shu, J.; Yang, H.; Fang, X.; Yuan, J. Graphene nanohybrids: Excellent electromagnetic properties for the absorbing and shielding of electromagnetic waves. *J. Mater. Chem. C* **2018**, *6*, 4586–4602. [CrossRef]
3. Jiang, D.; Murugadoss, V.; Wang, Y.; Lin, J.; Ding, T.; Wang, Z.; Shao, Q.; Wang, C.; Liu, H.; Lu, N.; et al. Electromagnetic Interference Shielding Polymers and Nanocomposites–a Review. *Polym. Rev.* **2019**, *59*, 280–337. [CrossRef]
4. Li, Q.; Zhang, Z.; Qi, L.; Liao, Q.; Kang, Z.; Zhang, Y. Toward the application of high frequency electromagnetic wave absorption by carbon nanostructures. *Adv. Sci.* **2019**, *6*, 1801057. [CrossRef] [PubMed]
5. Gupta, S.; Tai, N. Carbon materials and their composites for electromagnetic interference shielding effectiveness in X-band. *Carbon* **2019**, *152*, 159–187. [CrossRef]
6. Zeng, X.; Cheng, X.; Yu, R.; Stucky, G.D. Electromagnetic microwave absorption theory and recent achievements in microwave absorbers. *Carbon* **2020**, *168*, 606–623. [CrossRef]
7. Wang, L.; Li, X.; Shi, X.; Huang, M.; Li, X.; Zeng, Q.; Che, R. Recent progress of microwave absorption microspheres by magnetic–dielectric synergy. *Nanoscale* **2021**, *13*, 2136–2156. [CrossRef] [PubMed]
8. Wu, N.; Hu, Q.; Wei, R.; Mai, X.; Naik, N.; Pan, D.; Guo, Z.; Shi, Z. Review on the electromagnetic interference shielding properties of carbon based materials and their novel composites: Recent progress, challenges and prospects. *Carbon* **2021**, *176*, 88–105. [CrossRef]
9. Oh, J.; Oh, K.; Kim, C.; Hong, C. Design of radar absorbing structures using glass/epoxy composite containing carbon black in X-band frequency ranges. *Compos. Part. B* **2004**, *35*, 49–56. [CrossRef]
10. Zhou, Q.; Yin, X.; Ye, F.; Liu, X.; Cheng, L.; Zhang, L. A novel two-layer periodic stepped structure for effective broadband radar electromagnetic absorption. *Mater. Des.* **2017**, *123*, 46–53. [CrossRef]
11. Han, M.; Liang, D.; Deng, L. Fabrication and electromagnetic wave absorption properties of amorphous $Fe_{79}Si_{16}B_5$ microwires. *Appl. Phys. Lett.* **2011**, *99*, 082503.
12. Yin, P.; Deng, Y.; Zhang, L.; Li, N.; Feng, X.; Wang, J.; Zhang, Y. Facile synthesis and microwave absorption investigation of activated carbon@Fe_3O_4 composites in the low frequency band. *RSC Adv.* **2018**, *8*, 23048–23057. [CrossRef]
13. Lv, H.; Yang, Z.; Wang, P.L.; Ji, G.; Song, J.; Zheng, L.; Zeng, H.; Xu, Z.J. A Voltage-Boosting strategy enabling a Low-Frequency, flexible electromagnetic wave absorption device. *Adv. Mater.* **2018**, *30*, 1706343. [CrossRef] [PubMed]
14. Jia, Z.R.; Lan, D.; Lin, K.J.; Qin, M.; Kou, K.C.; Wu, G.L.; Wu, H.J. Progress in low-frequency microwave absorbing materials. *J. Mater. Sci. Mater. Electron.* **2018**, *29*, 17122–17136. [CrossRef]
15. Smith, F.C.; Scarpa, F. Design of honeycomb-like composites for electromagnetic and structural applications. *IEE Proc. Sci. Meas. Technol.* **2004**, *151*, 9–15. [CrossRef]

16. Panwar, R.; Lee, J.R. Recent advances in thin and broadband layered microwave absorbing and shielding structures for commercial and defense applications. *Funct. Compos. Struct.* **2019**, *1*, 032001. [CrossRef]
17. Li, W.; Xu, L.; Zhang, X.; Gong, Y.; Ying, Y.; Yu, J.; Zheng, J.; Qiao, L.; Che, S. Investigating the effect of honeycomb structure composite on microwave absorption properties. *Compos. Comm.* **2020**, *19*, 182–188. [CrossRef]
18. Chen, H.; Shen, R.; Han, L.; Zhou, Y.; Li, F.; Lu, H.; Weng, X.; Xie, J.; Li, X.; Deng, L. Closed-form representation for equivalent electromagnetic parameters of biaxial anisotropic honeycomb absorbing materials. *Mater. Res. Express* **2019**, *6*, 085804. [CrossRef]
19. Choi, W.; Kim, C. Broadband microwave-absorbing honeycomb structure with novel design concept. *Compos. Part. B* **2015**, *83*, 14–20. [CrossRef]
20. Liu, L.; Fan, C.Z.; Zhu, N.B.; Zhao, Z.Y.; Liu, R.P. Effective electromagnetic properties of honeycomb substrate coated with dielectric or magnetic layer. *Appl. Phys. A Mater. Sci. Process.* **2014**, *116*, 901–905. [CrossRef]
21. Pang, H.; Duan, Y.; Dai, X.; Huang, L.; Yang, X.; Zhang, T.; Liu, X. The electromagnetic response of composition-regulated honeycomb structural materials used for broadband microwave absorption. *J. Mater. Sci. Technol.* **2021**, *88*, 203–214. [CrossRef]
22. Zhou, P.H.; Huang, L.R.; Xie, J.L.; Liang, D.F.; Lu, H.P.; Deng, L.J. Prediction of microwave absorption behavior of grading honeycomb composites based on effective permittivity formulas. *IEEE Trans. Antennas Propag.* **2015**, *63*, 3496–3501. [CrossRef]
23. Buehler, F.U.; Seferis, J.C.; Zeng, S.Y. Consistency evaluation of a qualified glass fiber prepreg system. *J. Adv. Mater.* **2001**, *33*, 41–50.
24. He, Y.; Gong, R.; Cao, H.; Wang, X.; Zheng, Y. Preparation and microwave absorption properties of metal magnetic micropowder-coated honeycomb sandwich structures. *Smart Mater. Struct.* **2007**, *16*, 1501–1505. [CrossRef]
25. Feng, J.; Zhang, Y.; Wang, P.; Fan, H. Oblique incidence performance of radar absorbing honeycombs. *Compos. Part. B* **2016**, *99*, 465–471. [CrossRef]
26. Gao, S.; An, Q.; Xiao, Z.; Zhai, S.; Shi, Z. Significant promotion of porous architecture and magnetic Fe_3O_4 NPs inside honeycomb-like carbonaceous composites for enhanced microwave absorption. *RSC Adv.* **2018**, *8*, 19011–19023. [CrossRef]
27. He, L.; Zhao, Y.; Xing, L.; Liu, P.; Wang, Z.; Zhang, Y.; Wang, Y.; Du, Y. Preparation of reduced graphene oxide coated flaky carbonyl iron composites and their excellent microwave absorption properties. *RSC Adv.* **2018**, *8*, 2971–2977. [CrossRef]
28. Luo, H.; Chen, F.; Wang, X.; Dai, W.; Xiong, Y.; Yang, J.; Gong, R. A novel two-layer honeycomb sandwich structure absorber with high-performance microwave absorption. *Compos. Part. A* **2019**, *119*, 1–7. [CrossRef]
29. Shabanpour, J.; Beyraghi, S.; Oraizi, H. Reconfigurable honeycomb metamaterial absorber having incident angular stability. *Sci. Rep.* **2020**, *10*, 14920. [CrossRef]
30. Xu, J.; Zhang, X.; Zhao, Z.; Hu, H.; Li, B.; Zhu, C.; Zhang, X.; Chen, Y. Lightweight, Fire-Retardant, and Anti-Compressed Honeycombed-Like carbon aerogels for thermal management and High-Efficiency electromagnetic absorbing properties. *Small* **2021**, *17*, 2102032. [CrossRef]
31. Huang, Y.; Wu, D.; Chen, M.; Zhang, K.; Fang, D. Evolutionary optimization design of honeycomb metastructure with effective mechanical resistance and broadband microwave absorption. *Carbon* **2021**, *177*, 79–89. [CrossRef]
32. Rozanov, K.N. Ultimate thickness to bandwidth ratio of radar absorbers. *IEEE Trans. Antennas Propag.* **2000**, *48*, 1230–1234. [CrossRef]
33. Staude, I.; Schilling, J. Metamaterial-inspired silicon nanophotonics. *Nat. Photonics* **2017**, *11*, 274–284. [CrossRef]
34. Xu, W.; Xie, L.; Ying, Y. Mechanisms and applications of terahertz metamaterial sensing: A review. *Nanoscale* **2017**, *9*, 13864–13878. [CrossRef] [PubMed]
35. Sun, S.; He, Q.; Hao, J.; Xiao, S.; Zhou, L. Electromagnetic metasurfaces: Physics and applications. *Adv. Opt. Photonics* **2019**, *11*, 380–479. [CrossRef]
36. Zhang, L.; Mei, S.T.; Huang, K.; Qiu, C.W. Advances in full control of electromagnetic waves with metasurfaces. *Adv. Opt. Mater.* **2016**, *4*, 818–833. [CrossRef]
37. Dai, J.Y.; Zhao, J.; Cheng, Q.; Cui, T.J. Independent control of harmonic amplitudes and phases via a time-domain digital coding metasurface. *Light: Sci. Appl.* **2018**, *7*, 90. [CrossRef]
38. Chen, W.; Zhan, J.; Zhou, Y.; Chen, R.; Wang, Y.; Ma, Y. Microwave metamaterial absorbers with controllable luminescence features. *ACS Appl. Mater. Inter.* **2021**, *13*, 54497–54502. [CrossRef]
39. Landy, N.I.; Sajuyigbe, S.; Mock, J.J.; Smith, D.R.; Padilla, W.J. Perfect metamaterial absorber. *Phys. Rev. Lett.* **2008**, *100*, 207402. [CrossRef]
40. Watts, C.M.; Liu, X.L.; Padilla, W.J. Metamaterial electromagnetic wave absorbers. *Adv. Mater.* **2012**, *24*, OP98–OP120. [CrossRef]
41. Ding, F.; Cui, Y.; Ge, X.; Jin, Y.; He, S. Ultra-broadband microwave metamaterial absorber. *Appl. Phys. Lett.* **2012**, *100*, 103506. [CrossRef]
42. Yu, N.; Genevet, P.; Kats, M.A.; Aieta, F.; Tetienne, J.; Capasso, F.; Gaburro, Z. Light propagation with phase discontinuities: Generalized laws of reflection and refraction. *Science* **2011**, *334*, 333–337. [CrossRef] [PubMed]
43. Sun, S.; He, Q.; Xiao, S.; Xu, Q.; Li, X.; Zhou, L. Gradient-index meta-surfaces as a bridge linking propagating waves and surface waves. *Nat. Mater.* **2012**, *11*, 426–431. [CrossRef]
44. Li, Y.; Zhang, J.; Qu, S.; Wang, J.; Chen, H.; Xu, Z.; Zhang, A. Wideband radar cross section reduction using two-dimensional phase gradient metasurfaces. *Appl. Phys. Lett.* **2014**, *104*, 221110. [CrossRef]
45. Cui, T.J.; Qi, M.Q.; Wan, X.; Zhao, J.; Cheng, Q. Coding metamaterials, digital metamaterials and programmable metamaterials. *Light Sci. Appl.* **2014**, *3*, e218. [CrossRef]

46. Feng, M.; Li, Y.; Zheng, Q.; Zhang, J.; Han, Y.; Wang, J.; Chen, H.; Sai, S.; Ma, H.; Qu, S. Two-dimensional coding phase gradient metasurface for RCS reduction. *J. Phys. D Appl. Phys.* **2018**, *51*, 375103. [CrossRef]
47. Fan, Y.; Wang, J.; Li, Y.; Pang, Y.; Yang, J.; Meng, Y.; Qu, S. Goos–Hänchen shift in metallic gratings assisted by phase gradient metasurfaces. *Mater. Res. Express* **2018**, *5*, 125802. [CrossRef]
48. Huang, D.; Kang, F.; Zhou, Z.; Cheng, H.; Ding, H. An 'H'-shape three-dimensional meta-material used in honeycomb structure absorbing material. *Appl. Phys. A Mater. Sci. Process.* **2015**, *118*, 1099–1106. [CrossRef]
49. Ghosh, S.; Lim, S. Perforated lightweight broadband metamaterial absorber based on 3-D printed honeycomb. *IEEE Antenn. Wirel. Pr.* **2018**, *17*, 2379–2383. [CrossRef]
50. Li, S.M.; Jiang, S.C.; Wang, Y.L.; Gu, J.X.; Xing, L.Y. Study on "Metamaterial" Structural Absorbing Composite Technology. *Cailiao Gongcheng* **2017**, *45*, 10–14.
51. Sun, J.M.; Wen, L.; Li, B.T.; Yang, Z.; Xing, L.Y. Study on influences of heat moisture treatment on process and mechanical properties of cyanate ester adhesives. *China Adhes.* **2017**, *26*, 17–20.
52. Bakerjarvis, J.; Vanzura, E.J.; Kissick, W.A. Improved technique for determining complex permittivity with the transmission reflection method. *IEEE Trans. Microw. Theory Tech.* **1990**, *38*, 1096–1103. [CrossRef]
53. He, S.; Hu, J.; Zhang, C.; Wang, J.; Chen, L.; Bian, X.; Lin, J.; Du, X. Performance improvement in nano-alumina filled silicone rubber composites by using vinyl tri-methoxysilane. *Polym. Test.* **2018**, *67*, 295–301. [CrossRef]
54. Liao, Y.; Weng, Y.; Wang, J.; Zhou, H.; Lin, J.; He, S. Silicone rubber composites with high breakdown strength and low dielectric loss based on polydopamine coated mica. *Polymers* **2019**, *11*, 2030. [CrossRef] [PubMed]

Article

Bio-Based *Eucommia ulmoides* Gum Composites with High Electromagnetic Interference Shielding Performance

Hailan Kang [1,2], Sen Luo [1,2], Hongyang Du [1,2], Lishuo Han [1,2], Donghan Li [1,2], Long Li [1,2,*] and Qinghong Fang [1,2,*]

[1] College of Materials Science and Engineering, Shenyang University of Chemical Technology, Shenyang 110142, China; kanghailan@syuct.edu.cn (H.K.); luosen971012@163.com (S.L.); 13940204549@163.com (H.D.); hanls0022@163.com (L.H.); lidonghansyuct@126.com (D.L.)
[2] Key Laboratory for Rubber Elastomer of Liaoning Province, Shenyang University of Chemical Technology, Shenyang 110142, China
* Correspondence: lilong@syuct.edu.cn (L.L.); fqh80@126.com (Q.F.); Tel.: +86-189-0092-6770 (L.L.); +86-138-4010-2035 (Q.F.)

Abstract: Herein, high-performance electromagnetic interference (EMI) shielding bio-based composites were prepared by using EUG (*Eucommia ulmoides* gum) with a crystalline structure as the matrix and carbon nanotube (CNT)/graphene nanoplatelet (GNP) hybrids as the conductive fillers. The morphology of the CNT/GNP hybrids in the CNT/GNP/EUG composites showed the uniform distribution of CNTs and GNPs in EUG, forming a denser filler network, which afforded improved conductivity and EMI shielding effect compared with pure EUG. Accordingly, EMI shielding effectiveness values of the CNT/GNP/EUG composites reached 42 dB in the X-band frequency range, meeting the EMI shielding requirements for commercial products. Electromagnetic waves were mainly absorbed via conduction losses, multiple reflections from interfaces and interfacial dipole relaxation losses. Moreover, the CNT/GNP/EUG composites exhibited attractive mechanical properties and high thermal stability. The combination of excellent EMI shielding performance and attractive mechanical properties render the as-prepared CNT/GNP/EUG composites attractive candidates for various applications.

Keywords: *Eucommia ulmoides* gum; carbon nanotubes; graphene; electromagnetic shielding

Citation: Kang, H.; Luo, S.; Du, H.; Han, L.; Li, D.; Li, L.; Fang, Q. Bio-Based *Eucommia ulmoides* Gum Composites with High Electromagnetic Interference Shielding Performance. *Polymers* **2022**, *14*, 970. https://doi.org/10.3390/polym14050970

Academic Editor: Ahmed I. A. Abd El-Mageed

Received: 14 January 2022
Accepted: 26 February 2022
Published: 28 February 2022

Publisher's Note: MDPI stays neutral with regard to jurisdictional claims in published maps and institutional affiliations.

Copyright: © 2022 by the authors. Licensee MDPI, Basel, Switzerland. This article is an open access article distributed under the terms and conditions of the Creative Commons Attribution (CC BY) license (https://creativecommons.org/licenses/by/4.0/).

1. Introduction

With the widespread application of advanced electronic devices, our lives are enhanced by ever greater convenience. However, the development of electronic devices has created considerable quantities of electromagnetic (EM) pollution due to electromagnetic radiation and EM interference; this pollution affects human health and the normal operation of other devices [1–6]. In order to avoid these issues, many novel materials with high EMI shielding performance were explored with the aim of addressing problems related to EMI and radiation [7]. Conductive polymer composites (CPCs) consisting of polymer and conductive fillers were explored as promising alternative EMI shielding materials; these materials are generally metal-based materials [8–12]. Research on CPCs was motivated by their low weight, flexibility, resistance to corrosion and good processability [13–16].

Numerous efforts were devoted to the design and construction of CPCs using compounding polymers with carbon-based nanofillers, such as carbon black (CB), carbon fibers (CFs), carbon nanotubes (CNTs) and graphene [17–23]. For example, Cheng et al. [24] prepared CPCs using an ultrahigh-molecular-weight polyethylene/polypropylene (PP) blend with conductive CB, which exhibits an absorption-dominated EMI shielding effectiveness (EMI SE) as high as 27.29 dB over the X-band frequency range. Lin et al. [25] developed CF/PP composites with an EMI SE of ~20 dB via adding 15 wt% CFs. Since both CNTs and graphene have exceptional mechanical, electric transport properties and extremely high aspect ratios, they make excellent conducting fillers in CPCs for EMI shielding applications [26–31]. Wu et al. [32] prepared CNT/PP composites with an EMI SE of 43.1 dB by adding 5 wt% CNTs. Abraham et al. [33] produced CNT/styrene-butadiene rubber composites with ionic liquids, and the EMI SE of the composites was found to reach 35.06 dB at 18 GHz when adding 10 wt% CNTs. Lu et al. [34] fabricated graphene nanoplatelet (GnP)/ethylene propylene diene monomer rubber (EPDM) composites via a combination process that included mixing, ultrasonication and compression. The GnP/EPDM composites exhibit a high EMI SE of 33 dB in the range 8.2–12.4 GHz and 35 dB in the range 12.4–18 GHz.

The EMI shielding performance of CPCs is heavily dependent on the conductivity, dispersion and connectivity of fillers in the matrix [35,36]. Three-dimensional (3D) hybrid filler materials were successfully constructed via the hybridization of one-dimensional (1D) CNTs with two-dimensional (2D) graphene, which can overcome the problem of dispersion faced by the individual fillers. The π–π interaction between the CNTs and the graphene is beneficial in the nanoscale separation of the individual CNTs and graphene sheets [37,38]. Moreover, the long CNTs were found to be entangled with the graphene nanosheet, which inhibited their aggregation and provided efficient pathways for electron transfer. Zhao et al. [39] produced a material with an improved EMI shielding performance using polydimethylsiloxane with CNT/graphene hybrids; this material achieved an EMI SE of 31 dB. Kim et al. [40] fabricated CNT/GNP/PP composites with an EMI SE of 36.5 dB at 1.25 GHz. Mohammed et al. [41] prepared PP/polyethylene composites using GNP/CNT hybrid nanofillers. Across this research, it can be seen that using CNTs and graphene as hybrid fillers yield an effective route to the fabrication of CPCs with excellent EMI shielding performance.

With concern regarding the price fluctuations of fossil fuel reserves and the increasing environmental awareness, eco-friendly polymeric materials have attracted increasing research attention. *Eucommia ulmoides* gum (trans-1,4-polyisoprene, EUG) is extracted from the *Eucommia ulmoides* Oliv and an isomer of natural rubber. Due to its unique dual nature combining properties of both plastics and rubber, EUG was used as a thermoplastic material, thermoelastic shape memory material and as a highly elastic material in many fields. To date, there is little research regarding the use of EUG as an EMI shielding matrix. In the present work, we used EUG with a crystalline structure as a matrix material and CNT/GNP hybrids as conductive fillers to construct a shielding composite with a high EMI shielding performance. The effect of the CNT/GNP hybrids on the morphology, electrical conductivity and EMI shielding properties was investigated. Furthermore, the EMI shielding mechanism of the CNT/GNP/EUG composites was analyzed.

2. Materials and Methods

2.1. Materials

Eucommia ulmoides gum (EUG) was supplied by Shandong Beilong Eucommia Chemical Industry Co., Ltd., Weifang, China. Carbon nanotubes (CNTs, with a diameter in the range of 8–15 nm and length of 1–10 μm) were purchased from the Shanghai Kajite Chemical Technology Co., Ltd., Shanghai, China. Graphene nanosheets (GNPs) were provided by Shenzhen Hongda Changjin Technology Co., Ltd., Shenzhen, China. Triton X-100 (polyethylene glycol tertoctylphenyl ether) used in this work was provided by Shanghai Youdao Aoba Chemical Co., Ltd., Shanghai, China.

2.2. Methods

2.2.1. Preparation

CNTs and GNPs were modified via the use of Triton X-100. An amount of 0.1 g of GNPs (or CNTs) was mixed with 1 mL of Triton X-100 and subsequently grounded manually in an agate bowl for 30 min. The product was then dispersed in deionized water and magnetically stirred in a glass beaker for 20 min. The suspension was ultrasonicated for 60 min in ice bath conditions, filtrated and vacuum dried to obtain *m*-GNPs (or *m*-CNTs).

The calculated amounts of m-GNPs (0 wt%, 0.80 wt%, 2.35 wt% and 3.85 wt%) and m-CNTs (9.62 wt%) were dispersed in toluene at 2 mg/mL under the action of mechanical stirring; this was followed by sonication for 60 min. The m-GNPs and m-CNTs suspensions were subsequently mechanically stirred for 10 min and ultrasonicated for 30 min. Then CNT/GNP suspension was added to the EUG toluene solution; the mixture was then mechanically stirred for 30 min and ultrasonicated for 2 h. The mixture was precipitated using ethanol and then dried at 40 °C. The dried mixture was mixed with curing agents (ZnO, 4 phr; SA, 2 phr; sulfur, 2 phr; NOBS, 1.2 phr) on a two-roll mill at 65 °C. Then, the blends were cured into 1-mm-thick sheets under 150 °C at the vulcanization time. The CNT/GNP/EUG composites are referred to as ECGx, where x represents the content of GNPs added to the composites.

2.2.2. Characterization

The morphology of the composites was investigated via scanning electron microscopy (SEM, SU8010, Hitachi Co., Ltd., Tokyo, Japan) and transmission electron microscopy (TEM, FEI Talos F200, Thermo Fisher Scientific Inc, Shanghai, China). The samples used for SEM were fractured in liquid nitrogen and had a surface coated with a thin gold layer. The TEM samples were ultramicrotomed at −100 °C to produce sections with a thickness in the range of 50–60 nm. Dynamic mechanical rheology measurements were performed using a rubber process analyzer (RPA-8000, Goteh Testing Machines Co., Ltd., Qingdao, China). A strain sweep was undertaken from 1% to 200% at 60 °C and 1 Hz.

The electrical conductivity of the composites was measured using a four-probe conductometer (RTS-9, Guangzhou Four-Probes Technology Co. Ltd., Guangzhou, China). The EMI shielding performance of the composites was measured in the frequency range of 8.2–12.4 GHz (X band) at room temperature, using a vector network analyzer (E5071C, Agilent Technologies, Santa Clara, CA, USA); the test samples were of length 22.86 mm, width 10.16 mm and thickness 2.00 mm. The EMI performance parameters, including the SE total (SE_T), SE reflection (SE_R) and SE absorption (SE_A), were calculated from the scattering parameters (S_{11} and S_{21}).

$$R = |S_{11}|^2 \tag{1}$$

$$T = |S_{21}|^2 \tag{2}$$

$$SE_R = -log_{10}(1-R) \tag{3}$$

$$SE_A = -log_{10}(T/(1-R)) \tag{4}$$

$$SE_T = SE_R + SE_A \tag{5}$$

The tensile tests were carried out using an INSTRON 3365 testing machine (Instron Co., Ltd., Norwood, MA, USA) according to the standard ASTM D412 using a crosshead speed of 500 mm/min at 23 ± 2 °C. Cut the EUG/CNTs/GNPs composite sample into strips with a section width of 6 mm and thickness of 2 mm with a dumbbell cutter, and clamp both ends of each sample. For each data point, at least five specimens were tested, and an average value was taken. The thermal stability of the samples was measured via thermogravimetric analysis (TGA, Q50, TA Instruments, Rutherfordton, NC, USA) from 40 °C to 600 °C using a heating rate of 10 °C/min in an N_2 atmosphere. Differential scanning calorimetric (DSC) measurements were obtained using a DSC-Q200 (TA Instruments, Rutherfordton, NC, USA) under N_2 from −70 °C to 100 °C at a heating/cooling rate of 10 °C/min. The crystallinity of the EUG was determined, as

$$X_c = \frac{\Delta H_m}{\Delta H_m^*} \times 100\% \tag{6}$$

where ΔH_m^* is the theoretical value of 100% crystallized EUG (125.6 J/g).

3. Results and Discussion

3.1. The Morphology of the CNT/GNP/EUG Composites

It is generally accepted that the properties of composite material are closely correlated with the dispersion of the filler material used. Figures 1 and 2 show the SEM and TEM images of CNT/GNP/EUG composites. As shown in Figures 1a and 2a, the CNTs are uniformly dispersed within the EUG matrix without obvious agglomeration or entanglement. As shown in Figures 1b–d and 2b, the CNTs and GNPs are uniformly dispersed across the matrix after the shear force was applied during the open mill process. CNTs were observed to show an individual distribution state and are seen to be closely attached to the GNPs, further demonstrating that the GNPs lead to an efficient dispersion of the CNTs and implying that there exists a high-strength binding interaction between the GNPs and CNTs. Because of the π–π interaction between the CNTs and GNPs, GNPs can improve the dispersibility of CNTs and inhibit their re-aggregation due to their larger surface area and steric hindrance; rod-shaped CNTs can also reduce stacking effects of the GNPs [42]. Furthermore, with an increase in hybrid loading, the GNPs exhibit increased agglomeration; these were caused by the effect of the π–π bond between the GNPs and the strong Van der Waals force. Thus, the composites form a more complete CNT/GNP conductive network, which provides a favorable condition for the electrical conductivity of the composites.

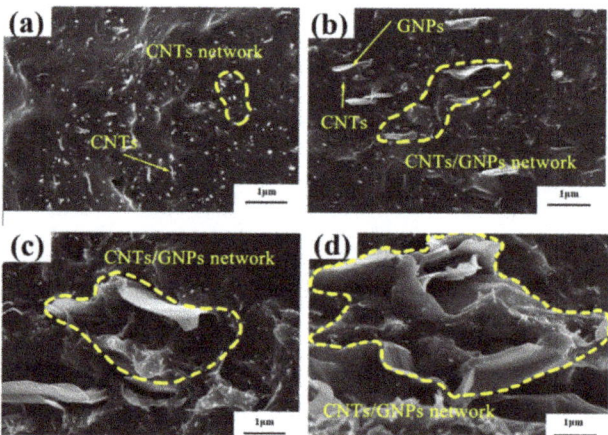

Figure 1. SEM images of CNT/GNP/EUG composites: (**a**) ECG0; (**b**) ECG1; (**c**) ECG2; (**d**) ECG3.

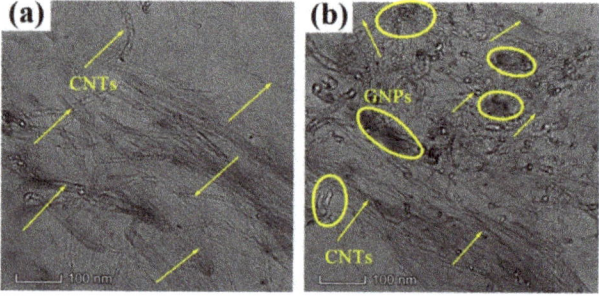

Figure 2. TEM images of CNT/GNP/EUG composites: (**a**) ECG0; (**b**) ECG3.

The Payne effect is widely used to characterize filler dispersion and filler networks. Figure 3 shows the storage modulus (G') as a function of the strain for the CNT/GNP/EUG composites. G' of the CNT/GNP/EUG composites is seen to decrease rapidly with increasing dynamic strain amplitudes; it is also seen to exhibit high sensitivity to strain, an indication of a typical Payne effect. The decrease in G' is due to the destruction of filler networks. G' increases significantly with increasing GNP loading of the CNT/GNP hybrids, demonstrating that denser filler networks are formed between CNTs and GNPs at higher CNT/GNP loading. The difference between G' at 200% and G' at 1% (referred to as ΔG') also increases with the increase in the GNP loading; this phenomenon indicates an increase in the Payne effect. In summary, with increasing GNP loading within CNT/GNP hybrids, the filler networks possess a higher modulus and a higher sensitivity to strain. The strong filler networks also demonstrate that the CNT/GNP hybrids are highly dispersed within the EUG matrix.

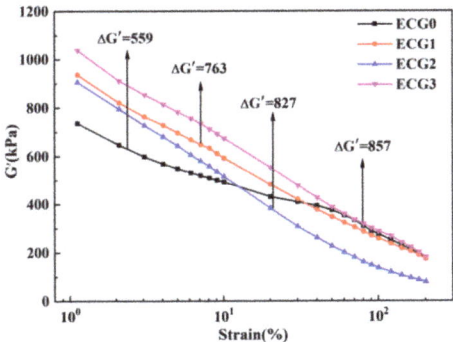

Figure 3. Storage modulus versus strain of CNT/GNP/EUG composites.

3.2. Electrical Conductivity and EMI Shielding Properties of the CNT/GNP/EUG Composites

Electrical conductivity is a key factor that determines the EMI shielding properties of a given composite [43]. Materials with high electrical conductivity usually exhibit good EMI shielding properties. The volume electrical conductivity values of the CNT/GNP/EUG composites fabricated here are shown in Figure 4. CNTs are observed to be non-directionally dispersed and distributed but in a directional distribution in local microbeams, which results in the composites having anisotropic characteristics across different micro-domains; these properties have an effect on the electrical conductivity of the rubber. The electrical conductivity of CNTs/EUG composites with 9.62 wt% CNTs is seen to be around 2.31 S/cm. This result indicates that the CNTs in the composites form a conductive network at this CNT loading. The electrical conductivity of the CNT/GNP/EUG composites slightly increases with increasing of GNP loading within CNT/GNP hybrids. As shown in Figure 2, the CNT/GNP hybrids were well dispersed within the EUG matrix, which is conducive to the formation of perfect conductive networks and leads to an improved EMI shielding performance.

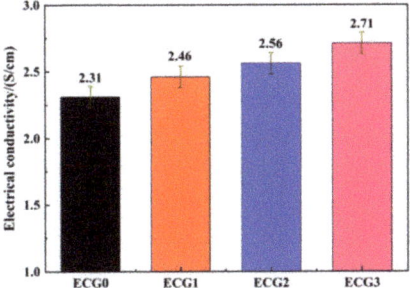

Figure 4. Electrical conductivity of CNT/GNP/EUG composites.

The outstanding electrical properties endow CNT/GNP/EUG composites with high EMI shielding properties [44]. The EMI shielding properties of the CNT/GNP/EUG composites were assessed by measuring the values of SE_T, SE_R and SE_A. Figure 5 shows the EMI SE_T, SE_R and SE_A of the CNT/GNP/EUG composites in the X-band frequency range. The pure EUG exhibits a very small SE_T value of less than 0.5 dB, indicating that EUG is almost completely transparent to the electromagnetic waves over the whole X-band frequency range. It can be seen from Figure 5a that the value of SE_T of the CNT/GNP/EUG composites gradually increased from 30.29 dB to 42.49 dB with the increase in the GNPs loading within the CNT/GNP hybrids. The enhanced EMI shielding properties are attributed to the increased conductive networks created by the synergistic effect of GNPs and CNTs.

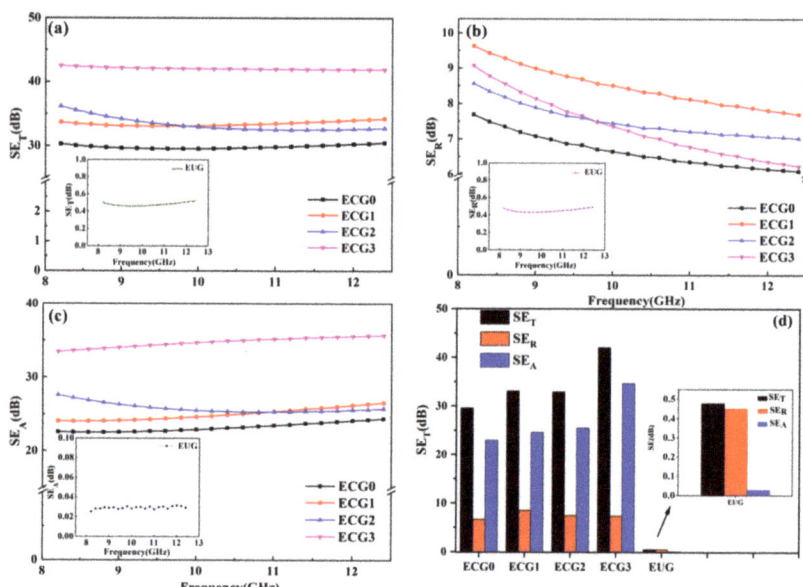

Figure 5. (a) EMI SE_T (b) EMI SE_R, (c) EMI SE_A and (d) EMI SE_T at 10 GHz of CNT/GNP/EUG composites.

In order to state the synergistic effect of 2D GNPs and 1D CNTs, the same weight loading of GNP and CNT composites were fabricated, and the comparative results of electrical conductivity and EMI shielding efficiency are summarized in Figure 6. The electrical conductivity and EMI SE_T of the CNT/GNP/EUG composite are all higher than CNT/EUG composite and GNP/EUG composite. As a 1D material, CNTs contact each other predominantly through point contacts. As a 2D material, GNPs contact each other via surface contacts. In CNT/GNP hybrids, GNPs function as the 'spacers' between the polymers, and the CNTs function as the 'wires' making up the conductive network. The combination of the two fillers creates much denser conductive networks than either filler used alone. Such perfect conductive networks permit the dissipation of electromagnetic waves through both reflection and absorption. Therefore, the CNT/GNP/EUG composites exhibit enhanced electrical conductivity and EMI shielding efficiency because of the synergistic effect of GNP and CNTs.

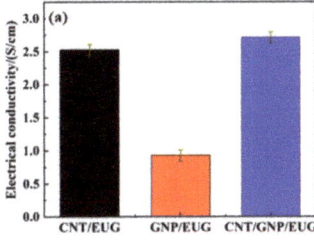 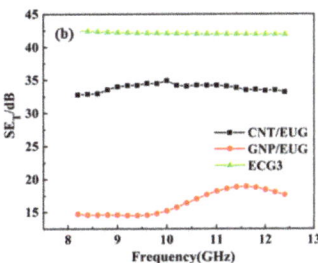

Figure 6. Comparison of (**a**) electrical conductivity and (**b**) EMI SE_T of CNT/EUG composite (13.47 wt% CNT), GNP/EUG composite (13.47 wt% GNP) and CNT/GNP/EUG composite (9.62 wt% CNT and 3.85 wt% GNP, ECG3).

The maximum EMI SE_T value of the CNT/GNP/EUG composite is seen to be as high as ~42 dB, a value that meets the EMI shielding requirement for commercial products. Table 1 compares the results obtained in this work with other EMI shielding polymeric composites reported in previous work [17,18,20,28,30–32]; it is seen that the CNT/GNP/EUG composites created in this work exhibit superior EMI SE_T compared with previously studied composites. The results indicate that this work provides a practical and effective method for the preparation of high-performance EUG-based EMI shielding composites.

Table 1. Comparison of EMI shielding properties of CNT/GNP/EUG composites with other CPCs.

Materials	Carbon Filler Contents	Thickness (mm)	SE (dB)	Frequency Range	Ref.
CB/Silicone Rubber	15 phr	2.0	20	0.01–10,000 MHz	[17]
CF/MWCNT/PBT/PolyASA	8 vol %+2 vol %	2.0	33.7	8.2–12.4 GHz	[18]
CNT/PE	10 wt%	3.0	35	0.5–1.5 GHz	[20]
RGO/CB/PDMS	15 wt%+17 wt%	1.5	28	8.2–18 GHz	[28]
graphene/SBR	27.81 wt%	3.0	45	8–12 GHz	[30]
GN/NBR	4 wt%	2.0	77	1–12 GHz	[31]
CNTs/RGO/Epoxy	0.83 wt%+1.2 wt%	3	42	8.2–12.4 GHz	[32]
CNTs/GNPs/EUG	9.62 wt%+3.85 wt%	2.0	42.5	8.2–12.4 GHz	this work

PBT: Polyb; PBT: Polybutylene terephthalate; ASA: Acrylonitrile Styrene acrylate copolymer; PDMS: Polydimethylsiloxane; SBR: 1,3-butadiene polymer.

3.3. EMI Shielding Mechanism of the CNT/GNP/EUG Composites

In order to fully elucidate the EMI shielding mechanism of the CNT/GNP/EUG composites, the values of SE_A, SE_R and the relative values at 10 GHz were calculated and plotted in Figure 5b,c,d, respectively. The values of SE_A and SE_R of the CNT/GNP/EUG composites all increase with increases in GNP loading of the CNT/GNP hybrids. More importantly, the SE_A values greatly increase with the GNP loading of the CNT/GNP hybrids. In all the CNT/GNP/EUG composites, the SE_A values are much greater than the SE_R values; this is because the conductive networks provide a sufficient number of interfaces that form multiple reflections and thus attenuate the incident electromagnetic waves [45]. For instance, the SE_T, SE_R and SE_A of ECG3 are 42 dB, 7.4 dB and 34.6 dB, respectively. Furthermore, the SE_A value that represents absorption makes up 82% of the SE_T value, whereas the SE_R value that represents reflection accounts for only 18% of the SE_T value. Thus, the results of this work indicate that absorption is the predominant shielding mechanism of the CNT/GNP/EUG composites.

The complex permittivity of the CNT/GNP/EUG composites was also measured in order to understand their EM absorption properties better. Figure 7 shows the real part (ε'), the imaginary part (ε'') and the dielectric loss tangent (tan $\delta_\varepsilon = \varepsilon''/\varepsilon'$) in the X-band frequency range. The values of ε' and ε'' represent the storage and loss capability of electric energy within a material, respectively, whereas tan δ_ε is used to evaluate the dielectric loss capacity of a material. Both ε' and ε'' decrease with increasing frequency; this decrease is related to the interface polarization [46]. It is noted that the values of ε′ and ε″ of the CNT/GNP/EUG composites increase with an increase in the GNP loading of the CNT/GNP hybrids. This is ascribed to the formation of perfect conducting networks by the CNTs and GNPs within the EUG matrix. The enhanced ε′ value is predominantly due to the creation of more dipoles within the composite. The uniform distribution of the CNT/GNP hybrids without agglomeration can create many interfaces and hence induce polarization at the interfaces of CNT/GNP, CNT/EUG and GNP/EUG. Moreover, the introduction of more layers of GNPs with a large specific surface area induces an increase in the interface polarization [47]. The increase in the value of ε″ with an increase in the GNP loading is attributed to the dielectric relaxation and electrical conductivity of the composites. The incorporation of GNPs enhances the electrical conductivity, resulting in high conduction loss. Furthermore, the presence of a high number of interfaces between the EUG matrix and the fillers results in further increases in the dielectric loss within the composite. The value of tan δ_ε is significantly increased by increases in GNP loading, indicating a superior dielectric loss capacity. The enhanced value of tan δ_ε represents significant dissipation of electrical energy and attenuation of wave absorption.

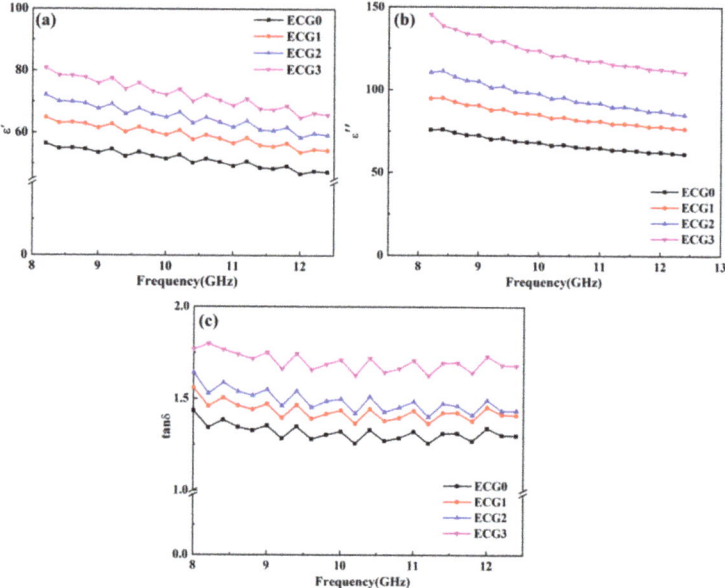

Figure 7. (a) ε′, (b) ε″ and (c) tan δ_ε of CNT/GNP/EUG composites.

Based on the preceding analysis, we established the shielding mechanism of the CNT/GNP/EUG composites, which is illustrated in Figure 8. When the EM waves reach the interfaces within the composite, a proportion of the EM waves are reflected at the surface of the CNT/GNP/EUG composite due to the impedance mismatch between the composites and the air, while a proportion of EM waves is absorbed by the CNT/GNP/EUG composite due to the conduction losses, multiple reflections from interfaces and the interfacial dipole relaxation losses. The CNT/GNP/EUG composites form more effective conductive networks due to the synergistic effect of 2D GNPs and 1D CNTs, thereby causing the attenuation of EM waves via Joule heating, which increases the EM absorption. Moreover, the numerous EUG/CNT, EUG/GNP and CNT/GNP interfaces facilitate substantial attenuation of the EM waves via multiple reflections and scattering. The interfacial polarizations also attenuate EM waves and enhance the EM wave absorption effect. Importantly, the boundary of crystalline–amorphous phases within EUG matrix can accumulate electrical charges, thus resulting in polarization loss, while the crystalline regions of the EUG matrix also increase the propagation of EM waves within the material and hence enhances EM absorption [48,49].

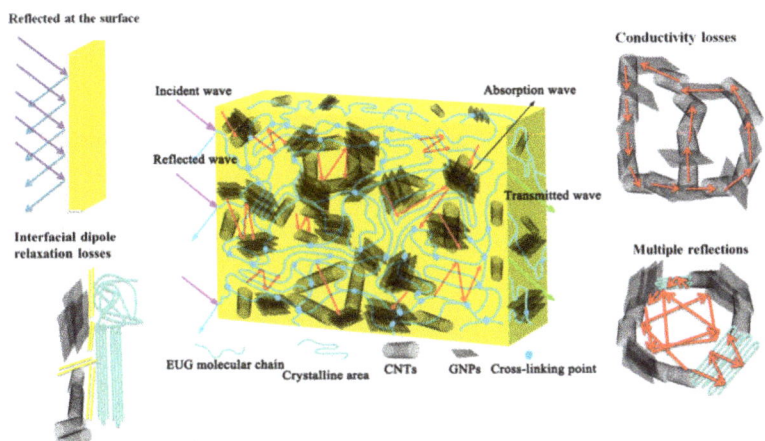

Figure 8. Proposed electromagnetic shielding mechanism for CNT/GNP/EUG composites.

3.4. Crystallinity of the CNT/GNP/EUG Composites

Thermal and crystalline behaviors were characterized by the DSC curves, and the crystallinity of EUG was calculated. As shown in Figure 9a, all samples exhibit one peak upon heating curves, which corresponds to the melting peak of crystalline EUG. Compared with pure EUG, the melting temperature (T_m) of the CNT/EUG composite shifts to a higher temperature, and the melting enthalpy (ΔH_m) increases from −23.95 to −35.48 J/g. Accordingly, the crystallinity (X_C) of the CNT/EUG composite also increases from 19.1% to 28.2%, implying that CNTs acted as the nucleating agent and promoted the crystallization of EUG segments. However, the T_m, ΔH_m and X_C of CNT/GNP/EUG composite all decrease with the increase in GNP loading of the CNT/GNP hybrids. These results indicate that the EUG segmental mobility was restricted by strong networks formed by the synergistic effect of CNTs and GNPs.

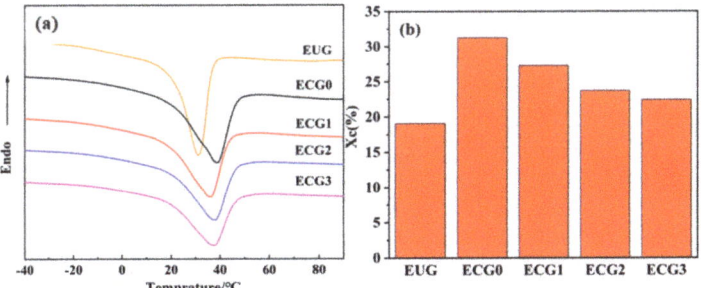

Figure 9. (**a**) DSC curves of CNT/GNP/EUG composites, (**b**) the crystallinity (Xc) of CNT/GNP/EUG composites.

3.5. Thermal and Mechanical Properties of the CNT/GNP/EUG Composites

Figure 10 depicts the TGA curves of the CNT/GNP/EUG composites; the related data are shown in Table 2. As shown in Figure 8, the thermal decomposition of CNT/GNP/EUG composites takes place in a single stage between 260 °C and 300 °C, predominantly due to the pyrolysis of the EUG macromolecules. The decomposition temperatures of the CNT/GNP/EUG composites with 5% and 30% weight loss (T_5, T_{30}) and the calculated heat resistance index (T_{HRI}) are all higher than 303 °C, 410 °C and 180 °C, respectively. The value of T_{HRI} of the CNT/GNP/EUG composites all increase with the increase in the GNP loading of the CNT/GNP hybrids. The value of T_{HRI} for the CNT/GNP/EUG is 191 °C, which is 11 °C higher than that of pure EUG (180 °C). The reason for this increase is due to the CNT and GNP fillers possessing high heat resistance, which improves the thermal stability of the EUG matrix.

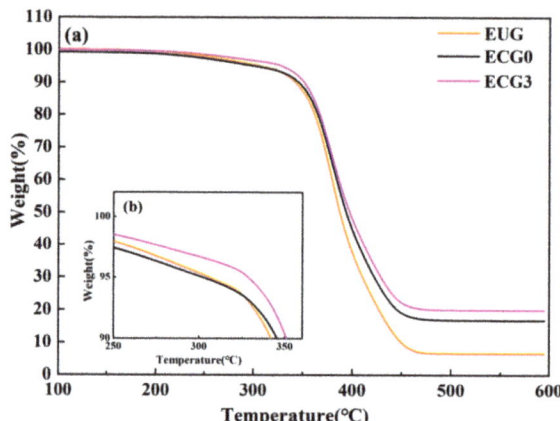

Figure 10. (a) Thermogravimetric curves of CNT/GNP/EUG composites. (b) Enlarged view of thermogravimetry at 100–90%.

Table 2. Thermal characteristics data of CNT/GNP/EUG composites.

Samples	Weight Loss Temperature (°C)		$T_{\text{Heat-resistance index}}$ (°C)
	$T_{d,5\%}$	$T_{d,30\%}$	
EUG	305	410	180
ECG0	303	423	184
ECG3	328	429	191

T_5 and T_{30} correspond to the decomposition temperature of 5% and 30% weight loss, respectively.; $T_{\text{Heat-resistance index}} = 0.49 * [T_5 + 0.6 * (T_{30} - T_5)]$.

The mechanical properties of the CNT/GNP/EUG composites are shown in Figure 11. The tensile strength of the CNT/GNP/EUG composites is seen to increase with an increase in the GNP loading of the CNT/GNP hybrids. The elongation at break initially increases and subsequently decreases with increasing GNP loading of the composite. The incorporation of GNPs in the CNT/GNP/EUG composite improves the dispersion of CNTs, resulting in enhanced tensile strength and elongation at break. On the other hand, the introduction of more CNT/GNP hybrids increases the filler–filler and filler–polymer networks, which further reinforce the composites and resists the removal of polymer chains; this results in improved tensile strength and decreased elongation at break. The Young's modulus of ECG0, ECG1, ECG2 and ECG3 are 0.05 GPa, 0.051 GPa, 0.052 GPa and 0.057 GPa, respectively. Besides, the shore A hardness of ECG0, ECG1, ECG2 and ECG3 are 79, 81, 82 and 84. The Young's modulus and the hardness of the CNT/GNP/EUG composite show a slight increase with the increase in the CNT/GNP hybrids, an indication of the reinforcement of CNT/GNP hybrids. ECG3 exhibits a tensile strength of 20.9 MPa and elongation at a break of 304%. Therefore, the combination of the excellent electromagnetic shielding performance and high mechanical properties makes the as-prepared CNT/GNP/EUG composite appropriate for a wide range of applications.

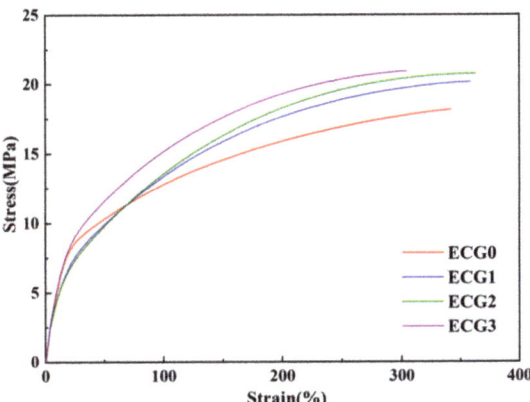

Figure 11. Stress-strain curves of CNT/GNP/EUG composites.

4. Conclusions

In summary, we have successfully fabricated bio-based composites with a high EMI SE and attractive mechanical properties by combining a matrix of EUG with a crystalline structure and CNT/GNP hybrid structures used as conductive fillers. The synergistic effect of the CNT/GNP hybrid originates from the bridging of CNTs between GNPs, which facilitates the enhanced filler networks, electrical conductivity, EMI shielding properties and mechanical properties of the CNT/GNP/EUG composites. The maximum electrical conductivity and EMI SE values of CNT/GNP/EUG composites are observed to be up to 2.71 S/cm and 42.49 dB, respectively. Furthermore, the tensile strength of the CNT/GNP/EUG composites was found to increase with an increase in the GNP loading of the CNT/GNP hybrids. These results indicate that the method studied here provides a practical and effective route towards the creation of high-performance EUG-based EMI shielding composites.

Author Contributions: Conceptualization, H.K., Q.F. and L.L.; methodology, L.L., S.L., H.D. and L.H.; formal analysis, D.L.; investigation, S.L. and H.D.; resources, H.K.; data curation, S.L. and L.H.; writing—original draft preparation, L.L.; writing—review and editing, D.L. and H.K.; supervision, Q.F.; project administration, H.K.; funding acquisition, H.K. and Q.F. All authors have read and agreed to the published version of the manuscript.

Funding: This work was supported by the National Natural Science Foundation of China (grant number 52073178); Natural Science Foundation of Liaoning, China (grant number 2019-MS-263); and Outstanding Young Talent Projects of the Shenyang University of Chemical Technology, China (grant number 2019YQ003).

Informed Consent Statement: Informed consent was obtained from all subjects involved in the study.

Conflicts of Interest: The authors declare no conflict of interest.

References

1. Cheng, H.; Wei, S.; Ji, Y.; Zhai, J.; Zhang, X.; Chen, J.; Shen, C. Synergetic effect of Fe_3O_4 nanoparticles and carbon on flexible poly (vinylidence fluoride) based films with higher heat dissipation to improve electromagnetic shielding. *Compos. Part A Appl. Sci. Manuf.* **2019**, *121*, 139–148. [CrossRef]
2. Wu, J.L.; Zhang, Y.R.; Gong, Y.Z.; Wang, K.; Chen, Y.; Song, X.P.; Lin, J.; Shen, B.Y.; He, S.J.; Bian, X.M. Analysis of the electrical and thermal properties for magnetic Fe_3O_4 coated SiC filled epoxy composites. *Polymers* **2021**, *13*, 3028. [CrossRef] [PubMed]
3. Araby, S.; Meng, Q.; Zhang, L.; Kang, H.; Majewski, P.; Tang, Y.; Ma, J. Electrically and thermally conductive elastomer/graphene nanocomposites by solution mixing. *Polymer* **2014**, *55*, 201–210. [CrossRef]
4. Li, X.H.; Li, X.; Liao, K.N.; Min, P.; Liu, T.; Dasari, A.; Yu, Z.Z. Thermally Annealed Anisotropic Graphene Aerogels and Their Electrically Conductive Epoxy Composites with Excellent Electromagnetic Interference Shielding Efficiencies. *ACS Appl. Mater. Interfaces* **2016**, *8*, 33230–33239. [CrossRef]

5. Zhai, S.X.; Dai, W.X.; Lin, J.; He, S.J.; Zhang, B.; Chen, L. Enhanced proton conductivity in sulfonated poly(ether ether ketone) membranes by incorporating sodium dodecyl benzene sulfonate. *Polymers* **2019**, *11*, 203. [CrossRef]
6. Zhan, Y.; Wang, J.; Zhang, K.; Meng, Y.; Yan, N.; Wei, W.; Peng, F.; Xia, H. Fabrication of a flexible electromagnetic interference shielding Fe_3O_4@reduced graphene oxide/natural rubber composite with segregated network. *Chem. Eng. J.* **2018**, *344*, 184–193. [CrossRef]
7. Li, Y.; Shen, B.; Pei, X.; Zhang, Y.; Yi, D.; Zhai, W.; Zhang, L.; Wei, X.; Zheng, W. Ultrathin carbon foams for effective electromagnetic interference shielding. *Carbon* **2016**, *100*, 375–385. [CrossRef]
8. Sun, R.; Zhang, H.B.; Liu, J.; Xie, X.; Yang, R.; Li, Y.; Hong, S.; Yu, Z.Z. Highly Conductive Transition Metal Carbide/Carbonitride(MXene)@polystyrene Nanocomposites Fabricated by Electrostatic Assembly for Highly Efficient Electromagnetic Interference Shielding. *Adv. Funct. Mater.* **2017**, *27*, 1702807–1702818. [CrossRef]
9. He, S.J.; Dai, W.X.; Yang, W.; Liu, S.X.; Bian, X.M.; Zhang, C.; Lin, J. Nanocomposite proton exchange membranes based on phosphotungstic acid immobilized by polydopamine-coated halloysite nanotubes. *Polym. Test.* **2019**, *73*, 242–249. [CrossRef]
10. He, S.J.; Wang, J.Q.; Yu, M.X.; Xue, Y.; Hu, J.B.; Lin, J. Structure and mechanical performance of poly(vinyl alcohol) nanocomposite by incorporating graphitic carbon nitride nanosheets. *Polymers* **2019**, *11*, 610. [CrossRef]
11. Guo, Y.; Xu, G.; Yang, X.; Ruan, K.; Ma, T.; Zhang, Q.; Gu, J.; Wu, Y.; Liu, H.; Guo, Z. Significantly enhanced and precisely modeled thermal conductivity in polyimide nanocomposites with chemically modified graphene via in situ polymerization and electrospinning-hot press technology. *J. Mater. Chem. C* **2018**, *6*, 3004–3015. [CrossRef]
12. Liao, Y.F.; Weng, Y.X.; Wang, J.Q.; Zhou, H.F.; Lin, J.; He, S.J. Silicone rubber composites with high breakdown strength and low dielectric loss based on polydopamine coated mica. *Polymers* **2019**, *11*, 2030. [CrossRef] [PubMed]
13. Liu, H.; Xu, Y.; Cao, J.; Han, D.; Yang, Q.; Li, R.; Zhao, F. Skin structured Silver/Three-dimensional Graphene/Polydimethylsiloxane Composites with Exceptional Electromagnetic Interference Shielding Effectiveness. *Compos. Part A Appl. Sci. Manuf.* **2021**, *148*, 106476–106507. [CrossRef]
14. Arjmand, M.; Chizari, K.; Krause, B.; Petra, P.; Uttandaraman, S. Effect of synthesis catalyst on structure of nitrogen-doped carbon nanotubes and electrical conductivity and electromagnetic interference shielding of their polymeric nanocomposites. *Carbon* **2016**, *98*, 358–372. [CrossRef]
15. He, S.J.; Wang, J.Q.; Hu, J.B.; Zhou, H.F.; Nguyen, H.; Luo, C.M.; Jun Lin, J. Silicone rubber composites incorporating graphitic carbon nitride and modified by vinyl tri-methoxysilane. *Polym. Test.* **2019**, *79*, 106005. [CrossRef]
16. Bian, X.M.; Tuo, R.; Yang, W.; Zhang, Y.R.; Xie, Q.; Zha, J.W.; Lin, J.; He, S.J. Mechanical, thermal, and electrical properties of BN-epoxy composites modified with carboxyl-terminated butadiene nitrile liquid rubber. *Polymers* **2019**, *11*, 1548. [CrossRef] [PubMed]
17. Zhang, J.; Zhang, H.; Wang, H.; Chen, F.; Zhao, Y. Extruded conductive silicone rubber with high compression recovery and good aging-resistance for electromagnetic shielding applications. *Polym. Compos.* **2019**, *40*, 1078–1086. [CrossRef]
18. Park, D.H.; Lee, Y.K.; Park, S.S.; Lee, C.S.; Kim, S.H.; Kin, W.N. Effects of hybrid fillers on the electrical conductivity and EMI shielding efficiency of polypropylene/conductive filler composites. *Macromol. Res.* **2013**, *21*, 905–910. [CrossRef]
19. Avi, B.; Eric, M.; Alan, T. Comparison of Experimental and Modeled EMI Shielding Properties of Periodic Porous xGNP/PLA Composites. *Polymers* **2019**, *11*, 1233.
20. Rahaman, M.; Chaki, T.K.; Khastgir, D. Development of high performance EMI shielding material from EVA, NBR, and their blends: Effect of carbon black structure. *J. Mater. Sci.* **2011**, *46*, 3989–3999. [CrossRef]
21. Maria, G.P.C.; Maxime, B.; Can, K.; Anastasios, C.M.; Nikolaos, K.; Gian, P.P.; Antonello, A.; Ernesto, D.M.; Costas, G. Thermoplastic polyurethane–graphene nanoplatelets microcellular foams for electromagnetic interference shielding. *Graphene Technol.* **2020**, *5*, 33–39.
22. Yim, Y.J.; Park, S.J. Electromagnetic interference shielding effectiveness of high-density polyethylene composites reinforced with multi-walled carbon nanotubes. *J. Ind. Eng. Chem.* **2015**, *21*, 155–157. [CrossRef]
23. Rollo, G.; Ronca, A.; Cerruti, P.; Gan, X.P.; Fei, G.X.; Xia, H.S.; Gorokhov, G.; Bychanok, D.; Kuzhir, P.; Lavorgna, M.; et al. On the Synergistic Effect of Multi-Walled Carbon Nanotubes and Graphene Nanoplatelets to Enhance the Functional Properties of SLS 3D-Printed Elastomeric Structures. *Polymers* **2020**, *12*, 1841. [CrossRef] [PubMed]
24. Cheng, H.; Cao, C.; Zhang, Q.; Wang, Y.; Liu, Y.; Huang, B.; Sun, X.L.; Guo, Y.; Xiao, L.; Chen, Q.; et al. Enhancement of Electromagnetic Interference Shielding Performance and Wear Resistance of the UHMWPE/PP Blend by Constructing a Segregated Hybrid Conductive Carbon Black-Polymer Network. *ACS Omega* **2021**, *6*, 15078–15088. [CrossRef]
25. Lin, C.W.; Lou, C.W.; Huang, C.H.; Huang, C.L.; Lin, J.H. Electromagnetically shielding composite made from carbon fibers, glass fibers, and impact-resistant polypropylene: Manufacturing technique and evaluation of physical properties. *J. Thermoplast. Compos. Mater.* **2014**, *27*, 1451–1460. [CrossRef]
26. Wu, N.; Qiao, J.; Liu, J.; Du, W.; Xu, D.; Liu, W. Strengthened electromagnetic absorption performance derived from synergistic effect of carbon nanotube hybrid with Co@C beads. *Adv. Compos. Hybrid Mater.* **2017**, *1*, 149–159. [CrossRef]
27. Wang, L.; Wu, Y.; Wang, Y.; Li, H.; Jiang, N.; Niu, K. Laterally Compressed Graphene Foam/Acrylonitrile Butadiene Styrene Composites for Electromagnetic Interference Shielding. *Compos. Part A Appl. Sci. Manuf.* **2020**, *133*, 105887. [CrossRef]
28. Anooja, J.B.; Dijith, K.S.; Surendran, K.P.; Subodh, G. A simple strategy for flexible electromagnetic interference shielding: Hybrid rGO@CB-Reinforced polydimethylsiloxane. *J. Alloy. Compd.* **2019**, *807*, 151678. [CrossRef]

29. Li, Y.; Pei, X.; Shen, B.; Zhai, W.; Zhang, L.; Zhen, W. Polyimide/graphene composite foam sheets with ultrahigh thermostability for electromagnetic interference shielding. *RSC Adv.* **2015**, *5*, 24342–24351. [CrossRef]
30. Li, Y.; Xu, F.; Lin, Z.; Sun, X.; Peng, Q.; Yuan, Y.; Wang, S.; Yang, Z.; He, X.; Li, Y. Electrically and thermally conductive underwater acoustically absorptive graphene/rubber nanocomposites for multifunctional applications. *Nanoscale* **2017**, *9*, 14476–14485. [CrossRef]
31. Al-Ghamdi, A.A.; Al-Ghamdi, A.A.; Al-Turki, Y.; El-Tantawy, F. Electromagnetic shielding properties of graphene/acrylonitrile butadiene rubber nanocomposites for portable and flexible electronic devices. *Compos. Part B Eng.* **2016**, *88*, 212–219. [CrossRef]
32. Wu, H.Y.; Zhang, Y.P.; Jia, L.C.; Yan, D.X.; Gao, J.F.; Li, Z.M. Injection Molded Segregated Carbon Nanotube/Polypropylene Composite for Efficient Electromagnetic Interference Shielding. *Ind. Eng. Chem. Res.* **2018**, *57*, 12378–12385. [CrossRef]
33. Abraham, J.; Arif, P.M.; Xavier, P.; Bose, S.; George, S.C.; Kalarikkal, N.; Thomas, S. Investigation into dielectric behaviour and electromagnetic interference shielding effectiveness of conducting styrene butadiene rubber composites containing ionic liquid modified MWCNT. *Polymer* **2017**, *112*, 102–115. [CrossRef]
34. Lu, S.; Bai, Y.; Wang, J.; Chen, D.; Ma, K.; Meng, Q.; Liu, X. Flexible GnPs/EPDM with Excellent Thermal Conductivity and Electromagnetic Interference Shielding Properties. *Nano* **2019**, *14*, 1950075. [CrossRef]
35. Gupta, S.; Tai, N.H. Carbon materials and their composites for electromagnetic interference shielding effectiveness in X-band. *Carbon* **2019**, *152*, 159–187. [CrossRef]
36. Guo, Y.; Pan, L.; Yang, X.; Ruan, K.; Han, Y.; Kong, J.; Gu, J. Simultaneous improvement of thermal conductivities and electromagnetic interference shielding performances in polystyrene composites via constructing interconnection oriented networks based on electrospinning technology. *Compos. Part A Appl. Sci. Manuf.* **2019**, *124*, 105484–105522. [CrossRef]
37. Sun, H.; Chen, D.; Ye, C.; Li, X.; Dai, D.; Yuan, Q.; Chee, K.; Zhao, P.; Jiang, N.; Lin, C.T. Large-area self-assembled reduced graphene oxide/electrochemically exfoliated graphene hybrid films for transparent electrothermal heaters. *Appl. Surf. Sci.* **2018**, *435*, 809–814. [CrossRef]
38. Huangfu, Y.; Ruan, K.; Qiu, H.; Lu, Y.; Liang, C.; Kong, J.; Gu, J. Fabrication and investigation on the PANI/MWCNT/thermally annealed graphene aerogel/epoxy electromagnetic interference shielding nanocomposites. *Compos. Part A Appl. Sci. Manuf.* **2019**, *121*, 265–272. [CrossRef]
39. Zhao, S.; Yan, Y.; Gao, A.; Zhao, S.; Cui, J.; Zhang, G. Flexible Polydimethylsilane Nanocomposites Enhanced with a Three-Dimensional Graphene/Carbon Nanotube Bicontinuous Framework for High-Performance Electromagnetic Interference Shielding. *ACS Appl. Mater. Interfaces* **2018**, *10*, 26723–26732. [CrossRef]
40. Kim, M.S.; Yan, J.; Joo, K.H.; Pandey, J.K.; Kang, Y.J.; Ahn, S.H. Synergistic effects of carbon nanotubes and exfoliated graphite nanoplatelets for electromagnetic interference shielding and soundproofing. *J. Appl. Polym. Sci.* **2013**, *130*, 3947–3951. [CrossRef]
41. Mohammed, H.; Al, S. Electrical, EMI shielding and tensile properties of PP/PE blends filled with GNP:CNT hybrid nanofiller. *Synth. Met.* **2016**, *217*, 322–330.
42. Güler, Ö.; Başgöz, Ö.; Güler, S.H.; Canbay, C.A.; Açıkgöz, Ş.; Boyrazlı, M. The synergistic effect of GNPs+CNTs on properties of polyester: Comparison with polyester–CNTs nanocomposite. *J. Mater. Sci. Mater. Electron.* **2021**, *32*, 17436–17447. [CrossRef]
43. Wang, Y.; Wang, W.; Qi, Q.; Xu, N.; Yu, D. Layer-by-layer assembly of PDMS-coated nickel ferrite/multiwalled carbon nanotubes/cotton fabrics for robust and durable electromagnetic interference shielding. *Cellulose* **2020**, *27*, 2829–2845. [CrossRef]
44. Shen, B.; Li, Y.; Yi, D.; Zhai, W.; Wei, X.; Zheng, W. Strong flexible polymer/graphene composite films with 3D saw-tooth folding for enhanced and tunable electromagnetic shielding. *Carbon* **2017**, *113*, 55–62. [CrossRef]
45. Wang, G.; Liao, X.; Yang, J.; Tang, W.; Zhang, Y.; Jiang, Q.; Li, G. Frequency-selective and tunable electromagnetic shielding effectiveness via the sandwich structure of silicone rubber/graphene composite. *Compos. Sci. Technol.* **2019**, *184*, 107847. [CrossRef]
46. Liu, W.; Liu, J.; Yang, Z.; Ji, G. Extended working frequency of ferrites by synergistic attenuation through a controllable carbothermal route based on prussian blue shell. *ACS Appl. Mater. Interfaces* **2018**, *10*, 28887–28897. [CrossRef]
47. Vovchenko, L.; Matzui, L.; Oliynyk, V.; Launets, V.; Mamunya, Y.; Maruzhenko, O. Nanocarbon/polyethylene composites with segregated conductive network for electromagnetic interference shielding. *Mol. Cryst. Liq. Cryst.* **2019**, *672*, 186–198. [CrossRef]
48. Xia, T.; Zhang, C.; Oyler, N.A.; Chen, X. Hydrogenated TiO_2 nanocrystals: A novel microwave absorbing material. *Adv. Mater.* **2013**, *25*, 6905–6910. [CrossRef]
49. Dong, J.Y.; Ullal, R.; Han, J.; Wei, S.; Ouyang, X.; Dong, J.; Dong, J.Z.; Gao, W. Partially crystallized TiO_2 for microwave absorption. *J. Mater. Chem. A* **2015**, *3*, 5285–5288. [CrossRef]

Article

Streamer Propagation along the Insulator with the Different Curved Profiles of the Shed

Xiaobo Meng [1], Liming Wang [2,*], Hongwei Mei [2,*], Bin Cao [2] and Xingming Bian [3]

1. School of Mechanical and Electrical Engineering, Guangzhou University, Guangzhou 510006, China; mengxb@gzhu.edu.cn
2. Tsinghua Shenzhen International Graduate School, Tsinghua University, Shenzhen 518055, China; huhu0512@126.com
3. The State Key Laboratory of Alternate Electrical Power System with Renewable Energy Sources, North China Electric Power University, Beijing 102206, China; bianxingming@ncepu.edu.cn
* Correspondence: wanglm@sz.tsinghua.edu.cn (L.W.); mei.hongwei@sz.tsinghua.edu.cn (H.M.); Tel.: +86-0755-26036695 (L.W.)

Abstract: The flashover along the insulator endangers the reliable operation of the electrical power system. The reasonable curved profiles of the shed could improve the flashover voltage, which would reduce power system outages. The research on the influence of the curved profiles of the shed on the streamer propagation along the insulator made of polymer was presented in the paper. The streamer propagation "stability" field, path, and velocity affected by the curved profiles of the shed, were measured by ultraviolet camera, ICCD camera, and photomultipliers. The "surface" component of the streamer is stopped at the shed with the different curved profiles, while the "air" component could go round the shed and reach the cathode. The streamer propagation "stability" fields are inversely proportional to the curved profiles of the shed. The streamer propagation velocities are proportional to the curved profiles of the shed. The relationship between the streamer propagation and the flashover propagation was discussed in depth. The subsequent flashover propagation is greatly affected by the streamer propagation path and "stability" field. Furthermore, the influence of the material properties on the streamer propagation path was also discussed in depth.

Keywords: streamer discharge; curved profiles; streamer propagation "stability" field; streamer propagation path; streamer propagation velocity

1. Introduction

The flashover along the insulator surface causes many widespread power outages every year. It endangers the stability and safety of electrical power system, resulting in a large number of social and economic losses. The reasonable shed design might improve the flashover voltage, which would reduce power system outages. The test results of the flashover along the insulator with the different shed configurations have been used to optimize the design of the insulator by many researchers at home and abroad [1–4].

The flashover along the insulator was supported by many researchers [5–7]. However, the streamer discharge along the insulator with a shed were little investigated [8–10]. The streamer propagation along the insulator is the important physical process before the flashover. The inhibited streamer discharge along the insulator can lead to the subsequent flashover process to disappear. Hence, the research on the streamer discharge along the insulator with the different sheds is helpful for designing the shed configuration of insulator.

The streamer discharge along the smooth insulator have been investigated widely, but the characteristics of streamer discharge along the insulator with a shed were rarely researched [11–15]. In the paper [12,13], the influences of the shed on the streamer discharge were discussed, the "surface" component of streamer propagation along the insulator was blocked at the shed, but the "air" component could cross the shed and reach the cathode plate. In the paper [15], the influences of the shed configuration on the streamer discharge were discussed and summarized, such as the diameter, location, and combination of the shed. However, the influences of the curved profiles of the shed on the streamer discharge was not studied in depth. The photograph of streamer discharge along the profiled insulator surfaces was not shot by advanced high-definition camera [13]. The relation between the streamer discharge and flashover along the insulator with the shed was still not very clear [16].

In the paper, the streamer propagation "stability" field, path, and velocity affected by the curved profiles of the shed, were measured by ultraviolet camera, ICCD camera, and photomultipliers. The streamer energy loss at the shed was estimated by a method, which was used to interpret physical phenomenon in the test. Furthermore, the relationship between the characteristics of the streamer and the flashover was discussed in depth. It provided theoretical and experimental basis for optimization design of the curved profiles of the shed.

2. Experimental Arrangement and Measurement System

The streamer propagation "stability" field, path, and velocity affected by the curved profiles of the shed, were measured by ultraviolet camera, ICCD camera, and photomultipliers in a three-electrode arrangement. The experiment arrangement and measurement system could be found in our past literature [15], as shown in Figure 1. The streamer discharge was triggered by a square pulse voltage applied to the needle electrode. A negative DC voltage was applied to two flat electrodes to produce a uniform electric field. The photographs of the streamer discharge paths were shot by the UV imaging detector (Ofil Corporation, Tel Aviv-Yafo, Israel). The micromorphology of streamer discharge was taken by an ICCD camera (Princeton Instruments, Trenton, NJ, USA). The photons radiated from the streamer discharge could be acquired by the photomultiplier (ET Enterprises Limited, Uxbridge, UK), so three photomultipliers were adopted to monitor the development process of the streamer discharge.

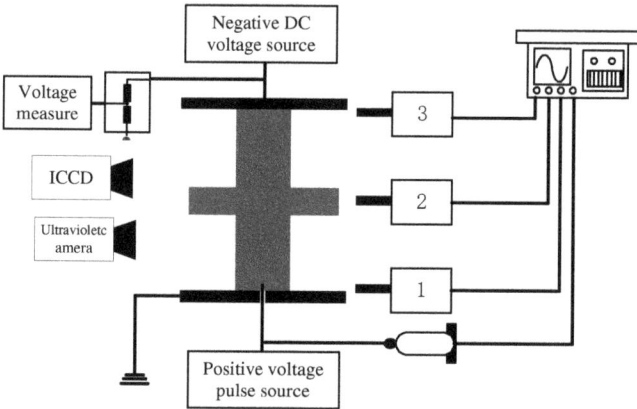

Figure 1. Experimental equipment.

The experiment was taken under standard atmospheric pressure, and the relative humidity was about 60% and temperature was about 20 °C. The insulators with the different curved profiles of the shed were made of polymer (nylon). The diameters of the sheds were 70 mm. The fillet diameters of the curved profiles of the sheds were 25 mm, 16 mm, and 5 mm respectively, as shown in Figure 2. The surface resistivity of the polymer (nylon) is 6.8×10^9 Ω under the test condition.

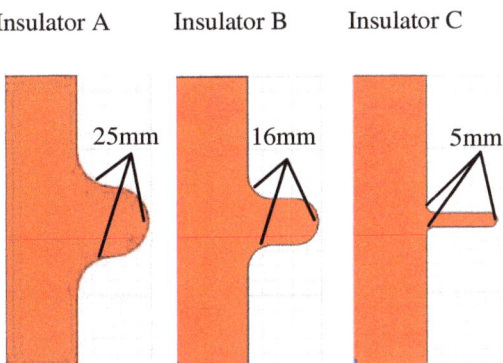

Figure 2. Insulators with the different fillet diameters of the curved profiles of the sheds.

3. Experimental Results

3.1. Streamer Propagation Fields

The streamer propagation probability under the different electric fields is an important parameter in the development of streamer, which is useful to reveal streamer characteristics. The detailed measurement method of the streamer propagation probability was based on our past literature [17]. The streamer propagation probability distribution is shown in Figure 3.

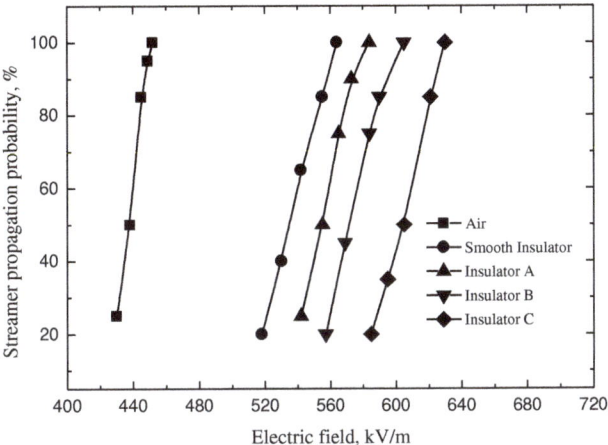

Figure 3. Streamer propagation probability.

The formula of Gaussian distribution could be used to calculate the streamer "stability" propagation fields E_{st} corresponding to streamer propagation probability of 97.5% [15]. Figure 4 shows that the relation between the pulse amplitude and the streamer "stability" propagation fields is linear. There is inverse proportional relationship between the streamer "stability" propagation fields and the pulse amplitude, as same as the result in the literature [18]. The streamer "stability" propagation fields for the insulator with a shed are larger than that for the smooth insulation surface and the air alone. Specifically, the streamer "stability" propagation fields are inversely proportional to the fillet diameters of the curved profiles of the sheds.

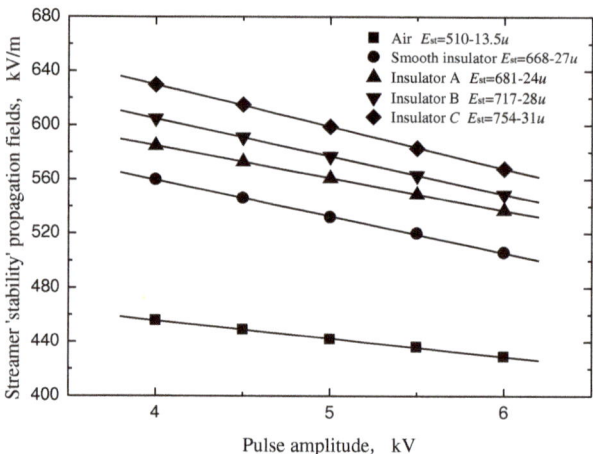

Figure 4. Stability fields for streamer propagation.

3.2. Light Emission

The photomultipliers were used to measure the streamer discharge along the insulators with the different curved profiles of the sheds. The double light peaks were detected by the photomultiplier 2 at the shed, whereas single light peak at the cathode were detected by the photomultiplier 3. The test results measured by the photomultiplier have been discussed in depth in the literature [15]. The "surface" component of streamer propagation along the insulator is blocked at the shed, but the "air" component could cross the shed and reach the cathode plate.

The photographs of the streamer discharge along the insulators were shot by the ultraviolet camera (Figures 5–7). The shooting method of the ultraviolet camera was listed in the literature [17]. Consistent with the measurement result of the photomultipliers, the "air" propagation path bypasses the shed and reaches the cathode, while the "surface" propagation path stops at the shed.

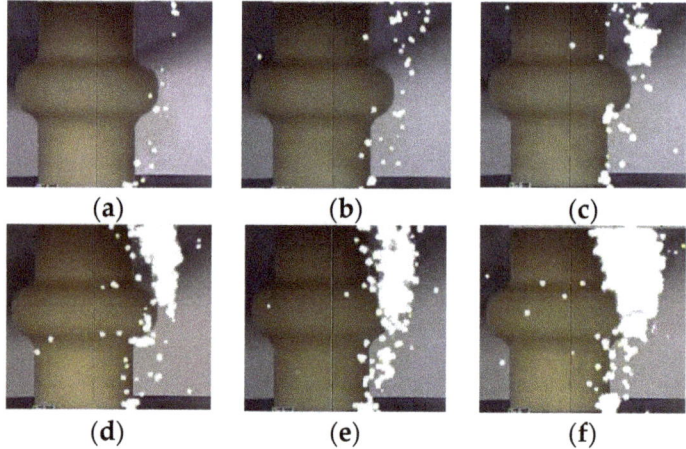

Figure 5. Streamer discharge photographs (Insulator A). (**a**) 570 kV/m, (**b**) 590 kV/m, (**c**) 630 kV/m, (**d**) 680 kV/m, (**e**) 710 kV/m, (**f**) 740 kV/m.

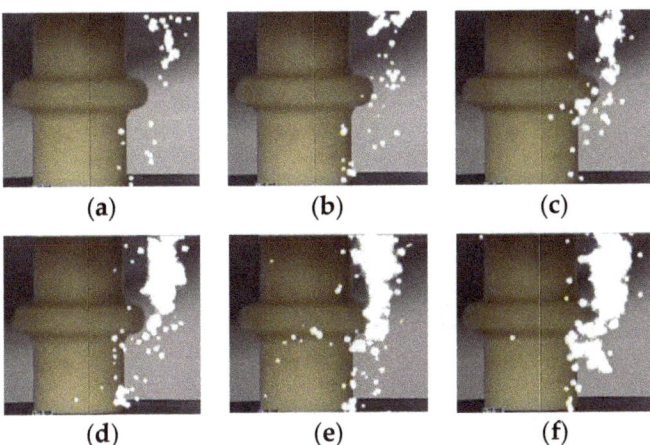

Figure 6. Streamer discharge photographs (Insulator B). (**a**) 610 kV/m, (**b**) 630 kV/m, (**c**) 650 kV/m, (**d**) 690 kV/m, (**e**) 710 kV/m, (**f**) 750 kV/m.

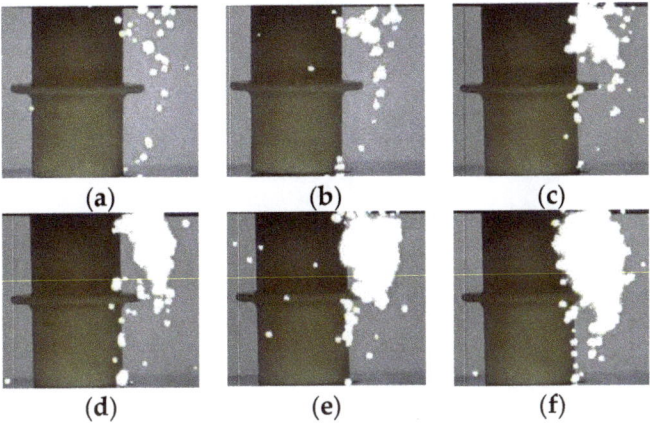

Figure 7. Streamer discharge photographs (Insulator C). (**a**) 630 kV/m, (**b**) 650 kV/m, (**c**) 690 kV/m, (**d**) 720 kV/m, (**e**) 750 kV/m, (**f**) 780 kV/m.

The ICCD camera (PI MAX3) made by Princeton Instruments was used to take photographs of the micromorphology of the streamer propagation along the insulators. Due to the weaknesses of the streamer discharge, each photograph recorded the three-times process of streamer discharges in order to make the photograph clearer. Figures 8–10 shows the streamer discharge photographs. It is found that there are also two streamer paths: the "air" propagation path and the "surface" propagation path. The "air" propagation path crosses the shed and arrives at the cathode. Different from the measurement results of the ultraviolet camera, the "surface" propagation path also crosses the shed and arrives at the cathode. However, the photomultiplier 3 detected single light peak at the cathode, which meant that the "surface" component was blocked by the shed, and only the "air" component crossed the shed. The reason to explain this contradiction is that the "surface" component is blocked by the shed. When the "air" component of streamer bypasses the shed, sometimes it might propagate along the "surface" propagation path due to the attraction of the surface charge [19–21], or sometimes it propagated along the "air" propagation path, or both cases existed in the one streamer discharge. Due to the larger fillet diameters of the curved profiles of the insulators A and B, the air component is more easily attracted by the surface charge and develops along the insulating material surface. This situation is not obvious on the surface of the insulator C due to the smallest fillet diameters of the curved profiles. In the literature [15], the "air" component of streamer might propagate along either the "surface" propagation path or the "air" propagation path. Therefore, the "surface" propagation path and the "air" propagation path behind the shed both belong to the "air" component of streamer due to the attraction of the surface charge. The greater the fillet diameters of the curved profiles, the easier it is for the "air" component of streamer to develop along the surface of the insulating material. The streamer propagation paths along the insulator with a shed are described in Figure 11.

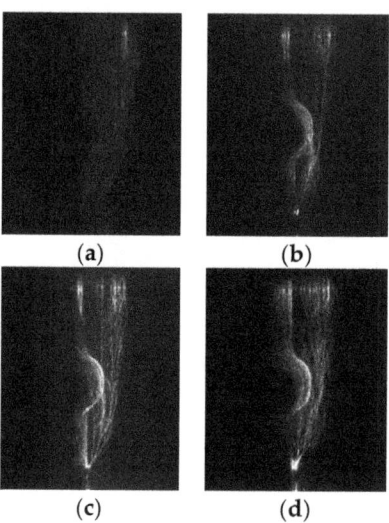

Figure 8. Streamer propagation photographs measured by ICCD camera (Insulator A). (**a**) 618 kV/m (**b**) 643 kV/m, (**c**) 690 kV/m, (**d**) 720 kV/m.

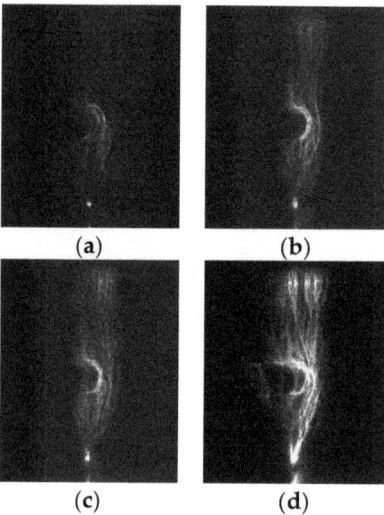

Figure 9. Streamer propagation photographs measured by ICCD camera (Insulator B). (**a**) 620 kV/m, (**b**) 660 kV/m, (**c**) 700 kV/m, (**d**) 740 kV/m.

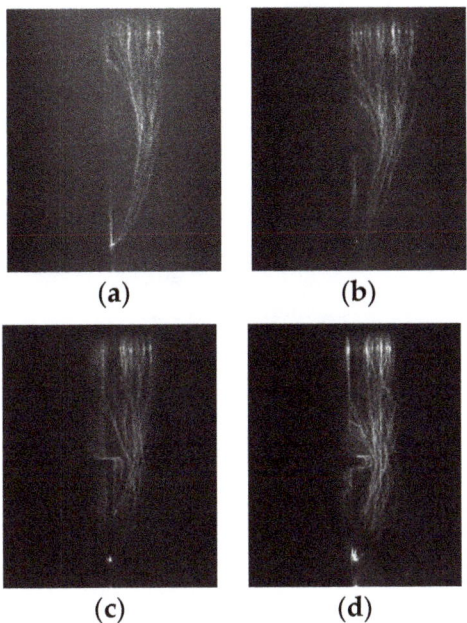

Figure 10. Streamer propagation photographs measured by ICCD camera (Insulator C). (**a**) 640 kV/m (**b**) 680 kV/m, (**c**) 720 kV/m (**d**) 760 kV/m.

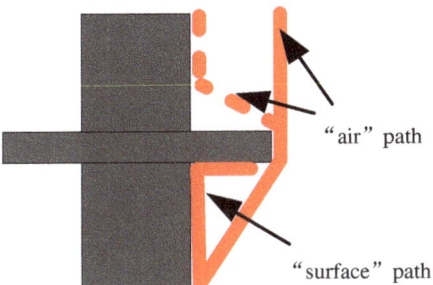

Figure 11. Streamer propagation paths along the insulators.

3.3. Streamer Propagation Velocity

The streamer "stability" propagation velocities V_{st} were defined as the velocities of streamer propagation at the stability fields [22]. The velocities of streamers propagating along the insulators with a shed decreases linearly but slowly with the pulse amplitude in Figure 12. The velocities of the "air" component along the insulators with a shed are higher than that along the smooth insulator in Figure 13. The reason could be found in the literature [15]. It is obvious that the "air" component velocities are proportional to the fillet diameters of the curved profiles of the sheds.

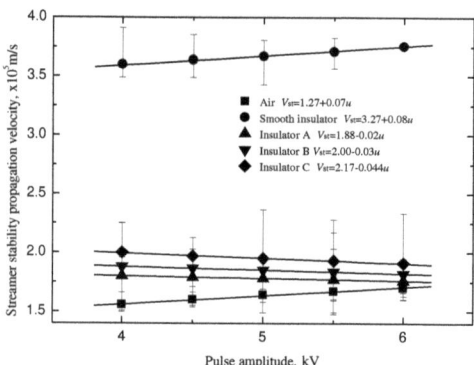

Figure 12. Streamer stability propagation velocity.

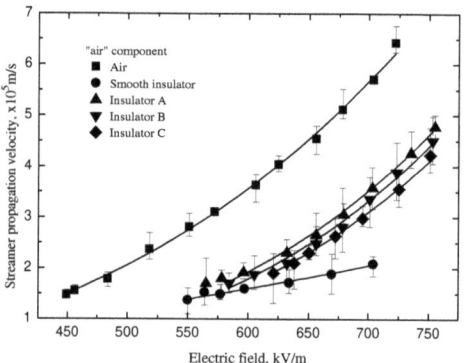

Figure 13. Streamer propagation velocity ("air" component).

3.4. Evolution of Streamer to Flashover

The characteristics of flashover along the insulator were investigated through improving the applied voltage between the plates. The flashover "stability" propagation fields E_{50} corresponding to flashover propagation probability of 50% could be acquired by the method of acquiring the streamer "stability" propagation fields E_{st}. Table 1 shows that the flashover "stability" propagation fields E_{50} increase with streamer "stability" propagation fields E_{50}. The flashover "stability" propagation fields are also inversely proportional to the fillet diameters of the curved profiles of the sheds. Hence, the applied electric field has the same effect on the streamer propagation and the subsequent flashover propagation.

Table 1. The comparison of streamer "stability" propagation fields Est and flashover "stability" propagation fields E50.

Shed Configuration	E_{st} (kV/m)	E_{50} (kV/m)
Insulator A	585	824
Insulator B	605	842
Insulator C	630	863

The photographs of the flashover path were acquired by a high-speed camera in Figure 14. It shows that the flashover propagates along the insulator with only one propagation path. In front of the shed, the flashover propagates along either the "surface" or "air" propagation path of the previous streamer. The flashover propagated along the "air" propagation path of the previous streamer behind the shed due to the interruption of the "surface" propagation path at the shed. Hence, the streamer channel has a great influence on the subsequent flashover. The reasonably optimized curved profiles of the sheds could improve the electric fields for the streamer propagation, thereby, the electric fields for the flashover propagation are also improved. The perspective of restraining the streamer discharge to prevent the flashover discharge would be a new idea to design the curved profiles of the shed.

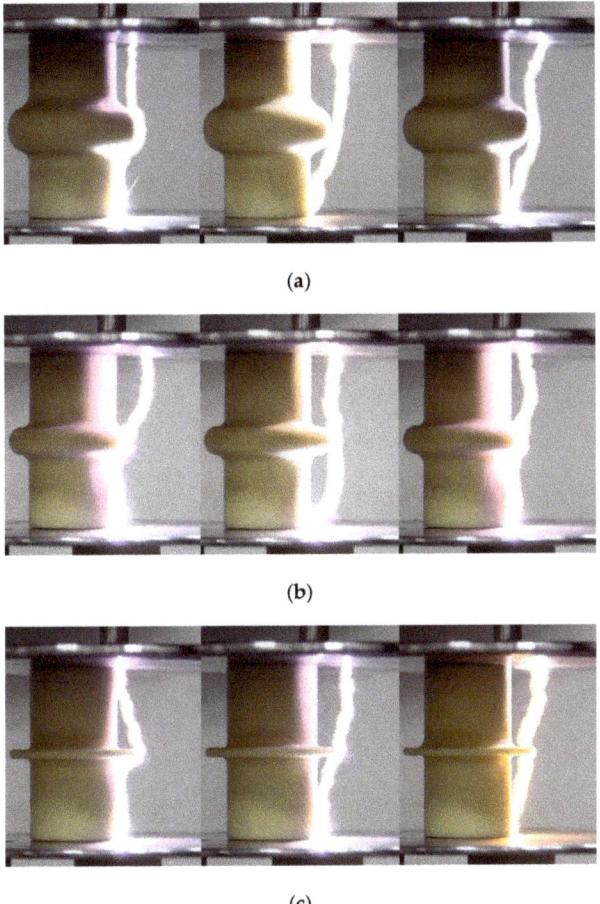

Figure 14. Photographs of flashover along the insulator with the shed. (**a**) Insulator A. (**b**) Insulator B. (**c**) Insulator C.

4. Discussion

4.1. Tangential Electric Field along Streamer Propagation Path

Ansoft Maxwell was adopted to calculate the electric fields along the insulator. Figure 15 shows the tangential electric field along the whole insulator surface. The tangential electric field is small and even negative in the regions of the curved profiles of the shed. The reason why the "surface" component of streamer would not cross this small electric field region is that the energy from this small electric field could not maintain the streamer propagation. Hence, the "surface" propagation path fails to reach the cathode.

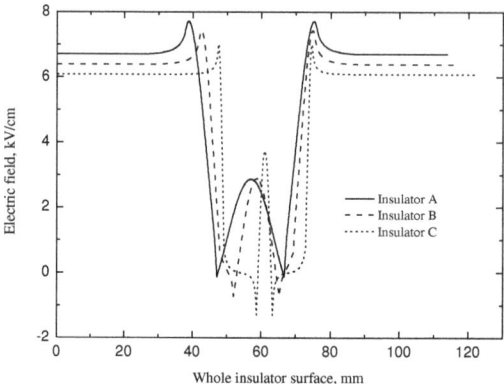

Figure 15. Tangential electric field along the whole insulator surface.

However, the "air" component could cross the shed and reach the cathode plate. Hence, the streamer propagation characteristic parameters, such as the streamer "stability" propagation fields and the streamer "stability" propagation velocities, are all determined by the "air" component of streamer. The distribution of the tangential electric field along the "air" path is important to explain the streamer propagation characteristic parameters. Figure 16 shows that the tangential electric field along the "air" path before and behind the shed is proportional to the fillet diameters of the curved profile of the shed; whereas, the tangential electric field at the shed is inversely proportional to the fillet diameters.

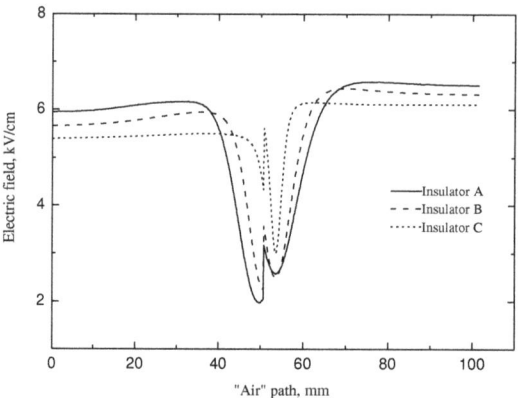

Figure 16. Tangential electric field along the "air" path.

The streamer has not yet obtained much energy from the electric field in front of the shed, so the applied electric field has the greatest impact on it. When the streamer propagates along the insulator with the largest fillet diameters of the curved profile of the shed, the tangential electric field along the "air" path is the largest, so it is easy for the streamer to develop and to cross the shed where the tangential electric field is greatly reduced. Therefore, lower electric field is required for the streamer to propagate along the insulator with the larger fillet diameters of the curved profile of the shed. Under the same electric field, the "air" component velocities along the insulator are proportional to the fillet diameters of the curved profiles of the sheds.

4.2. Streamer Propagation Energy Loss at the Shed

The streamer discharge would lose much energy at the shed due to the low tangential electric field and the charges in the streamer attaching to the shed [23–26]. The method in the literature [15] was used to estimate the energy loss at the shed. The calculated energy loss at the shed (L_{shc}) differed by an error (Q_{ste}) from the real energy loss (L_{shc}) as shown in Equation (1).

$$L_{sh} = L_{shfc} + L_{shsc} - (Q_{stfe} - Q_{stse}) = L_{shc} - Q_{ste} \tag{1}$$

In Table 2, E_{st1} is the stability fields for streamer propagation along insulators with a shed. v_1/v_2 is the "air" component velocity before/after the shed at E_{st1}. Q_{Ef1}/Q_{Es1} is the energy obtained from ambient electric field before/after the shed. E_{f2} is the ambient field corresponding to the same value of v_1 when streamer propagates along smooth insulators during the first half part of air gap. E_{s2} is the ambient field corresponding to the same value of v_2 when streamer propagates in air alone during the second half part of air gap. Q_{Ef2}/Q_{Es2} is the energy obtained from ambient electric field E_{f2}/E_{s2} in its case.

Table 2. Energy loss at the shed.

Shed Configuration	Insulator A	Insulator B	Insulator C
E_{st1} (kV/m)	585	605	630
v_1 (10^5 m/s)	1.4	1.45	1.52
v_2 (10^5 m/s)	2.48	2.68	2.84
Q_{Ef1} (10^4 J/C)	2.93	3.03	3.15
Q_{Es1} (10^4 J/C)	2.93	3.03	3.15
E_{f2} (kV/m)	580	595	610
E_{s2} (kV/m)	527	535	550
Q_{Ef2} (10^4 J/C)	2.9	2.975	3.05
Q_{Es2} (10^4 J/C)	2.635	2.675	2.75
L_{shc} (10^4 J/C)	0.315	0.4	0.5

Table 2 showed the energy loss at the shed (L_{shc}) is inversely proportional to the fillet diameters of the curved profile of the shed. From the perspective of the energy loss, the streamer propagation energy loss at the shed is inversely proportional to the fillet diameters of the curved profile of the shed. Therefore, the streamer "stability" propagation fields are inversely proportional to the fillet diameters, and the streamer propagation velocities are proportional to the fillet diameters under the same electric field.

4.3. Influence of the Material Properties

In the literature [12,13], the authors found that the "air" component of streamer discharge could bypass the shed and reach the cathode, the "surface" component of streamer discharge could also bypass the shed with the large fillet diameters of the curved profile and reach the cathode. However, we used more advanced experimental equipment and experimental testing methods in this paper, such as the ICCD camera and the UV imaging detector was adopted to shoot the streamer discharge photographs. The different test conclusion was obtained that the "surface" component of streamer discharge could not bypass the shed, even the fillet diameters of the curved profile of the shed is much large. The "surface" propagation path behind the shed does not belong to the "surface" component of streamer discharge, but belongs to the "air" component of streamer discharge.

Our conclusion is closer to the actual situation, in addition to our more advanced measuring equipment, it is more important that we consider the influence of material properties on the streamer discharge. There are much microporous defects (physical defect and chemical defect) on the surface of the polymer in Figure 17 measured by the scanning electron microscope. The surface adsorbed charges on the polymer increase with the increase of the microporous defects [27]. Therefore, the surface of the polymer would adsorb a lot of negative charge with the negative DC voltage applied on the gap.

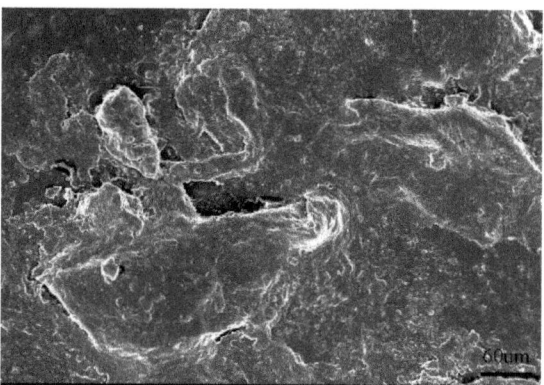

Figure 17. Scanning electron microcopy images.

The method of the thermally stimulated current was used to measure the trap charges (nC) on the surface of the polymer. The trap charges on the polymer could be calculated by the result of the thermally stimulated current in Figure 18. The trap charge on the polymer used in the test is 1879 nC, which is greater than other materials. The more trap charges means that the more negative charge is accumulated or deposited on the polymer surface. The result of the thermally stimulated current proves the result of the scanning electron microscope.

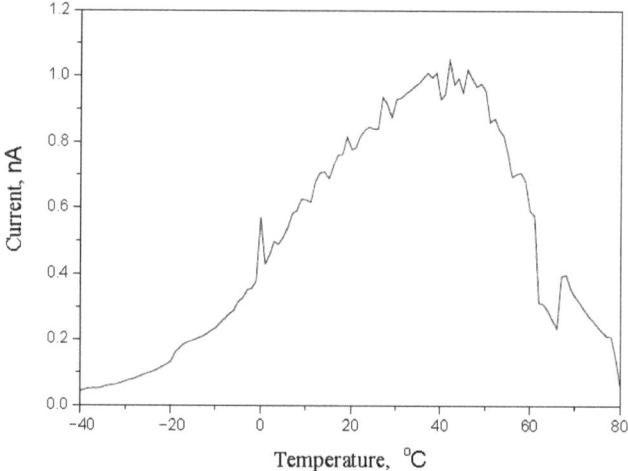

Figure 18. The result of the thermally stimulated current.

As shown in Figure 19, the vector of electric fields is parallel to the surface of the polymer with no surface charge. However, the vector of electric fields points to the surface of the polymer with negative surface charge. Therefore, the air component of streamer discharge behind the shed is more easily attracted by the polymer surface and develops along the "surface" propagation path. It could explain reasonably that why the "surface" propagation path and the "air" propagation path behind the shed both belong to the "air" component of streamer in the Section 3.2.

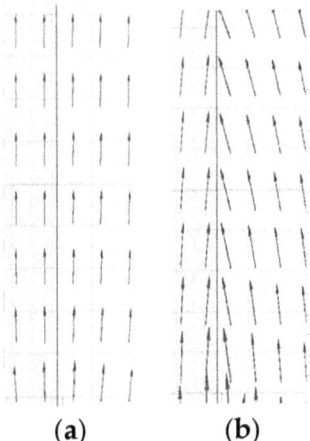

Figure 19. Vector plots of the electric field distribution along the surface of the polymer. (**a**) no surface charge, (**b**) $-10\ \mu C/m^2$ surface charge. The darker part is polymer material, the white part is air.

This contradiction of the measurement result from the ICCD camera and the UV imaging detector has given us great enlightenment. It is easy to produce confusing results or conclusions for a single test equipment or measurement method to study the physical process of the temporal and spatial evolution of streamer discharge. Combining a variety of test equipment or measurement methods, and then comprehensively analyzing the test results, can accurately reveal the physical development process of streamer discharge.

The test results tell us that the smaller fillet diameters of the curved profile of the shed would be adopted to obtain the larger flashover voltage in electrical power system. However, the outdoor insulators of the transmission lines are prone to accumulating pollution on the surface in the atmospheric environment, and the larger fillet diameters of the curved profile of the shed would remove pollutants easily through rain and wind. Hence, the fillet diameters of the curved profile of the shed not only affect the flashover voltage of the insulator, but also the removal of pollutants through rain and wind. In engineering practice, the fillet diameters of the curved profile of the shed should be chosen to consider the impact on both sides, taking into account both flashover voltage and cleaning capacity. Sometimes in order to obtain high cleaning ability, the flashover voltage can be appropriately reduced to obtain better environmental adaptability. These results provide a theoretical basis for promoting the shape of the shed.

5. Conclusions

From typical photomultiplier signals and the streamer discharge photographs, the "surface" component of streamer discharge was blocked at the shed. Only the "air" component crossed the shed, and the "air" component behind the shed might propagate along either the "surface" propagation path or the "air" propagation path.

The streamer "stability" propagation fields were inversely proportional to the fillet diameters of the curved profiles of the sheds. The streamer propagation velocities were proportional to the fillet diameters.

The applied electric field had the same effect on the streamer propagation and the subsequent flashover propagation. Furthermore, the streamer channel had a great influence on the development path of the subsequent flashover. The perspective of restraining the streamer discharge to prevent the flashover discharge would be a new idea to design the curved profiles of the shed.

The tangential electric field along the streamer propagation path and the streamer propagation energy loss at the shed could be used to explain the characteristics of the streamer discharge. From the perspective of the tangential electric field, when the streamer propagates along the insulator with the largest fillet diameters of the curved profile of the shed, the tangential electric field along the "air" propagation path is the largest, so it is easy for the streamer to develop and to cross the shed where the tangential electric field is greatly reduced. From the perspective of the streamer propagation energy loss, the streamer propagation energy loss at the shed is inversely proportional to the fillet diameters of the curved profile of the shed. Therefore, lower electric field is required for the streamer to propagate along the insulator with the larger fillet diameters of the curved profile of the shed.

The surface of the polymer would adsorb a lot of negative charge with the negative DC voltage applied on the gap. The vector of electric fields points to the surface of the polymer with negative surface charge. Hence, the air component of streamer discharge behind the shed is more easily attracted by the polymer surface and develops along the "surface" propagation path. It could explain reasonably that why the "surface" propagation path and the "air" propagation path behind the shed both belong to the "air" component of the streamer.

Author Contributions: Conceptualization, L.W.; methodology, L.W.; formal analysis, X.B.; software, B.C.; data curation, X.M.; writing—original draft preparation, X.M.; writing—review and editing, B.C.; supervision, X.B.; project administration, H.M.; funding acquisition, L.W. All authors have read and agreed to the published version of the manuscript.

Funding: This study was supported by the State Key Laboratory of Alternate Electrical Power System with Renewable Energy Sources (Grant No. LAPS20004), the GuangDong Basic and Applied Basic Research Foundation (2021A1515110568).

Institutional Review Board Statement: Not applicable.

Informed Consent Statement: Not applicable.

Data Availability Statement: Not applicable.

Conflicts of Interest: The authors declare no conflict of interest. The funders had no role in the design of the study; in the collection, analyses, or interpretation of the data; in the writing of the manuscript; nor in the decision to publish the results.

References

1. Jin, L.; Ai, J.; Han, S.; Zhou, G. Probability Calculation of Pollution Flashover on Insulators and Analysis of Environmental Factors. *IEEE Trans. Power Deliv.* **2021**, *36*, 3714–3723. [CrossRef]
2. Ghayedi, M.; Shariatinasab, R.; Mirzaie, M. AC flashover dynamic model suggestion and insulation level selection under fan-shaped pollution. *Int. J. Electr. Power Energy Syst.* **2022**, *134*, 107438. [CrossRef]
3. Salem, A.A.; Lau, K.Y.; Rahiman, W.; Al-Gailani, S.A.; Abdul-Malek, Z.; Rahman, R.A.; Al-Ameri, S.M.; Sheikh, U.U. Pollution Flashover Characteristics of Coated Insulators under Different Profiles of Coating Damage. *Coatings* **2021**, *11*, 1194. [CrossRef]
4. Ghayedi, M.; Shariatinasab, R.; Mirzaie, M. AC flashover dynamic theoretical and experimental model under fan-shaped and longitudinal pollution on silicone rubber insulator. *IET Sci. Meas. Technol.* **2021**, *15*, 719–729. [CrossRef]
5. Chrzan, K.L.; Brzezinski, H.M. Anomalous flashovers of silicone rubber insulators under the artificial rain test. *Arch. Electr. Eng.* **2021**, *70*, 835–844.
6. He, S.; Wang, J.; Hu, J.; Zhou, H.; Nguyen, H.; Luo, C.; Lin, J. Silicone rubber composites incorporating graphitic carbon nitride and modified by vinyl tri-methoxysilane. *Polym. Test.* **2019**, *79*. [CrossRef]
7. Liao, Y.; Weng, Y.; Wang, J.; Zhou, H.; Lin, J.; He, S. Silicone Rubber Composites with High Breakdown Strength and Low Dielectric Loss Based on Polydopamine Coated Mica. *Polymers* **2019**, *11*, 2030. [CrossRef]
8. Perez, A.; Beroual, A.; Jacquier, F.; Girodet, A. Measurement of Streamer Propagation Velocity over Solid Insulator Surface in a C4F7N/CO2/O-2 Mixture under Lightning Impulse Voltages. *IEEE Trans. Dielectr. Electr. Insul.* **2021**, *28*, 485–491. [CrossRef]
9. Li, X.; Sun, A.; Zhang, G.; Teunissen, J. A computational study of positive streamers interacting with dielectrics. *Plasma Sources Sci. Technol.* **2020**, *29*, 065004. [CrossRef]
10. Li, X.; Sun, A.; Zhang, G.; Teunissen, J. A computational study of negative surface discharges: Characteristics of surface streamers and surface charges. *IEEE Trans. Dielectr. Electr. Insul.* **2020**, *27*, 1178–1186. [CrossRef]
11. Shanmugam, G.; Karakkad, S. Influence of the insulator geometry on the streamer propagation characteristics in polymeric insulators under positive polarity lightning impulse voltages. *IET Sci. Meas. Technol.* **2018**, *12*, 1082–1088. [CrossRef]

2. Pritchard, L.; Allen, N. Streamer propagation along profiled insulator surfaces. *IEEE Trans. Dielectr. Electr. Insul.* **2002**, *9*, 371–380. [CrossRef]
3. Allen, N.; Mikropoulos, P. Surface profile effect on streamer propagation and breakdown in air. *IEEE Trans. Dielectr. Electr. Insul.* **2001**, *8*, 812–817. [CrossRef]
4. Allen, N.L.; Hashem, A.; Rodrigo, H.; Tan, B.H. Streamer development on silicone-rubber insulator surfaces. *IEE Proc. Sci. Meas. Technol.* **2004**, *151*, 31–38. [CrossRef]
5. Meng, X.; Mei, H.; Wang, L.; Guan, Z.; Zhou, J. Characteristics of streamer propagation along insulation surface: Influence of shed configuration. *IEEE Trans. Dielectr. Electr. Insul.* **2016**, *23*, 2145–2155. [CrossRef]
6. Geng, J.; Chen, Y.; Lv, F.; Wang, P.; Ding, Y. Photo electric properties during streamer-to-leader transition in a long positive sphere–plane gap. *Phys. Plasmas* **2020**, *27*, 083509. [CrossRef]
7. Meng, X.; Mei, H.; Chen, C.; Wang, L.; Guan, Z.; Zhou, J. Characteristics of streamer propagation along the insulation surface: Influence of dielectric material. *IEEE Trans. Dielectr. Electr. Insul.* **2015**, *22*, 1193–1203. [CrossRef]
8. Meng, X.; Wang, L.; Mei, H.; Zhang, C. Influence of Silicone Rubber Coating on the Characteristics of Surface Streamer Discharge. *Polymers* **2021**, *13*, 3784. [CrossRef]
9. Tumiran, M.; Maeyama, H.; Kobayashi, S.; Saito, Y. Flashover from surface charge distribution on alumina insulators in vacuum. *IEEE Trans. Dielectr. Electr. Insul.* **1997**, *4*, 400–406. [CrossRef]
20. Yuan, M.; Zou, L.; Li, Z.; Pang, L.; Zhao, T.; Zhang, L.; Zhou, J.; Xiao, P.; Akram, S.; Wang, Z.; et al. A review on factors that affect surface charge accumulation and charge-induced surface flashover. *Nanotechnology* **2021**, *32*, 262001. [CrossRef]
21. Jun, X.; Chalmers, I.D. The influences of surface charge upon flash-over of particle-contaminated insulators in SF6 under impulsevoltage conditions. *J. Phys. D: Appl. Phys.* **1997**, *30*, 1055–1063. [CrossRef]
22. Allen, N.L.; Mikropoulos, P.N. Streamer propagation along insulating surfaces. *IEEE Trans. Dielectr. Electr. Insul.* **1999**, *6*, 357–362. [CrossRef]
23. Allen, N.; Tan, B.; Rodrigo, H. Progression of Positive Corona on Cylindrical Insulating Surfaces Part II: Effects of Profile on Corona. *IEEE Trans. Dielectr. Electr. Insul.* **2008**, *15*, 390–398. [CrossRef]
24. Meyer, H.K.; Mauseth, F.; Marskar, R.; Pedersen, A.; Blaszczyk, A. Streamer and surface charge dynamics in nonuniform air gaps with a dielectric barrier. *IEEE Trans. Dielectr. Electr. Insul.* **2019**, *26*, 1163–1171. [CrossRef]
25. Shao, T.; Kong, F.; Lin, H.; Ma, Y.; Xie, Q.; Zhang, C. Correlation between surface charge and DC surface flashover of plasma treated epoxy resin. *IEEE Trans. Dielectr. Electr. Insul.* **2018**, *25*, 1267–1274. [CrossRef]
26. Liu, Y.; Wu, G.; Gao, G.; Xue, J.; Kang, Y.; Shi, C. Surface charge accumulation behavior and its influence on surface flashover performance of Al2O3-filled epoxy resin insulators under DC voltages. *Plasma Sci. Technol.* **2019**, *21*, 055501. [CrossRef]
27. Yamano, Y.; Kasuga, K.; Kobayashi, S.; Saito, Y. Surface flashover and charging characteristics on various kinds of alumina under nonuniform electric field in vacuum. In Proceedings of the 20th International Symposium on Discharges and Electrical Insulation in Vacuum, Tours, France, 1–5 July 2002.

Article

The Dynamic Behaviour of Multi-Phase Flow on a Polymeric Surface with Various Hydrophobicity and Electric Field Strength

Qi Li [1,*], Rui Liu [1], Li Li [2], Xiaofan Song [3], Yifan Wang [2] and Xingliang Jiang [1]

1. State Key Laboratory of Power Transmission Equipment and Systerm Security and New Technology, College of Electrical Engineering, Chongqing University, Chongqing 400044, China; 201911021010@cqu.edu.cn (R.L.); profjiang2018@gmail.com (X.J.)
2. Electric Power Research Institute of Guangdong Power Grid Co., Ltd., Guangzhou 510080, China; liligz@163.com (L.L.); haomafan2006@126.com (Y.W.)
3. State Grid Henan Economic Research Institute, Zhengzhou 450032, China; sxf810118@sina.com
* Correspondence: qi.li@cqu.edu.cn; Tel.: +86-185-2345-0925

Abstract: The dynamic behaviour of rain droplets on the insulator surface is a key measure to its reliability and performance. This is due to the fact that the presence and motion of rain droplets cause intensive discharge activities, such as corona and low current arcing, which accelerate the ageing process and flashovers. This article aims to investigate and characterize the movement of a rain droplet placed on an inclined insulator surface subject to an intensive electric field. The rain droplets' movement on hydrophobic surfaces in the absence of an electric field is investigated. A high speed camera is used to capture the footage and finite element method (FEM) is used to simulate the multi-physics phenomenon on two polymeric surfaces, namely, silicon rubber (SiR) and PTFE (polytetrafluoroethylene). A 'creepage' motion was observed. The inception of motion and the movement speed are analysed in correlation with various surface conditions. Models are established to estimate the moisture and potential discharge characteristics on the inclined polymeric surfaces. They are further utilized to analyse the actual insulators subject to wet conditions.

Keywords: droplet vibration; high voltage insulator; polymeric surface; corona discharge; arcing; creepage distance

1. Introduction

The distortion and movement of rain drops in an electric field has been the subject of a number of experimental research studies. High voltage AC outdoor insulators can be distinct by their bulk insulation material as ceramic insulators (glass or porcelain) or composite insulators (polymeric or non-ceramic insulators). Composite insulators consist of pultruded glass cores and polymeric shed sheaths are fast becoming the main choice of materials because of their improved wetting behaviour and greater degree of hydrophobicity [1,2]. The discharge activity on the surface of polymeric insulator surfaces is one of the ageing mechanisms leading to insulator failure. Such activity between the rain droplets causes the production of radicals that chemically react with the surface of the insulator, thereby altering the properties of the insulator material. The practical aspect of the subject seems to be an important factor in justifying the extensive work carried out on electrical discharges from rain drops on polymeric surfaces.

The surface of operational polymeric insulator experiences hydrophobicity weakness and even loss due to the surface discharge of the rain droplet. Due to the reduced hydrophobicity, the rain droplets coalesce and form 'rivulets' of conductivity, which bridge the insulation path on the insulator surface leading to the flashover between HV and ground. Most experimentation revolves around discharges between rain drops on the top of flat horizontal surfaces. As the weather-proof sheds of insulators are neither horizontal nor planar, it is worth studying the behaviour of rain droplets in more complex situations.

This article aims to investigate the behaviour of a rain droplet on an inclined insulator surface under the influence of intense AC electric field. This unique experiment and its results are necessary to further understand the behaviour of moisture [3] and ageing process for polymeric surfaces, especially their application within power systems for insulators and bushings.

2. Literature Review

The study of thunderstorms inspired the early work on the impact of electrical fields on rain drops and much of the activity was concerned with the distortion and break-up of individual drops in high fields. The discharges on the surface of polymeric insulators are important in the ageing mechanisms of insulators. Since the introduction of hydrophobic polymeric insulators, a considerable amount of work has been conducted addressing the electrical discharges from rain droplets on polymeric surfaces. Rowland [4] highlighted the critical role played by the local ageing of the polymers' surface in the formation of 'wet fingers' of conductivity. Karady et al., for example, describes the formation of the filaments between rain droplets on aged commercial silicon rubber insulators and a mechanism by which this led to flashover of single sheds [5–7].

The vibration and distortion of rain drops in the presence of an alternating field (ac) is well established [8–11]. A rain droplet in a strong electric field deforms as a result of the interaction of electrostatic force and the surface tension of the rain droplet [12]. The movement of rain droplet in alternating field (ac) is characterized by periodic vibration or fluctuation. Higashiyama and Yamada investigated in detail the behaviour of a rain droplet placed on the surface of the hydrophobic polymeric sheet in presence of an ac field. They demonstrated the change in droplet shape during vibration, showing that the droplet is deformed and synchronized with the ac field [8]. Their study revealed that the frequency and the volume of rain droplet greatly affect the amplitude of vibration.

An electric field alone can force raindrops up an incline in order to move away from a high-stress region pointing out the importance of gravity on inclined surfaces [13]. Phillips et al. and Cheng et al. have shown elongation of single raindrops [14,15]. They observed that raindrops do not change their shape gradually, but in a series of steps, and do not necessarily recover their shape after the field is removed. Krivda and Birtwhistle [13] explained that natural vibrations of a rain drop change its shape during the ac cycle and hence can reduce the insulation path. This increases the risk of flashover. It was shown that when several raindrops coalesce they can bridge a significant distance of insulation and hence discharges appear between them. The corona discharge takes place at the triple junction area of the rain–air–solid interface as the electric field at this point is intensified due to the difference in permittivity of silicon rubber, air and the rain droplet [16].

The corona discharge destroys the hydrophobicity of the insulator and thus a long chain of elongated drops is formed. Under the influence of electric field, this can lead to the flow of currents over the surface of the insulation [5].

The discharge activity between rain droplets on the insulator surface plays an important role in its ageing mechanisms. It is also critical in the processes leading up to the flashover of these insulators in highly polluted or marine environments. However, limited research has been conducted on the impact of surface shapes and dimensions on surface hydrophobicity and water sliding behaviour.

3. Methodology

To address this gap in the literature, this article aimsu to design a specific experimental set up to investigate the factors affecting the rain droplets' dynamic movement on an inclined insulation surface. To simplify, a flat inclined surface was used with the capability to vary its inclined angle. The first step was to investigate the behaviour of the rain droplets on an inclined plane in the absence of electric field [17]. This will facilitate the design process while identifying the key variables that need to be controlled. In the second step, the experiments were conducted in a high voltage environment. The aims and objectives of the study can change depending upon need and setup availability/feasibility.

3.1. In the Absence of an Electric Field

Silicon rubbers with different additives are materials of high interest for high voltage composite insulators. A number of other materials were also chosen for investigation in order to examine the role of surface energy or hydrophobicity. These materials included glass, PTFE and a commercial silicon rubber-based material (SiR). A drop of de-ionised water was placed on the sample surface by a pipette and the angle of inclination gradually increased until the droplet started to move. The experiment was repeated with three different sizes of droplet (50, 90 and 140 µL) and material types (SiR, PTFE and Glass). The diagram of the test setup is shown in Figure 1.

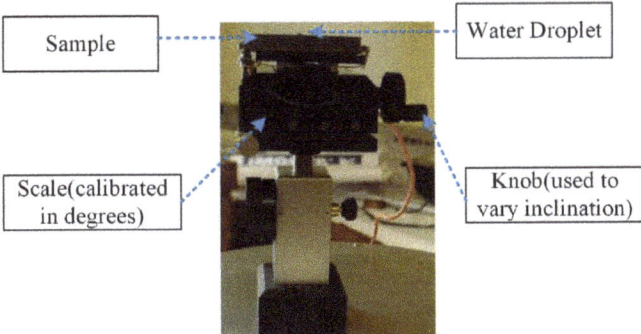

Figure 1. Test up to measure inclination angle.

The hydrophobicity of each material was determined by measuring the contact angle with a 100 µL drop of de-ionised water. The geometry of the point of contact between the liquid, solid and air is the key to controlling the process, so this is a particularly useful measurement in that context. Figure 2 shows the diagram of the syringe unit of the optical contact angle measuring device.

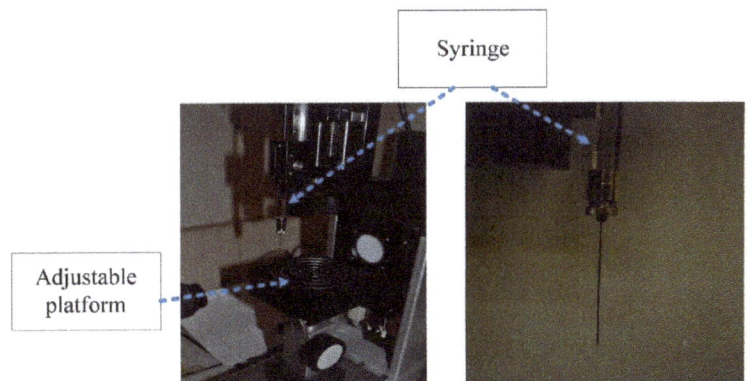

Figure 2. Syringe unit of the optical contact angle measuring device (Data Physics OCA Series).

3.2. In the Presence of an Electric Field

The basic understanding developed through investigating rain droplet behaviour on different materials (without E-field) helped to design the experiment. The experimental setup consists of two large cylindrical metal electrodes with rounded edges. The electrode surfaces were smooth with little or no asperities in order to obtain uniform electric field.

A silicon rubber sample with an inclination of 30 was used as a test material. The metal electrodes were separated by about 50 mm. A 50 µL drop of de-ionised water was placed in the central region on top of the insulator sample. The droplet size (volume) was controlled by the electronic syringe unit of the optical contact angle measuring device. A high speed camera (imaging rate 3902 fps) was used to observe and record the movement of the rain droplet under the influence of high AC electric field. A diagram of the experimental setup is shown in Figure 3.

Figure 3. Diagrammatic overview of the experimental setup.

The electrical diagram of the same experiment is shown in Figure 4. The components include a high voltage transformer, the rated voltage of which is 80 kV AC (frequency 50 Hz), and a 125 kΩ current limiting resistor. The circuit also contains a surge protection device and a voltage divider. A digital storage oscilloscope was used to monitor the voltage waveforms.

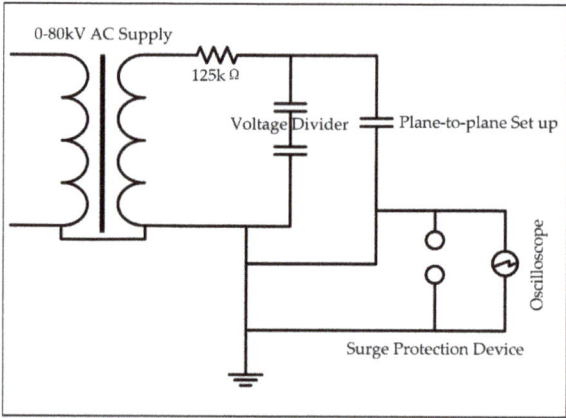

Figure 4. Diagram of the test circuit.

A 'break down test' was conducted in order to find the maximum voltage that can be safely applied to the test setup before the dielectric (air between electrodes) breaks down. A fairly large rain droplet with the volume 100 µL was applied to the sample surface and the voltage was gradually increased. The breakdown occurred at 55 kV and hence it was decided to perform all experimentation below the 35 kV voltage level.

The silicon rubber used in the experiment was designed using basic geometry. The thickness of the material was 7 mm. A 30° inclined silicon rubber sample was made using a sharp cutting tool as shown in Figure 5. The dimensions of the sample are shown in Table 1.

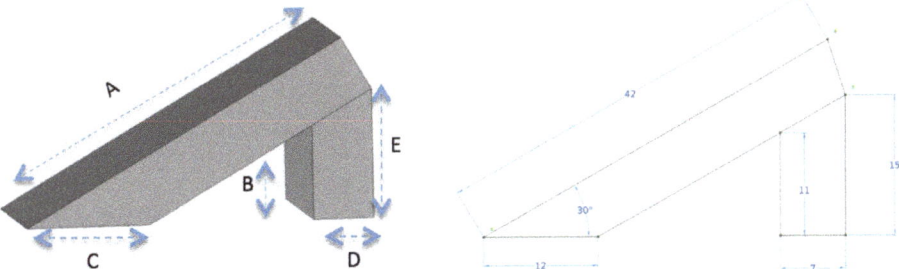

Figure 5. Silicon rubber sample with 30 degrees inclination.

Table 1. Sample dimensions.

A	B	C	D	E
42 mm	11 mm	12 mm	7 mm	15 mm

3.3. Numerical Simulation Using the Finite Element Method (FEM)

When the rain droplet is under the electric field, the electric field force drives the movement and deformation of the rain droplet, which leads to the distortion of the field strength and changes the size of the electric field force [18,19]. The interaction between the two is a coupling problem of electricity and fluid dynamics, which is defined as electrohydrodynamics. After a sensitivity study of various simplified versions of the simulation in comparison with the measurements, the FEM was identified as the most effective tool to understand the physical mechanism behind the droplets' dynamics.

4. Results and Discussion

4.1. Contact Angles and Inclination Angle of Samples

The experimentation conducted in absence of electric field was quite useful in developing a basic understanding about key concepts, such as surface hydrophobicity and water contact angles. The contact angles between rain drops on different surfaces are shown in Table 2; each value given is an average of three measurements.

Table 2. Contact angles between water droplets on different materials.

Materials	SiR	PTFE	Glass
Contact Angles (°)	108	100	38

Table 3 shows the inclination angle at run-off for different materials and droplet sizes (volume). In the experiments carried out, a fresh sample of material was used for all readings. The line of sight was kept perpendicular to the scale while taking the readings in order to avoid a parallax error.

The results in Table 2 show that the highest contact angles were seen for silicon rubber sample. This can be explained by the excellent hydrophobicity of silicon rubber. Water molecules were seen to form discrete droplets on the surface of the insulation when the material is highly hydrophobic.

Table 3. Inclination angle at run-off.

Sample Type	Volume 50 μL	90 μL	140 μL
SiR	52°	40°	32°
PTFE	49°	37°	28°
Glass	33°	21°	16°

Table 3 shows that, for a given volume, a greater degree of inclination was required to run off a water droplet from the SiR surface compared to PTFE and glass. The surface properties of these materials, such as surface roughness and hydrophobicity, are important in water droplet movement. The results were important in the design of the experiment conducted in presence of high electric field (AC). It provided the range of the droplet sizes and materials that can be used to meet the research specifications.

4.2. Creepage Phenomenon on the Insulator Surface

The main highlight of the experiment was the identification of a 'Creepage Phenomenon', which has not been observed earlier in the literature. The vibration and distortion of water drops on horizontal surfaces under the influence of an alternating field (AC) is well established. It was observed that, if the water droplet placed on an inclined insulator surface is subjected to an ac field, it tends to creep along the surface. Additionally, the movement is affected by numbers of factors, including the strength of field, volume of water droplet, type of the insulating material and the roughness of insulator surface. This dynamic behaviour of water droplet can play an important role in the ageing mechanisms of composite systems.

The voltage was uniformly increased from zero kV in steps of 5 kV. The 'Creepage Phenomenon' was seen at 30 kV (AC) using the experimental setup shown in Figure 3. The high-speed camera captured the movement and distortion of the water droplet placed on the inclined surface, which is shown in Figure 6. It can be clearly seen that the droplet creeps down the insulator surface during the course of the experiment.

Figure 6 shows the movement of the water droplets corresponding to footage of 8 s. It was found that:

The water droplet in strong electric field deforms as a result of the interaction of electrostatic force and the surface tension of the water droplet. Additionally, it vibrates with many vibration modes throughout the creepage movement. The deformation and vibration modes of the water droplet depend on a number of factors. The factors include the magnitude of the applied field, the surface tension, droplet size (volume) and the water density of the water droplet [13,20,21]. Additionally, the actual insulator surface has a non-uniform electric field distribution. This means that it is possible to find many vibration modes occurring simultaneously on its surface.

Upon careful observation of the high-speed camera footage, the water droplet was seen to elongate in the action of electric field. The change of water droplet shape would have a significant influence on the insulation performance of insulators.

4.3. Speed of Creeping Movement

The speed of the water droplet creeping down the insulator surface was calculated using MATLAB. This helped to quantify the experiment. The relationship between time and the speed of movement is shown in Figure 7. The graph expresses the variation of instantaneous speed against time. The average speed was also calculated, which is seen to increase with time.

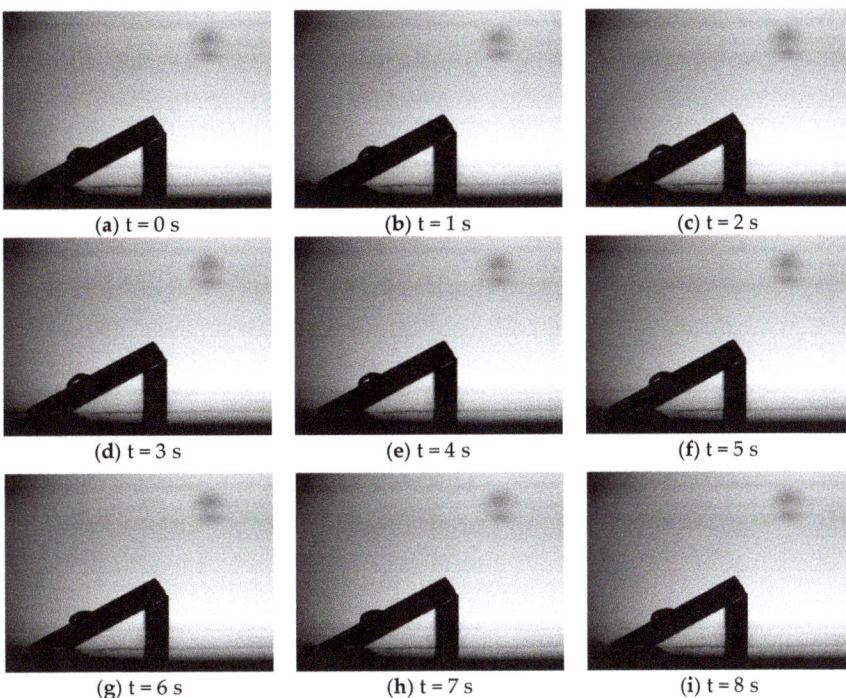

Figure 6. Still pictures from high-speed camera footage.

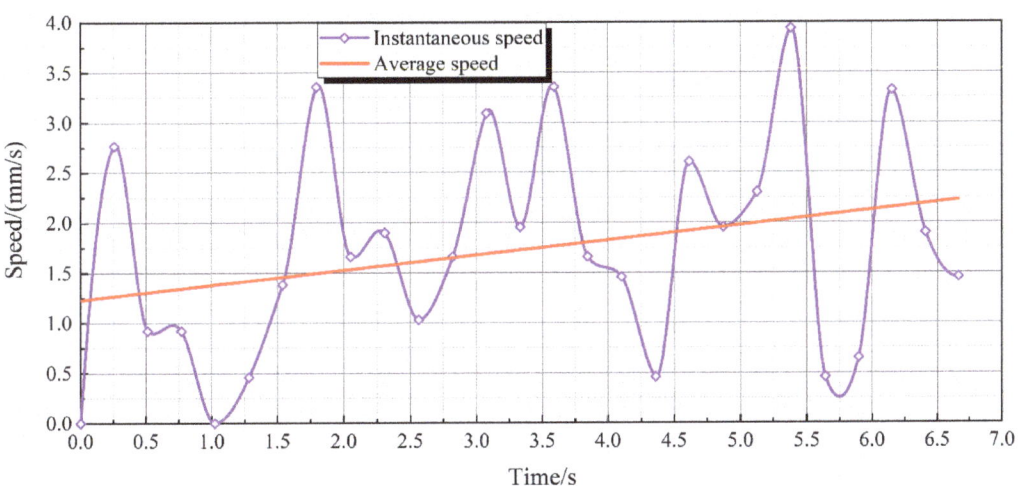

Figure 7. Speed of rain droplet movement with time.

Figure 7 shows that:

The maximum instantaneous speed is recorded as 3.94 mm/s at 5.38 s. The variation in instantaneous speed can be caused due to a number of factors, especially hydrophobicity and surface roughness.

Overall, the average speed of water droplet motion increases with time. The reason for this is that the combined forces on the water droplet are the driving forces.

It was seen that the water droplet placed on the SiR sample under the influence of ac electric field vibrates. This results in the deformation of the water droplet shape. The change in shape can cause the enhancement of a local electric field in the triple junction area of water–air–solid. Cheng [15] points out that the surface of composite insulators can experience hydrophobicity loss as a result of local electric field enhancement. Corona discharge may appear at the triple junction as the electric field is most intensified in that area [22]. This can facilitate the movement of the water droplet on the sample surface and thereby an increase in instantaneous speed is observed.

The energy transfer mechanism can also play an important role in justifying the speed variation. The energy (surface tension) of vibrating water droplet is transferred to the kinetic energy of the moving droplet and vice versa. After each subsequent movement, the dynamics change, resulting in different conditions needed to initiate the movement again.

4.4. The Occurance of the Creeping Phenomenon

It was quite interesting to observe that the phenomenon did not recur, despite many attempts to reproduce it. Both old and new silicon rubber samples were employed to see the reproducibility of the experiment. Later, the angle of inclination was also increased to 40 degrees to see the effect of level of inclination on water droplet movement. This had no impact on the water droplet movement as the creepage phenomenon was not observed at all. An increase in the electric field resulted only in severe vibration and distortion of the water droplet, but no movement. The same experiment was repeated on PTFE (another hydrophobic insulator) in order to examine the effect of surface energy or hydrophobicity. This experiment again did not see any creeping of the water droplet subject to same experimental conditions.

The last stage involved changing the droplet size (volume) to investigate its effect on the water droplet movement. In this case, the 'Creepage Phenomenon' was successfully repeated four times with a water droplet of about 75 µL. This indicates two main possibilities: the water droplet size measured as 50 µL in the initial experiment was inaccurately dispensed by the electronic syringe system, or there are some other unrevealed factors governing the movement of water droplets on inclined surfaces.

4.5. Electric Field with an Insulator

The experimental set up was designed in a way to simulate the electric stress of real insulators in service. This was quite important in developing a useful experiment whose relevance can be directly related to the practical world of overhead line composite insulators. The finite element method was employed as an effective tool to compute the electric field. Figure 8a shows the electric field distribution of an insulator in service. It can be seen that the electric field distribution on a real insulator varies from 0 to 9.754 kV/cm along the insulator length.

The electric field distribution on the silicon rubber sample surface is shown in Figure 8b. The top electrode is connected to the high voltage end, so the electric field intensity increases in magnitude from bottom to top on the sample surface. The graph shown in Figure 8 confirms that the electric field on the sample surface simulates the electric stress experienced by real insulators. This shows that a great proportion of sample surface experiences the same electric stress as that of real high voltage composite system insulator.

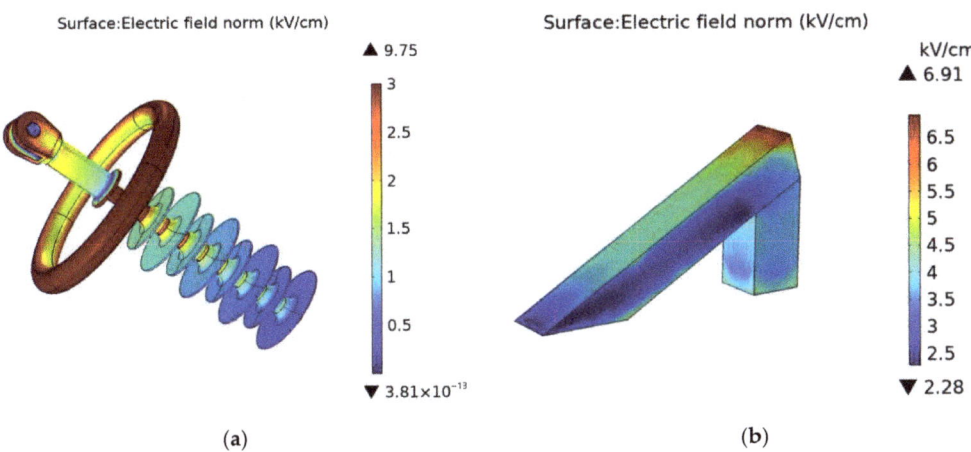

Figure 8. (a) Electric field distribution on a real insulator; (b) Electric field distribution on a sample (SiR).

5. Conclusions

It can be concluded that the water droplet placed on an inclined insulator surface vibrates with various modes. This is due to the reason that the insulators in service experience different electric field distribution on its surface and form wide range of droplet sizes. A 'Creepage Phenomenon' was observed where the water droplet (placed on an inclined polymeric surface) tends to creep along its surface in presence of high AC electric field. The experiment proved that this phenomenon is water-droplet-size dependent. A water droplet of about 75 µL is needed to observe the creepage movement of the water droplet on a 30 degrees inclined silicon rubber sample. The average speed of movement is seen to increase with time. The effect of hydrophobicity (for fixed volumes) did not have an important impact on the creepage movement. The outcomes from the indicated that this new phenomenon, so called 'creepage', plays a significant role in the ageing process of high voltage composite insulators.

Author Contributions: Conceptualization, Methodology, Writing—review and editing, Q.L. Writing—original draft preparation, Formal analysis, R.L. Validation, Resources, L.L. Visualization, Software, X.S. Investigation, Data curation, Y.W. Supervision, X.J. All authors have read and agreed to the published version of the manuscript.

Funding: This work was supported by the Natural Science Foundation of China (51637002).

Institutional Review Board Statement: Not applicable.

Informed Consent Statement: Not applicable.

Data Availability Statement: Data sharing not applicable.

Conflicts of Interest: The authors declare no conflict of interest.

References

1. Bhadra, P.; Siu, S.W.I. Effect of Concentration, Chain Length, Hydrophobicity, and an External Electric Field on the Growth of Mixed Alkanethiol Self-Assembled Monolayers: A Molecular Dynamics Study. *Langmuir* **2021**, *37*, 1913–1924. [CrossRef] [PubMed]
2. Phuong, P.T.; Oliver, S.; He, J.; Wong, E.H.H.; Mathers, R.T.; Boyer, C. Effect of Hydrophobic Groups on Antimicrobial and Hemolytic Activity: Developing a Predictive Tool for Ternary Antimicrobial Polymers. *Biomacromolecules* **2020**, *21*, 5241–5255. [CrossRef] [PubMed]
3. Klesse, G.; Tucker, S.J.; Sansom, M.S.P. Electric Field Induced Wetting of a Hydrophobic Gate in a Model Nanopore Based on the 5-HT3 Receptor Channel. *ACS Nano* **2020**, *14*, 10480–10491. [CrossRef] [PubMed]

4. Rowland, S.M.; Lin, F.C. Stability of alternating current discharges between water drops on insulation surfaces. *J. Phys. D Appl. Phys.* **2006**, *39*, 3067–3076. [CrossRef]
5. Karady, G.G. Flashover mechanism of non-ceramic insulators. *IEEE Trans. Dielectr. Electr. Insul.* **1999**, *6*, 718–723. [CrossRef]
6. Karady, G.G.; Minesh Shah, R.L. Brown Flashover mechanism of silicone rubber insulators used for outdoor insulation-I. *IEEE Trans. Power Deliv.* **1995**, *10*, 1965–1971. [CrossRef]
7. Brown, R.L.; Karady, G.G. Flashover mechanism of silicone rubber insulators used for outdoor insulation-II. *IEEE Trans. Power Deliv.* **1995**, *10*, 1972–1978. [CrossRef]
8. Yoshio, H.; Yamada, T.; Sugimoto, T. Vibration of water droplet located on a hydrophobic sheet under the tangential AC field. In Proceedings of the IEEE IAS Annual Meeting 1999, Phoenix, AZ, USA, 3–7 October 1999; pp. 1825–1830.
9. Higashiyama, Y.; Takada, T.I.; Sugimoto, T. Resonant phenomena of a water droplet located on a hydrophobic sheet under ac field. *J. Electrostat.* **2003**, *39*, 59–65. [CrossRef]
10. Higashiyama, Y.; Yamada, T.; Sugimoto, T. Effect of Resonance of a Water Droplet Located on a Hydrophobic Sheet on AC Flashover. In Proceedings of the IEEE Industry Applications Society, Pittsburgh, PA, USA, 13–18 October 2002; IEEE: Piscataway, NJ, USA, 2002; Volume 3, pp. 2198–2203.
11. Schutte, T.; Hornfeldt, S. Dynamics of electrically stressed water drops on insulating surfaces. In Proceedings of the 1990 IEEE International Symposium on Electrical Insulation, Toronto, ON, Canada, 3–6 June 1990; pp. 202–207.
12. Hara, M.; Ishibe, S.; Akazaki, M. Corona discharge and electrical charge on water drops dripping from D.C. transmission conductors-an experimental study in laboratory. *J. Electrostat.* **1979**, *6*, 235–257. [CrossRef]
13. Krivda, A.; Birtwhistle, D. Breakdown between water drops on wet polymer surfaces. In Proceedings of the Conference on Electrical Insulation and Dielectric Phenomena (CEIDP), Annual Report, Kitchener, ON, Canada, 14–17 October 2001; pp. 572–580.
14. Philips, A.J.; Childs, D.J.; Schneider, H.M. Water Drop Corona Effects on Full-Scale 500 kV Non-Ceramic Insulators. *IEEE Trans. Power Deliv.* **1999**, *14*, 258–265. [CrossRef]
15. Cheng, Z.X.; Liang, X.D.; Zhou, Y.X.; Wang, S.W.; Guan, Z.C. Observation of Corona and Flashover on the Surface of Composite Insulators. In Proceedings of the 2003 IEEE Bologna PowerTech Conference, Bologna, Italy, 23–26 June 2003; IEEE: Piscataway, NJ, USA, 2002; Volume 2, pp. 822–827.
16. Sarathi, R.; Nagesh, G. Classification of Discharges Initiated by Liquid Droplet on Insulation Material under AC Voltages Adopting UHF Technique. *J. Phys. D Appl. Phys.* **2008**, *41*, 177–182. [CrossRef]
17. Zhang, J.; Tan, J.; Pei, R.; Ye, S.; Luo, Y. Ordered Water Layer on the Macroscopically Hydrophobic Fluorinated Polymer Surface and Its Ultrafast Vibrational Dynamics. *J. Am. Chem. Soc.* **2021**, *143*, 13074–13081. [CrossRef] [PubMed]
18. Li, Q.; Rowland, S.M.; Dupere, I.; Morris, R.S. The impact of water droplet vibration on corona inception on conductors under 50 Hz AC fields. *IEEE Trans. Power Deliv.* **2018**, *33*, 2428–2436. [CrossRef]
19. Li, Q.; Shuttleworth, R.; Dupere, I.; Zhang, G.; Rowland, S.M.; Morris, R.S. FEA modelling of a water droplet vibrating in an electric field. In Proceedings of the Conference Record of IEEE International Symposium on Electrical Insulation, San Juan, PR, USA, 10–13 June 2012; pp. 449–453.
20. Li, Q.; Rowland, S.M. The Vibration Characteristics of Rain Droplet on the Surface of the Overhead Line Conductors. In Proceedings of the 2018 IEEE 2nd International Electrical and Energy Conference (CIEEC), Beijing, China, 4–6 November 2018; pp. 633–636.
21. Li, Q.; Rowland, S.M. Wet conductor surfaces and the onset of corona discharges. In Proceedings of the Annual Report Conference on Electrical Insulation and Dielectric Phenomena, Ann Arbor, MI, USA, 18–21 October 2015; pp. 245–248.
22. Roxburgh, J. Disintegration of Pairs of Water Drops in an Electric Field. *Proc. R. Soc. Lond.* **1966**, *295*, 84–97.

Article

Investigation of the Compatibility and Damping Performance of Graphene Oxide Grafted Antioxidant/Nitrile-Butadiene Rubber Composite: Insights from Experiment and Molecular Simulation

Meng Song [1,*], Xiulin Yue [1], Chaokang Chang [1], Fengyi Cao [1], Guomin Yu [1] and Xiujuan Wang [2,*]

[1] School of Materials and Chemical Engineering, Zhongyuan University of Technology, Zhengzhou 450007, China; yuelll333@126.com (X.Y.); changchaokang@126.com (C.C.); caofengyi0513@126.com (F.C.); yuguomin1129@126.com (G.Y.)
[2] Key Laboratory of Rubber-Plastics, Ministry of Education/Shandong Provincial Key Laboratory of Rubber-Plastics, Qingdao University of Science & Technology, Qingdao 266042, China
* Correspondence: chengsimengyin@126.com (M.S.); wangxj@qust.edu.cn (X.W.)

Abstract: Rubber damping materials are widely used in electronics, electrical and other fields because of their unique viscoelasticity. How to prepare high-damping materials and prevent small molecule migration has attracted much attention. Antioxidant 4010NA was successfully grafted onto graphene oxide (GO) to prepare an anti-migration antioxidant (GO-4010NA). A combined molecular dynamics (MD) simulation and experimental study is presented to investigate the effects of small molecules 4010NA, GO, and GO-4010NA on the compatibility and damping properties of nitrile-butadiene rubber (NBR) composites. Differential scanning calorimetry (DSC) results showed that both 4010NA and GO-4010NA had good compatibility with the NBR matrix, and the T_g of GO-4010NA/NBR composite was improved. Dynamic mechanical analysis (DMA) data showed that the addition of GO-4010NA increased the damping performance of NBR than that of the addition of 4010NA. Molecular dynamics (MD) simulation results show GO-4010NA/NBR composites have the smallest free volume fraction (FFV) and the largest binding energy. GO-4010NA has a strong interaction with NBR due to the forming of hydrogen bonds (H-bonds). Grafting 4010NA onto GO not only inhibits the migration of 4010NA but also improves the damping property of NBR matrixes. This study provides new insights into GO grafted small molecules and the design of high-damping composites.

Keywords: molecular dynamics simulation; damping performance; nitrile-butadiene rubber; graphene oxide; antioxidant 4010NA

1. Introduction

Rubber damping materials are widely used in electronics, electrical, aerospace, and automobiles for vibration and noise reduction due to their unique viscoelastic properties [1–3]. As some special fields have higher requirements for rubber damping materials, the preparation of high-performance damping materials has become a research hotspot. In recent years, the preparation, and research of organic hybrid damping materials have attracted much attention in the field of rubber damping [4,5]. Rubber damping materials with a dynamic hydrogen bond (H-bond) network can be prepared by adding small molecular compounds with polar functional groups, such as hydroxyl and amino into the polar rubber matrix, which can significantly improve the damping properties of the material [6,7]. In the previous studies, researchers often used nitrile-butadiene rubber (NBR) as the polar rubber matrix and added hindered phenol antioxidants (such as AO-80 or AO-60, etc.) to prepare organic hybrid damping materials with high-damping performance [8–10].

NBR has good damping performance due to the strong polarity of the -CN functional groups. However, NBR molecular chain contains a large number of unsaturated carbon-

carbon double bonds, which is easy to age in heat, oxygen, ozone, light, and other external conditions, thus shortening the service life and affecting the damping effect. At present, adding amine antioxidants into a rubber matrix is one of the easiest ways to prevent rubber aging. For most rubbers, amine antioxidants are more effective in preventing long-term oxidative degradation [11]. Many researchers improve the phenomenon of rubber aging easily by adding a small molecular antioxidant, but the antioxidant tends to migrate from the rubber matrix due to their low molecular weight, thus affecting the damping effect, resulting in shortened service life and environmental pollution [12–14]. At present, there are two solutions: the first is to improve the molecular weight of antioxidants [15,16]; the second is to graft antioxidants to the polymer chain or filler surface [17–21], which is an effective method for anti-migration of antioxidants.

Graphene oxide (GO) is a two-dimensional carbon material, arranged in a honeycomb shape, with an extremely high specific surface area and high stability. Compared with graphene, its surface contains a large number of oxygen-containing functional groups, such as hydroxyl (-OH), carboxyl (-COOH), and epoxy functional groups (-CH (O) CH-), which make GO have excellent mechanical, thermal, and electrical properties [22–25] and more practical application value and broad prospects. It can usually show a significant enhancement effect on some polar rubbers [26–29].

Grafting antioxidants on the surface of GO is an effective method to reduce the migration of antioxidants. At present, experimental studies have been carried out to graft antioxidants onto GO and other fillers. For example, Zhong et al. [30] modified GO with the antioxidant p-phenylenediamine (PPD), and added the modified GO into the NBR matrix, which significantly improved the thermal stability of the rubber matrix. Yao et al. [31] functionalized GO nanosheets through an acyl chloride reaction and then reacted the product with hindered phenol (HP) to obtain GO-g-HP with excellent anti-aging effects. Zhong et al. [32] grafted antioxidant 4020 to the surface of GO, which showed better anti-migration performance in styrene-butadiene rubber (SBR) than free antioxidants. By grafting the antioxidant onto the GO surface, the antioxidant can be fixed to prevent the migration in the rubber matrix, and the dispersity in the rubber can be improved. Generally, amine antioxidants contain polar functional groups, the addition of antioxidants can also form an H-bonds network with a polar rubber matrix, which can improve the damping and aging properties of rubber.

At present, there are few studies on the microstructure of GO grafted antioxidants from the molecular level, and the microscopic mechanism of adding GO grafted antioxidants into rubber to prevent migration and improve the damping performance of the rubber matrix. Molecular dynamics (MD) simulation, as a novel, practical, and powerful theoretical tool, can not only explain phenomena and processes that are difficult to be considered in traditional experiments but also predict experimental results. MD simulation has been widely used to study the relationship between microstructure and properties of materials [8,10,33].

In this study, antioxidant N-isopropyl-N'-phenyl-p-phenylenediamine (4010NA) was selected as a damping additive. NBR was selected as a polar rubber matrix. Small molecule 4010NA contains imino, which is easy to form H-bonds with NBR. Firstly, antioxidant 4010NA was grafted onto GO to prepare an anti-migration antioxidant (GO-4010NA). Then, the microstructure, compatibility, and damping properties of GO-4010NA/NBR composites were investigated by combining the experiment and MD simulation. For comparative analysis, control systems including NBR, GO/NBR, 4010NA/NBR, GO/4010NA/NBR were also studied. We expect to establish correlations between the microstructures and the damping properties.

2. Experimental Section
2.1. Materials

NBR with an acrylonitrile mass fraction of 41% (N220S) was purchased from Japan synthetic rubber co., Ltd. (Tokyo, Japan). GO was provided by Sixth Element Materials Technology Co., Ltd. (Changzhou, China). Antioxidant N-isopropyl-N′-phenyl-p-phenylenediamine (4010NA) was obtained from Shangshun Chemical (Heze, China). N, N-dimethylformamide (anhydrous grade, 99.8%), and sulfoxide chloride were purchased from Aladdin (Shanghai, China). All other raw materials are commercially available industrial products.

2.2. Preparation of NBR Composites

2.2.1. Synthesis of Anti-Migration Antioxidant GO-4010NA

The steps for the synthesis of anti-migration antioxidants are shown in Figure 1. First, 1 g GO was weighed and dissolved in 150 mL sulfoxide chloride ($SOCl_2$). The reaction was performed at room temperature for 1 h under ultrasound, and then at 85 °C for 6 h, after which $SOCl_2$ was removed by rotary steaming. Then it was added to 100 mL DMF for an ultrasound for 1 h, 5 g 4010NA was dissolved in 200 mL DMF and added to GO solution for reflux reaction at 120 °C for 48 h. Then the reaction solution was pumped and filtered, the solid product was poured into anhydrous ethanol for soaking for 12 h. During the extraction and filtration process, anhydrous ethanol was used for washing 4–6 times. Finally, the solid product was dried in a 40 °C vacuum oven for 12 h to obtain GO-4010NA.

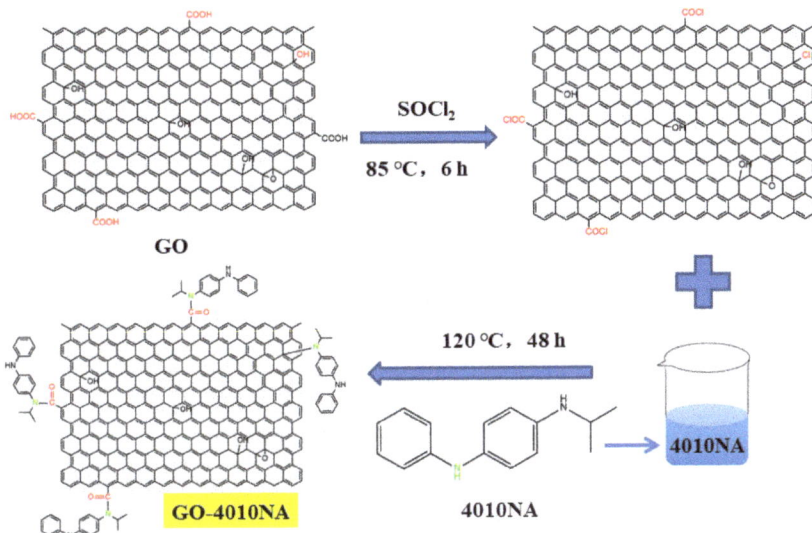

Figure 1. The synthesis route of damping additives.

2.2.2. Preparation of NBR Composites

The experimental formulae are shown in Table 1. First, 100 g NBR was plasticized at room temperature for 3 min on a two-roll mill. Then 4010NA (0.1 g), GO (1 g), GO (1 g) and 4010NA (0.1 g), GO-4010NA (1 g) were added to the above NBR, respectively. Then, the pure NBR and the four groups of samples were mixed at room temperature for 5 min. Rubber additives are added to the above samples, including 5 phr of zinc oxide (ZnO), 1 phr of stearic acid (SA), 0.5 phr of promoter D (diphenyl guanidine), 0.5 phr of promoter DM (dibenzothiazole disulfide), 0.2 phr of promoter TMTD (tetramethyl thiuram disulfide), 2 phr of sulfur hit the triangle package and cutting material, mixing evenly, mixing for 10 min. Finally, the samples were hot-pressed and vulcanized for 15 min at 15 MPa and 150 °C, and then cooled naturally to room temperature.

Table 1. The experimental formulae of NBR composites [a].

Sample	Ingredients (phr)			
	NBR	GO	4010NA	GO-4010NA
NBR	100			
GO/NBR	100	1		
4010NA/NBR	100		0.1	
4010NA/GO/NBR	100	1	0.1	
GO-4010NA/NBR	100			1

[a] Other rubber additives: ZnO, 5 phr; SA, 1 phr; D, 0.5 phr; DM, 0.5 phr; TMTD, 0.2 phr; S, 2 phr.

2.3. Characterization

Fourier transform infrared spectra (FTIR) were obtained from a Nicolet iS50 spectrometer made by Thermo Scientific Inc. (Waltham, MA, USA) within the 4000–400 cm^{-1}. The potassium bromide pellet technique and attenuated total reflection (ATR) technique were applied to powder antioxidants and NBR composites, respectively.

X-ray photoelectron spectroscopy (XPS) spectra were determined by an ESCALAB 250 Xi made by Thermo Fischer Inc. (Waltham, MA, USA). Excitation source was Al-Ka ray (HV = 1486.6 eV), operating voltage was 12.5 kV.

Differential scanning calorimetry (DSC) measurements were performed using a Netzsch DSC 200F3 calorimeter made by NETZSCH Scientific Instruments Trading Ltd. (Germany) under a nitrogen atmosphere. Samples were heated at a rate of 20 °C/min from room temperature to 100 °C, kept at 100 °C for 5 min, then cooled to −80 °C at a rate of 20 °C/min, and then heated to 180 °C at a rate of 10 °C/min.

Dynamic mechanical analysis (DMA) measurements were obtained by a Q800 dynamic mechanical analyzer made by TA instruments Inc. (New Castle, DE, USA). The samples' length, width, and thickness were 20 mm, 10 mm, and approximately 2 mm, respectively. The temperature dependence of the loss factor (tan δ) was measured from −50 °C to 150 °C at a constant frequency of 10 Hz and a heating rate of 3 °C/min in tension mode.

3. Model and Simulation Details

Materials Studio (MS) 7.0 software was used to reveal the effects of different fillers on NBR composites from the microscopic aspect, providing theoretical guidance for the experimental study of GO graft antioxidants.

The MD simulation was carried out using the Forcite and Amorphous Cell modules. During the simulation, the Andersen thermostat is used for temperature control, and the Berendsen barostat is used for pressure control. In the MD model, COMPASS force field is adopted. In COMPASS force field, the total energy E_T of the system is the sum of bond energy and non-bond energy, and the calculation formula is as follows [8].

$$E_T = E_b + E_\theta + E_\phi + E_\chi + E_{cross} + E_{ele} + E_{vdW} \quad (1)$$

In the formula, E_b is the bond stretching energy, E_θ is the bond Angle bending energy, E_φ is the dihedral Angle torsion energy, E_χ is the out-of-plane energy, E_{cross} is the cross term interaction energy, E_{ele} is electrostatic interaction energy and E_{vdW} is van der Waals interaction energy. The sum of the first five energies is the bond energy, and the sum of the last two energies is the non-bond energy.

3.1. Construction of GO Model

The GO model adopts the classic Lerf–Klinowski model: $C_{10}O_1(OH)_1(COOH)_{0.5}$ [34–36], which represents the results of the standard oxidation process. GO model is $C_{297}O_{28}(OH)_{28}(COOH)_{14}$, hydroxyl and epoxy groups are randomly distributed on the surface of GO, carboxyl groups are distributed on the edge of GO, and the oxidation degree is 21.27%. The length and width of the GO model are 34.295 Å and 22.811 Å, as shown in Figure 2.

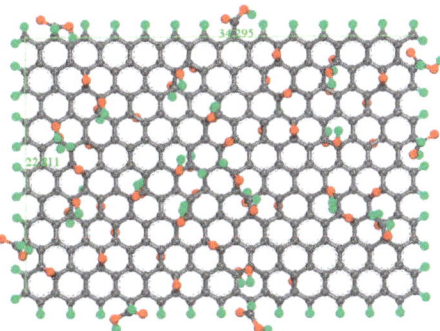

Figure 2. Molecular models for GO. The green atom is H, the gray atom is C, and the red atom is O.

3.2. Construction of Composite System Model

NBR is a polar rubber, because the molecular chain contains polar group nitrile (-NH) and the double bond structure has good performance and is widely used. Therefore, we selected NBR and constructed pure NBR (NBR molecular chain consists of 50 repeating units, 41% acrylonitrile content, and 59% butadiene content). 4010NA/NBR, GO/NBR, 4010NA/GO/NBR and GO-4010NA/NBR composite models were also constructed. The 4010NA/NBR composite has 4 NBR polymer chains and 2 4010NA small molecules, the GO/NBR composite has 4 NBR polymer chains and 1 GO, and the 4010NA/GO/NBR composite has 4 NBR polymer chains, 2 4010NA small molecules and 1 GO. The GO-4010NA/NBR composite has 4 NBR polymer chains and one GO-4010NA, as shown in Figure 3.

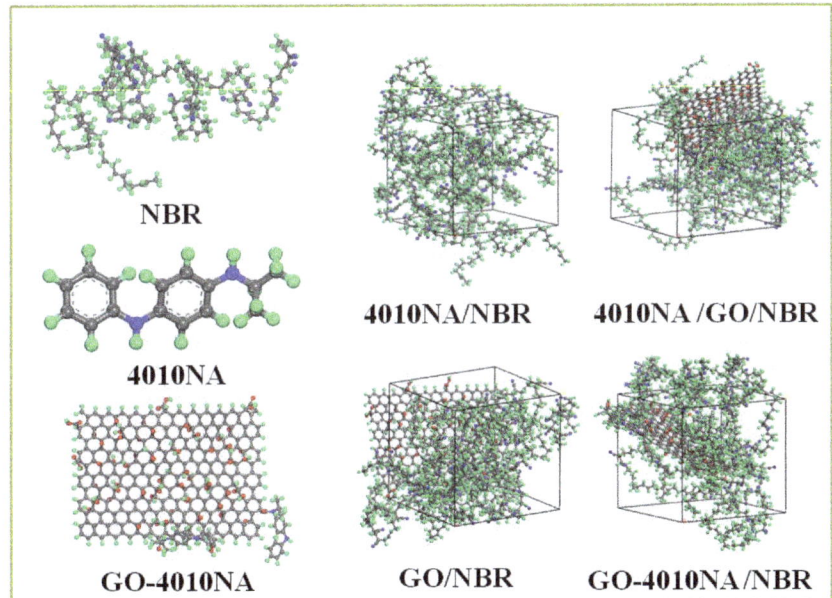

Figure 3. The construction of the 4010NA/NBR, GO/NBR, 4010NA/GO/NBR, and GO-4010NA/NBR composites amorphous cells (The green atom is H, the gray atom is C, the blue atom is N, and the red atom is O).

After the periodic cells are constructed, a series of simulations are needed in order to keep the system in equilibrium. First, the energy of amorphous cells is optimized by 2 million steps by geometric optimization with a convergence value of 1.0×10^{-5} kcal/mol/Å. The optimized cells are annealed from 200 K to 500 K with 200 annealing cycles. Then, the dynamics simulation was performed by an NVT ensemble (constant atomic number, constant volume, and constant temperature) at room temperature 298 K and the time length was set at 1000 ps. Finally, the dynamics simulation of the NPT ensemble (constant atomic number, constant pressure, and constant temperature) was carried out. The temperature is set at room temperature 298 K, the pressure is set at 0.1 MPa, and the time length is set at 1000 ps. The relevant physical parameters are calculated for the system reaching perfect equilibrium.

4. Results and Discussion

4.1. Structure Analysis of the Synthetic Antioxidant GO-4010NA

The structure of the anti-migration damping agent (GO-4010NA) was characterized by FTIR, as shown in Figure 4. From Figure 4a, the peak at a wavenumber of 3378 cm^{-1} is attributed to -NH- vibration. The peaks at wavenumber 1597 cm^{-1} and 1518 cm^{-1} are attributed to the benzene ring vibration [36]. In Figure 4b, pure GO has a broad absorption peak in the wavenumber range 2500–3750 cm^{-1}, which belongs to the -OH vibrations. The carbonyl group -C=O has a strong infrared absorption peak at 1736 cm^{-1} [10]. Compared to GO, the carbonyl peak is absent, and the hydroxyl peak is weakened in the GO-4010NA spectra. In addition, new peaks appear at 1550 cm^{-1} and 1492 cm^{-1}, which belong to the benzene ring vibration, indicating that 4010NA was grafted to GO.

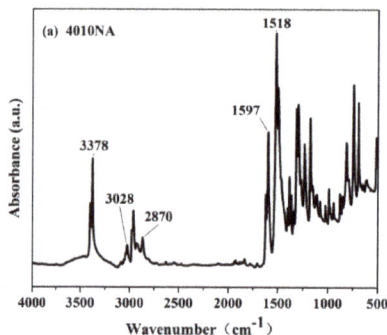 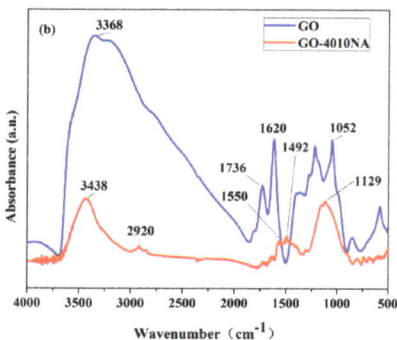

Figure 4. FTIR spectrum of (**a**) 4010NA and (**b**) GO and GO-4010NA.

To further illustrate whether 4010NA was grafted to GO, an XPS test was performed as shown in Figure 5. From Figure 5b, for GO, two peaks are detected at 284.8 eV (C1s) and 534.6 eV (O1s). For GO-4010NA from Figure 5c, in addition to strong signals of C1s and O1s, N1s signals were also found at 398.2 eV, indicating that 4010NA molecules were linked to the GO layer. All the above results indicate that the small molecule 4010NA was successfully grafted onto GO.

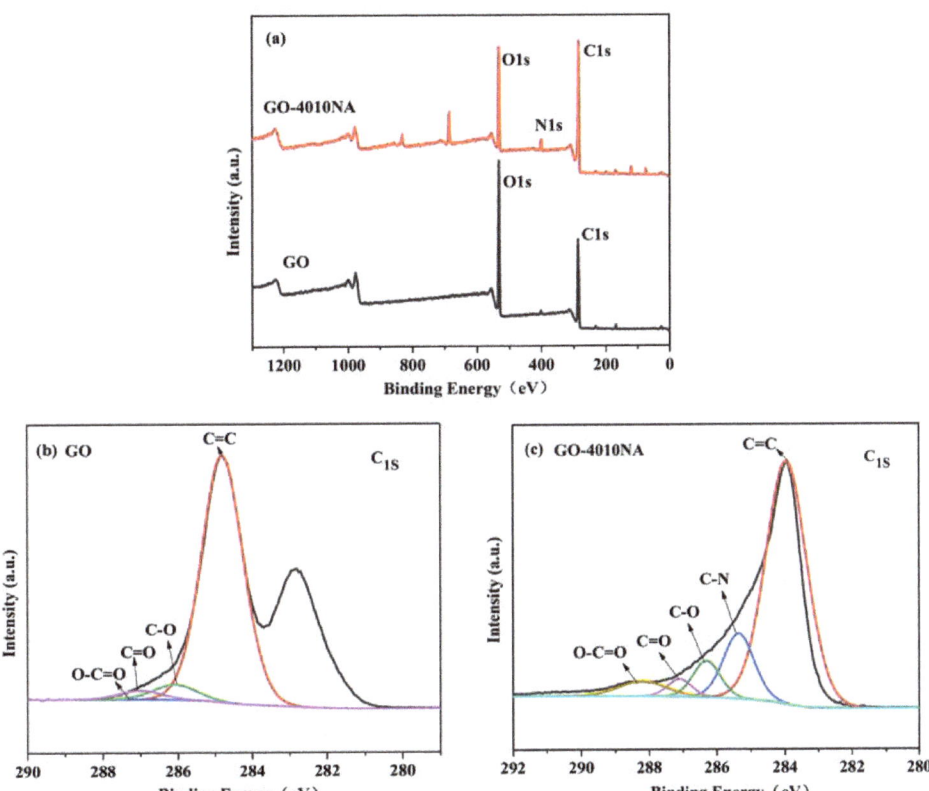

Figure 5. (**a**) XPS survey scans for GO and GO-4010NA, XPS C 1s core-level spectra for (**b**) GO and (**c**) GO-4010NA.

4.2. FTIR Analysis of NBR Composites

Figure 6 shows the FTIR spectrum of NBR composites with different contents. The telescopic vibration of -OH is generally at the wavenumber range 3125–3704 cm^{-1}. In Figure 6a, the spectrum of the neat NBR and 4010NA/NBR hardly reveals any absorbance band in the wavenumber range 3200–4000 cm^{-1} [8], whereas the other three systems containing GO showed significant peaks at 3400–3600 cm^{-1}. Compared with GO/NBR and 4010NA/GO/NBR composites, the -OH peak in GO-4010NA/NBR composites becomes stronger and shows a red shift, which is mainly attributed to the formation of hydrogen bonds in GO-4010NA/NBR composites. It can be seen from Figure 6b that the stretching vibration peak in the range of 2220–2260 cm^{-1} is -CN groups. After the addition of GO-4010NA, the -CN peak value is significantly weakened compared with pure NBR, indicating that many -CN groups in the composite are involved in the formation of hydrogen bonds.

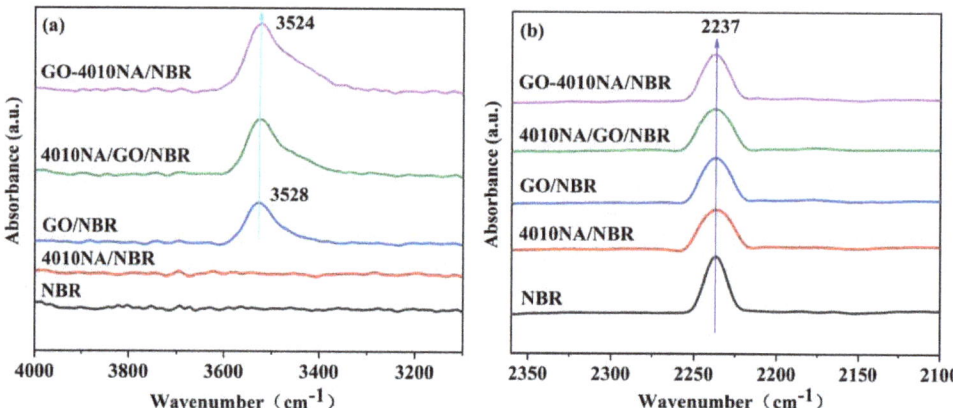

Figure 6. FTIR spectrum of (**a**) NBR composites with different fillers added illustrating the hydroxyl -OH stretching vibration region and (**b**) NBR composites with different fillers added illustrating the the nitrile -CN stretching vibration region.

4.3. DSC Analysis NBR Composites

Figure 7 shows the DSC curves and the T_g of different NBR composites. From Figure 7a, all the NBR composites have only one T_g, indicating good compatibility between filler and NBR matrix [6]. The glass transition temperature (T_g) of pure NBR was −8.3 °C, and the T_g increases after the addition of 4010NA or GO as shown in Figure 7b, indicating that the interactions of the composites are enhanced. Compared with 4010NA/NBR composite, the T_g of the GO-4010NA/NBR composite is increased to −7.5 °C. This increase is due to the H-bond network formed between GO-4010NA and NBR matrix [7,8]. The formation of H-bonds promotes the interaction of the composites and restricts the movement of NBR polymer chains, leading to an increase in the T_g.

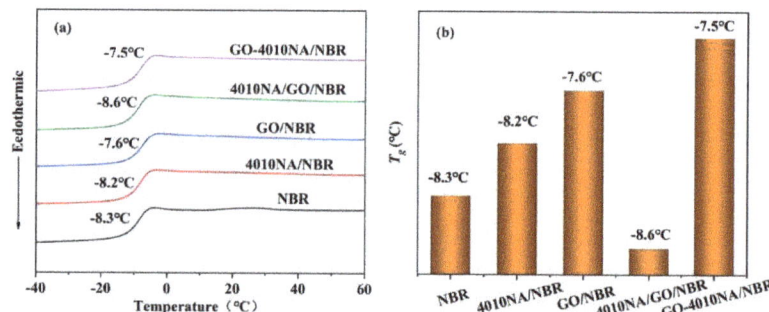

Figure 7. (**a**) DSC curves of different NBR composites and (**b**) the Tg of different NBR composites.

4.4. Dynamic Mechanical Properties of NBR Composites

Figure 8 shows the temperature dependence of the loss factor (tan δ) value and storage modulus (E') of NBR composites. Table 2 shows the damping parameters of NBR composites. When the small molecule 4010NA is added to the NBR matrix, the tan δ peak value (denote as tan δ_{max}) is 1.71 and the loss peak area TA is 29.58. When GO-4010NA was added, the tan δ_{max} increased to 1.73 and TA increased to 30.05. The results show that the addition of GO-4010NA can improve the damping performance of NBR composites more than the addition of 4010NA, which is caused by the strong interactions between GO-4010NA and NBR. It indicates that more hydrogen bonds may be formed between GO-4010NA and NBR, thus improving the damping performance of the NBR composite.

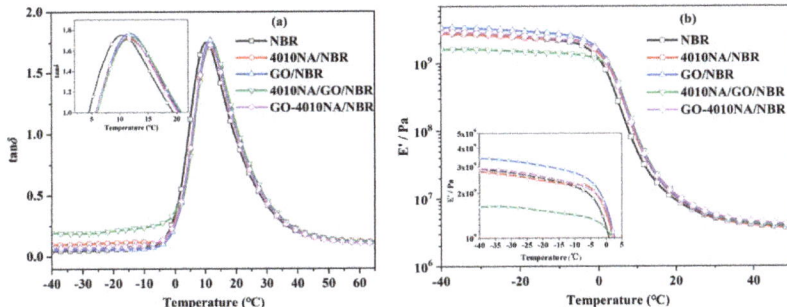

Figure 8. Temperature dependence of (**a**) the loss tangent (tan δ) values and (**b**) storage modulus for NBR composites.

Table 2. Damping properties of the NBR composites.

Sample	Tan δ Peak Position(°C)	Tan δ_{max}	Temperature Range > 0.3 (°C)			TA
			T_1	T_2	ΔT	
NBR	10.39	1.75	−0.56	34.04	34.60	31.12
4010NA/NBR	11.78	1.71	0.13	35.17	35.04	29.58
GO/NBR	11.73	1.77	0.55	34.98	34.43	31.30
4010NA/GO/NBR	11.95	1.74	−2.18	35.23	37.41	27.50
GO-4010NA/NBR	11.75	1.73	0.34	34.68	34.34	30.05

4.5. Molecular Simulation Data Analysis

4.5.1. Compatibility Analysis of 4010NA, GO, GO-4010NA and NBR

Improving the dispersity of filler is a key problem in the preparation of high-damping composites. The compatibility between filler and rubber can be evaluated by solubility parameters. The widely used solubility parameters are Hildebrand and Hansen solubility parameters [37].

$$\delta_T = \sqrt{CED} = \sqrt{\frac{E_{c,T}}{V}} \quad (2)$$

where δ_T is the Hildebrand solubility parameter, V is molar volume, and $E_{C,T}$ is cohesive energy. The non-bonding energy consists of three parts, namely dispersion, polarity, and hydrogen bonding. Hansen also proposed to divide the Hildebrand solubility parameter into three parts [38].

$$\frac{E_{C,T}}{V} = \frac{E_{C,D}}{V} + \frac{E_{C,P}}{V} + \frac{E_{C,H}}{V} \quad (3)$$

where $E_{C,D}$, $E_{C,P}$, and $E_{C,H}$ represent dispersive, polar, and hydrogen bond cohesive energy, respectively. Combined with Equations (2) and (3), δ_T can also be divided into three parts:

$$\delta_T = \delta_D{}^2 + \delta_P{}^2 + \delta_H{}^2 \quad (4)$$

δ_D, δ_P, and δ_H represent the dispersion, polar, and hydrogen bonding solubility parameters, known as Hansen solubility parameters. In the COMPASS force field, two energies are contained in E_{ele}.

$$E_{ele} = E_{C,P} + E_{C,H} \quad (5)$$

$$E_{vdW} = E_{C,D} \quad (6)$$

Then, Equation (4) can be expressed as:

$$\delta_T{}^2 = \delta_{vdW}{}^2 + \delta_{ele}{}^2 \quad (7)$$

Therefore, in the COMPASS force field, the three-component Hansen solubility parameter is converted into a two-component solubility parameter (δ_{vdW} and δ_{ele}). It can be expressed by the following formula [39,40]:

$$R = \sqrt{(\delta_{vdW,A} - \delta_{vdW,B})^2 + (\delta_{ele,A} - \delta_{ele,B})^2} \quad (8)$$

The compatibility of GO, 4010NA, and polymer is predicted by Cohesive Energy Density calculation R in the Forcite module. The physical significance of R is that the Hansen solubility of rubber is the distance from the spherical coordinate to the Hansen solubility parameter of filler. The smaller R is, the closer the solubility parameter is and the better the compatibility is [38,40]. The δ, δ_{vdW}, and δ_{ele} values of NBR, 4010NA, GO and GO-4010NA are calculated. Table 3 shows the δ_{vdW} and δ_{ele} values of NBR, 4010NA, GO, and GO-4010NA. Then, Equation (8) is used to calculate the value of R. The δ and R value of NBR, 4010NA, GO and GO-4010NA are shown in Figure 9. The results show that the R value increases first and then decreases, and the addition of GO made the R value increase, but the R value decreases after the addition of GO-4010NA, indicating that the GO graft 4010NA improves the compatibility with NBR, and there is good compatibility between NBR and GO-4010NA, which was consistent with the above DMA analysis.

Table 3. Solubility parameters of NBR, 4010NA, GO, and GO-4010NA by MD simulation.

Materials	δ_{vdW} (cal/cm^3)$^{0.5}$	Standard Error (cal/cm^3)$^{0.5}$	δ_{ele} (cal/cm^3)$^{0.5}$	Standard Eror (cal/cm^3)$^{0.5}$
NBR	8.36	0.006	3.84	0.005
4010NA	8.80	0.026	4.01	0.029
GO	10.32	0.009	5.98	0.016
GO-4010NA	7.53	0.009	4.38	0.012

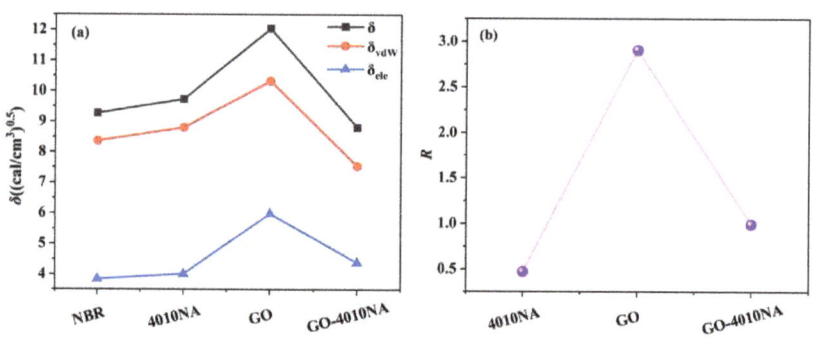

Figure 9. (a) Solubility parameter of NBR, 4010NA, GO, and GO-4010NA. (b) R value of NBR, 4010NA, GO, and GO-4010NA.

4.5.2. Charge Analysis of Atoms on Polar Functional Groups

Table 4 shows the atomic charges on the polar functional groups of NBR, 4010NA, and GO molecules obtained by MD simulation. The stronger the electronegativity, the stronger the H-bond is expected to form. According to the atomic charge distribution, it can be predicted that four types of H-bonds may be formed in the NBR composites as shown in Figure 10. The type a H-bond may be between the -CN groups of NBR molecular chains and the -NH- groups of small molecules 4010NA, expressed as (4010NA)-NH . . . NC-(NBR). The type b H-bond may be between -CN groups of NBR molecular chains and -OH of GO, expressed as (GO)-OH . . . NC-(NBR). The type c H-bond may be between -CN groups of NBR molecular chains and -COOH groups of GO, expressed as (GO)-COOH . . . NC-(NBR). The type d H-bond may be between -OH groups of GO and -C-O-C- groups of

GO, expressed as (GO)-O ... HO-(GO). The specific types and numbers of H-bonds need to be further calculated.

Table 4. Charges and forced field type of the atoms.

Polar Functional Groups	Atom	q(e)	Forced Field Type
-CN (NBR)	C	0.234	c2t
	N	−0.428	n1t
-NH (4010NA)	N	−0.373	n3h1
	H	0.353	h1n
-OH (GO)	H	0.410	h1o
	O	−0.570	o2h
-C-O-C (GO)	C	0.160	c44o
	O	−0.320	o2e
-CO$_1$O$_2$H (GO)	H	0.410	h1o
	O$_2$	−0.455	o2c
	O$_1$	−0.450	o1=
	C	0.495	c3'

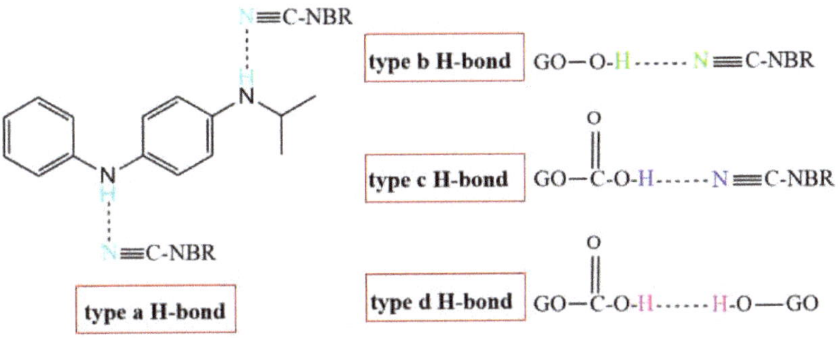

Figure 10. The four types of H-bonds might be formed in the NBR composites.

4.5.3. Accurate Statistics of Types and Number of H-Bonds

The types and numbers of H-bonds can be quantitatively calculated by MD simulation. As listed in Table 5, four types of H-bonds are formed in the NBR composites, including intermolecular H-bond interactions (such as H-bond types a, b, and c) and intramolecular H-bond interactions (such as H-bond type d). There is only one type of H-bond interaction in the 4010NA/NBR composite. With the addition of GO, it can be seen that the types and numbers of H-bonds increase significantly because there are more oxygen-containing functional groups on GO. The oxygen-containing functional groups on GO can not only form intermolecular H-bonds with NBR but also form intramolecular H-bonds between the oxygen-containing functional groups, so the addition of GO can significantly increase the types and numbers of H-bonds. Compared with the 4010NA/NBR composite, GO-4010NA/NBR composite has more H-bonds, indicating stronger interaction and better damping performance of the composite, which is consistent with the previous DMA experimental analysis.

Table 5. Numbers of H-bonds in NBR composites.

Sample	4010NA/NBR	GO/NBR	4010NA/GO/NBR	GO-4010NA/NBR
No. of type a H-bond	3	0	1	2
No. of type b H-bond	0	11	12	8
No. of type c H-bond	0	5	5	4
No. of type d H-bond	0	11	12	11
No. of total H-bonds	3	27	30	25

4.5.4. Binding Energy Analysis of NBR Composites

Binding energy ($E_{binding}$) is defined as the negative value of intermolecular energy (E_{inter}), which can reflect the mixing capacity and compatibility between each component. The larger $E_{binding}$ value is, the stronger the interface interaction is. If $E_{binding}$ is positive, the compatibility between two components is good. Formula (9) can be used to calculate [8]:

$$E_{binding} = -E_{inter} = -(E_{total} - E_{NBR} - E_{filler}) \qquad (9)$$

where E_{total}, E_{NBR}, and E_{filler} are the total energies of the NBR composite, NBR, and the corresponding filler of composite, respectively.

The $E_{binding}$ of different NBR composites is shown in Figure 11. Compared with 4010NA/NBR, the addition of GO obviously increases the $E_{binding}$ of NBR composites, indicating a stronger interaction between GO and NBR and corresponding to the above H-bonds analysis. In particular, GO-4010NA/NBR composite has the largest binding energy, which is caused by the strong H-bonds of the composites, indicating the better damping performance of the composites.

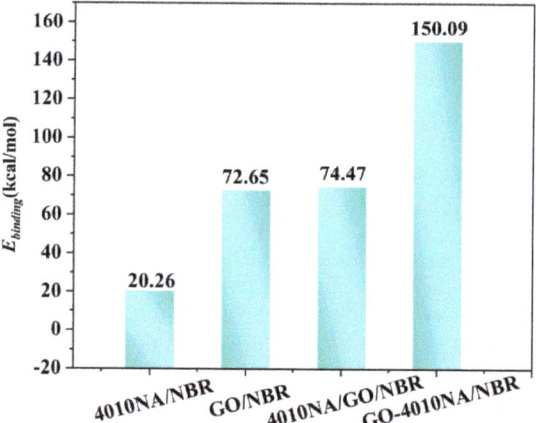

Figure 11. Binding energies of the different NBR composites.

4.5.5. Free Volume Fraction (FFV) of Different NBR Composites

Free volume fraction (FFV) represents the stacking degree of polymer and is the percentage of free volume V_f in the total volume V, which can be obtained from the expression below [39]:

$$\text{FFV} = \frac{V - V^*}{V} = \frac{V_f}{V} \qquad (10)$$

where V, V_f, and V^* represent the total volume, free volume, and occupied volume of the composite, respectively.

The FFV of different NBR composites is shown in Figure 12. Compared with pure NBR, the FFV of 4010NA/NBR composite decreases. This is due to the formation of H-bonds, which restricts the free movement of NBR molecular chains. This is consistent with the previous analysis of H-bonds above. When GO is added into NBR, the FFV becomes smaller, mainly because a large number of hydrogen bonds are generated between GO and NBR, which makes the NBR molecular chain pile closer and the FFV decreases.

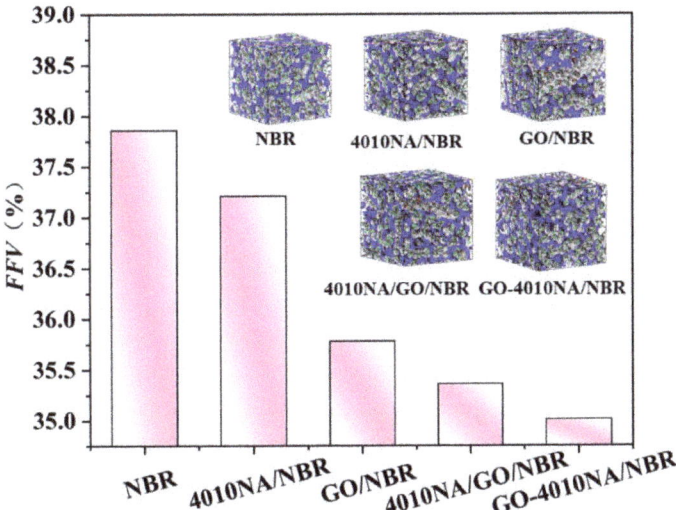

Figure 12. FFV of the different NBR composites by MD simulation, gray and blue areas represent the occupied and free volume, respectively.

4.5.6. Migration of Antioxidant 4010NA and Anti-Migration GO-4010NA

The mean square of the particle displacement in time t is called the mean square displacement (MSD). MSD can be used to describe the variation of molecular motion with time in the system [38]. Therefore, MSD can be used to study the migration characteristics of antioxidant molecules [39]. The equation for calculating MSD in MD simulation is [38]:

$$MSD(t) = \frac{1}{N} \sum_{i=1}^{N} \left[r_i(t) - r_i(0) \right]^2 \quad (11)$$

where, N is the total number of selected particles in the simulation system, and $r_i(t)$ and $r_i(0)$ are the final and initial position of atom i over the time interval t.

The movement of antioxidant 4010NA and GO-4010NA in 4010NA/NBR and GO-4010NA/NBR composites is simulated and analyzed at 298 K and 373 K, respectively. MSD curves of free and grafted 4010NA molecules are shown in Figure 13. The results show that MSD of both free 4010NA and grafted 4010NA increase with the increase of temperature, indicating that 4010NA was more mobile and migrated easily from the rubber matrix at a high temperature. At the same temperature, the MSD of free 4010NA is higher than that of GO-4010NA, indicating that the GO grafted 4010NA restricts the movement of 4010NA. The grafting method was effective to inhibit the migration of 4010NA.

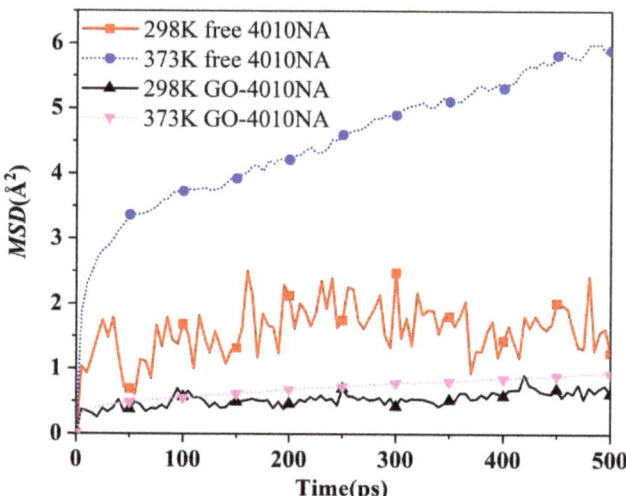

Figure 13. The MSD curves of 4010NA and GO-4010NA in NBR composites at different temperatures.

5. Conclusions

In this work, a new type of antioxidant was successfully prepared by chemical grafting 4010NA to the GO surface by $SOCl_2$. The effects of 4010NA and GO-4010NA on the compatibility and damping performance of NBR were studied by experimental and MD simulation methods. The main conclusions are as follows:

(1) DSC results show that NBR has good compatibility with 4010NA, GO, or GO-4010NA. Compared with 4010NA/NBR composite, the T_g of the GO-4010NA/NBR composite is increased, which is due to more H-bond networks formed between GO-4010NA and the NBR matrix. DMA results show that the addition of GO-4010NA can increase the damping performance of NBR more effectively than the addition of 4010NA.

(2) Through MD simulation, the two-component solubility parameters between 4010NA, GO, GO-4010NA, and NBR matrix is calculated. Compared with 4010NA, the addition of GO-4010NA significantly improved the compatibility with NBR. The MD simulation is used to calculate the H-bonds, binding energy, and FFV of the NBR composites with different fillers. Compared with 4010NA/NBR composite, the GO-4010NA/NBR composite has more H-bonds, larger binding energy, and smaller FFV, indicating GO-4010NA/NBR composite has the better damping performance, which is consistent with the DMA results.

(3) Grafting 4010NA onto GO not only inhibits the migration of 4010NA but also improves the damping property of the NBR matrix. GO-4010NA is expected to be a functional filler for the preparation of high-damping NBR composites.

Author Contributions: Conceptualization, M.S., and X.W.; methodology, M.S.; software, X.Y., and C.C.; investigation, X.Y., and C.C.; resources, G.Y.; data curation, F.C., and G.Y.; writing—original draft preparation, X.Y., and M.S.; writing—review and editing, M.S., and X.W.; funding acquisition, M.S. All authors have read and agreed to the published version of the manuscript.

Funding: This work was supported by the National Natural Science Foundation of China under Grant (Nos. 51603236, 52173057, 51703260), the Independent Innovation Application Research Project funded by the Basic Scientific Research Expenses of Zhongyuan University of Technology (No. K2020YY014), the Training Program for Young Backbone Teachers in Henan Colleges and Universities (No. 2021GGJS110), the Natural Science Foundation of Zhongyuan University of Technology (No. K2022MS005), and the China National Textile and Apparel Council (No. 2019042).

Institutional Review Board Statement: Not applicable.

Informed Consent Statement: Not applicable.

Data Availability Statement: The data presented in this study are available on request from the corresponding author.

Acknowledgments: The author would like to thank the National Natural Science Foundation of China under Grant (Nos. 51603236, 52173057, 51703260) for the financial support in completing this work. This work was supported by High Performance Computing Platform of BUCT.

Conflicts of Interest: The authors declare no conflict of interest.

References

1. Xu, K.M.; Hu, Q.M.; Wu, H.; Guo, S.Y.; Zhang, F.S. Designing a polymer-based hybrid with simultaneously improved mechanical and damping properties via a multilayer structure construction: Structure evolution and a damping mechanism. *Polymers* **2020**, *12*, 446. [CrossRef] [PubMed]
2. Wu, C.F.; Yamagishi, T.A.; Nakamoto, Y.; Ishida, S.I.; Nitta, K.H. Organic hybrid of chlorinated polyethylene and hindered phenol. I. Dynamic mechanical properties. *J. Polym. Sci. Pol. Phys.* **2000**, *38*, 2285–2295. [CrossRef]
3. Zhang, J.H.; Wang, L.F.; Zhao, Y.F. Fabrication of novel hindered phenol/phenol resin/nitrile butadiene rubber hybrids and their long-period damping properties. *Polym. Compos.* **2012**, *33*, 2125–2133. [CrossRef]
4. Xu, K.M.; Zhou, H.D.; Hu, Q.M.; Wang, J.H.; Huang, Y.; Chen, J.L. Molecular insights into chain length effects of hindered phenol on the molecular interactions and damping properties of polymer-based hybrid materials. *Polym. Eng. Sci.* **2020**, *3*, 446–454. [CrossRef]
5. Zhao, X.Y.; Xiao, D.L.; Wu, S.Z.; Feng, Y.P.; Zhang, L.Q.; Wang, W.M. Oriented distribution structure, interaction, and performance of thermoplastic polyurethane/selective hindered amine hybrids. *J. Appl. Polym. Sci.* **2011**, *120*, 906–913. [CrossRef]
6. Zhao, X.Y.; Xiang, P.; Tian, M.; Fong, H.; Jin, R.; Zhang, L.Q. Nitrile butadiene rubber/hindered phenol nanocomposites with improved strength and high damping performance. *Polymer* **2007**, *20*, 6056–6063. [CrossRef]
7. Song, M.; Yue, X.L.; Wang, X.J.; Cao, F.Y.; Li, Y.N.; Su, C.H.; Qin, Q. Effect of hindered phenol AO-80 on the damping properties for nitrile-Butadiene rubber/phenolic resin: Molecular simulation and experimental study. *Macromol. Mater. Eng.* **2020**, *305*, 2000222. [CrossRef]
8. Qiao, B.; Zhao, X.Y.; Yue, D.M.; Zhang, L.Q.; Wu, S.Z. A combined experiment and molecular dynamics simulation study of hydrogen bonds and free volume in nitrile-butadiene rubber/hindered phenol damping mixtures. *J. Mater. Chem.* **2012**, *22*, 12339–12348. [CrossRef]
9. Zhu, J.; Zhao, X.Y.; Liu, L.; Yang, R.N.; Song, M.; Wu, S.Z. Thermodynamic analyses of the hydrogen bond dissociation reaction and their effects on damping and compatibility capacities of polar small molecule/nitrile-butadiene rubber systems: Molecular simulation and experimental study. *Polymer* **2018**, *155*, 152–167. [CrossRef]
10. Song, M.; Yue, X.L.; Wang, X.J.; Huang, M.J.; Ma, M.X.; Pan, W.; Qin, Q. Improved high-temperature damping performance of nitrile-butadiene rubber/phenolic resin composites by introducing different hindered amine molecules. *e-Polymers* **2020**, *20*, 482–490. [CrossRef]
11. Komethi, M.; Othman, N.; Ismail, H.; Sasidharan, S. Comparative study on natural antioxidant as an aging retardant for natural rubber vulcanizates. *J. Appl. Polym. Sci.* **2012**, *124*, 1490–1500. [CrossRef]
12. Nawaz, S.; Hillborg, H.; Hedenqvist, M.S.; Gedde, U.W. Migration of a phenolic antioxidant from aluminium oxide-poly (ethylene-co-butyl acrylate) nanocomposites in aqueous media. *Polym. Degrad. Stabil.* **2013**, *98*, 475–480. [CrossRef]
13. Graciano-Verdugo, A.Z.; Soto-Valdez, H.; Peralta, E.; Cruz-Zárate, P.; Islas-Rubio, A.R.; Sánchez-Valdes, S.; Sánchez-Escalante, A.; González-Méndez, N.; González-Ríosb, H. Migration of α-tocopherol from LDPE films to corn oil and its effect on the oxidative stability. *Food Res. Int.* **2010**, *43*, 1073–1078. [CrossRef]
14. Wang, X.J.; Chen, X.H.; Song, M.; Wang, Q.F.; Zheng, W.; Song, H.J.; Fan, Z.H.; Thu, A.M. Effects of hindered phenol organic molecules on enhancing thermo-oxidative resistance and damping capacity for nitrile butadiene rubber: Insights from experiments and molecular simulation. *Ind. Eng. Chem. Res.* **2020**, *59*, 11494–11504. [CrossRef]
15. Buchmüller, Y.; Wokaun, A.; Gubler, L. Polymer-bound antioxidants in grafted membranes for fuel cells. *J. Mater. Chem. A* **2014**, *2*, 5870–5882. [CrossRef]
16. Wu, W.J.; Zeng, X.R.; Li, H.Q.; Lai, X.J.; Xie, H. Synthesis and antioxidative properties in natural rubber of novel macromolecular hindered phenol antioxidants containing thioether and urethane groups. *Polym. Degrad. Stabil.* **2015**, *111*, 232–238. [CrossRef]
17. Kim, T.H.; Oh, D.R. Melt grafting of maleimides having hindered phenol antioxidant onto low molecular weight polyethylene. *Polym. Degrad. Stabil.* **2004**, *84*, 499–503. [CrossRef]
18. Gao, X.W.; Meng, X.F.; Wang, H.T.; Wen, B.; Ding, Y.F.; Zhang, S.M.; Yang, M.S. Antioxidant behaviour of a nanosilica-immobilized antioxidant in polypropylene. *Polym. Degrad. Stabil.* **2008**, *93*, 1467–1471. [CrossRef]
19. Zhou, J.J.; Wei, L.Y.; Wei, H.T.; Zheng, J.; Huang, G.S. The synthesis of graphene-based antioxidants to promote anti-thermal properties of styrene-butadiene rubber. *RSC Adv.* **2017**, *7*, 53596–53603. [CrossRef]
20. Zhu, L.; Chen, X.; Shi, R.R.; Zhang, H.; Han, R.; Cheng, X.; Zhou, C.J. Tetraphenylphenyl-modified damping additives for silicone rubber: Experimental and molecular simulation investigation. *Mater. Design* **2021**, *202*, 109551. [CrossRef]

21. Yang, H.L.; Cai, F.; Luo, Y.L.; Ye, X.; Zhang, C.; Wu, S.Z. The interphase and thermal conductivity of graphene oxide/butadiene-styrene-vinyl pyridine rubber composites: A combined molecular simulation and experimental study. *Compos. Sci. Technol.* **2020**, *188*, 107971. [CrossRef]
22. Panahi, F.; Fareghi-Alamdari, R.; Khajeh Dangolani, S.; Khalafi-Nezhad, A.; Golestanzadeh, M. Graphene grafted N-methyl-4-pyridinamine(G-NMPA):an efficient heterogeneous organocatalyst for acetylation of alcohols. *ChemistrySelect* **2017**, *2*, 474–479. [CrossRef]
23. Lin, Y.L.; Chen, Y.Z.; Zeng, Z.K.; Zhu, J.R.; Wei, Y.; Li, F.C.; Liu, L. Effect of ZnO nanoparticles doped graphene on static and dynamic mechanical properties of natural rubber composites. *Compos. Part A Appl. Sci. Manuf.* **2015**, *70*, 35–44. [CrossRef]
24. Cao, R.; Chen, Z.; Wu, Y.H.; Tu, Y.F.; Wu, G.X.; Yang, X.M. Precisely controlled growth of poly(ethyl acrylate) chains on graphene oxide and the formation of layered structure with improved mechanical properties. *Compos. Part A Appl. Sci. Manuf.* **2017**, *93*, 100–106. [CrossRef]
25. Mohamed, M.A.; Yehia, A.M.; Banks, C.E.; Allam, N.K. Novel MWCNTs/graphene oxide/pyrogallol composite with enhanced sensitivity for biosensing applications. *Biosens. Bioelectron.* **2017**, *89*, 1034–1041. [CrossRef]
26. Smaoui, I.; Domatti, A.; Kharrat, M.; Dammak, M.; Monteil, G. Eco-friendly nanocomposites between carboxylated acrylonitrile-butadiene rubber (XNBR) and graphene oxide or graphene at low content with enhanced mechanical properties. *Fuller. Nanotub. Carbon Nanostructures* **2016**, *24*, 769–778. [CrossRef]
27. Yu, B.; Shi, Y.Q.; Yuan, B.H.; Qiu, S.; Xing, W.Y.; Hu, W.Z.; Song, L.; Lo, S.M.; Hu, Y. Enhanced thermal and flame retardant properties of flame-retardant-wrapped graphene/epoxy resin nanocomposites. *J. Mater. Chem. A* **2015**, *15*, 8034–8044. [CrossRef]
28. He, S.J.; Luo, C.M.; Zheng, Y.Z.; Xue, Y.; Song, X.P.; Lin, J. Improvement in the charge dissipation performance of epoxy resin composites by incorporating amino-modified boron nitride nanosheets. *Mater. Lett.* **2021**, *298*, 130009. [CrossRef]
29. He, S.J.; Wang, J.Q.; Hu, J.B.; Zhou, H.F.; Nguyen, H.; Luo, C.; Lin, J. Silicone rubber composites incorporating graphitic carbon nitride and modified by vinyl tri-methoxysilane. *Polym. Test.* **2019**, *79*, 106005. [CrossRef]
30. Zhong, R.; Zhang, Z.; Zhao, H.G.; He, X.R.; Wang, X.; Zhang, R. Improving thermo-oxidative stability of nitrile rubber composites by functional graphene oxide. *Materials* **2018**, *11*, 921. [CrossRef]
31. Yao, J.L.; Liu, S.X.; Huang, Y.J.; Ren, S.J.; Lv, Y.D.; Kong, M.Q.; Li, G.X. Acyl-chloride functionalized graphene oxide chemically grafted with hindered phenol and its application in anti-degradation of polypropylene. *Prog. Nat. Sci.* **2020**, *30*, 328–336. [CrossRef]
32. Zhong, B.C.; Dong, H.H.; Luo, Y.F.; Zhang, D.Q.; Jia, Z.X.; Jia, D.M.; Liu, F. Simultaneous reduction and functionalization of graphene oxide via antioxidant for highly aging resistant and thermal conductive elastomer composites. *Compos. Sci. Technol.* **2017**, *151*, 156–163. [CrossRef]
33. Luo, Y.L.; Li, T.T.; Li, B.; Chen, X.L.; Luo, Z.Y.; Gao, Y.Y.; Zhang, L.Q. Effect of the nanoparticle functionalization on the cavitation and crazing process in the polymer nanocomposites. *Chinese J. Polym. Sci.* **2021**, *39*, 249–257. [CrossRef]
34. Medhekar, N.V.; Ramasubramaniam, A.; Ruofr, R.S.; Shenoy, V.B. Hydrogen bond networks in graphene oxide composite paper: Structure and mechanical properties. *ACS Nano* **2010**, *4*, 2300–2306. [CrossRef] [PubMed]
35. Lerf, A.; He, H.; Forster, M.; Klinowski, J. Structure of graphite oxide revisited. *J. Phys. Chem. B* **1998**, *102*, 4477–4482. [CrossRef]
36. Cai, F.; You, G.H.; Luo, K.Q.; Zhang, H.; Zhao, X.Y.; Wu, S.Z. Click chemistry modified graphene oxide/styrene-butadiene rubber composites and molecular simulation study. *Compos. Sci. Technol.* **2020**, *190*, 108061. [CrossRef]
37. Hansen, C.M. *Hansen Solubility Parameters: A User's Handbook*; CRC Press: Boca Raton, FL, USA, 2007.
38. Luo, K.Q.; You, G.H.; Zhao, X.Y.; Lu, L.; Wang, W.C.; Wu, S.Z. Synergistic effects of antioxidant and silica on enhancing thermo-oxidative resistance of natural rubber: Insights from experiments and molecular simulations. *Mater. Design* **2019**, *181*, 107944. [CrossRef]
39. Wang, X.J.; Song, M.; Liu, S.T.; Wu, S.Z.; Thu, A.M. Analysis of phthalate plasticizer migration from PVDC packaging materials to food simulants using molecular dynamics simulations and artificial neural network. *Food Chem.* **2020**, *317*, 126465. [CrossRef]
40. Luo, Y.L.; Wang, R.G.; Wang, W.; Zhang, L.Q.; Wu, S.Z. Molecular dynamics simulation insight into two-component solubility parameters of graphene and thermodynamic compatibility of graphene and styrene butadiene rubber. *J. Phys. Chem. C* **2017**, *121*, 10163–10173. [CrossRef]

Article

Improved Dielectric Breakdown Strength of Polyimide by Incorporating Polydopamine-Coated Graphitic Carbon Nitride

Yinjie Dong, Zhaoyang Wang, Shouchao Huo, Jun Lin and Shaojian He *

State Key Laboratory of Alternate Electrical Power System with Renewable Energy Sources, North China Electric Power University, Beijing 102206, China; ddxdyj@126.com (Y.D.); wang94269264@163.com (Z.W.); huo472@foxmail.com (S.H.); jun.lin@ncepu.edu.cn (J.L.)
* Correspondence: heshaojian@ncepu.edu.cn

Abstract: Breakdown strength is an important parameter for polymer dielectric, and introducing inorganic filler into the polymer matrix is an efficient method to improve the breakdown strength. In this work, graphitic carbon nitride nanosheets (CNNS) were ultrasonically exfoliated and coated with polydopamine to obtain modified nanosheets (DCNNS), and then polyimide (PI) composite films with various CNNS and DCNNS were prepared and compared. Owing to the abundant hydroxyl groups of polydopamine, good filler-polymer compatibility and uniform filler dispersion were achieved for PI/DCNNS composites. Both breakdown strength and dielectric constant were improved with the addition of either CNNS or DCNNS. However, at the same filler content, the PI/DCNNS composites exhibited higher breakdown strength and dielectric constant than the PI/CNNS. The PI composite with 0.5 wt% DCNNS showed the highest breakdown strength of ~300 kV/mm, increased by 67.6% as compared to the pure PI, while the PI/CNNS composite with the same filler content only increased by 14.5%.

Keywords: polyimide; graphitic carbon nitride nanosheets; polydopamine; interfacial interaction; breakdown strength

1. Introduction

Polymer dielectrics with good dielectric properties and high-energy storage density play an important role in modern information and electronic industries connected to, for example, charge storage devices and embedded capacitors [1,2]. However, with the rapid development of the modern power industry, more requirements are proposed on the stability of polymer dielectrics and electronic devices operating under high electrical conditions. For the insulation dielectric materials, once the surface flashover or body breakdown phenomenon occurs, it can lead to insulation dielectric surface degradation and an equipment short circuit, ultimately threatening the operational reliability of the power equipment. What is more, according to the formula of energy storage density of the linear polymer, $U = \frac{1}{2}\varepsilon_0\varepsilon_r E_b^2$, the energy storage density (U) is proportional to the square of breakdown strength (E_b) tolerated by the film, so it is of great significance to improve the breakdown strength of polymer in the power industry [3,4].

Polyimide (PI), as a typical engineering polymer material, has been widely used in insulation materials, microelectronics, aerospace and other fields in recent years due to its high thermal stability, excellent electrical insulation, and mechanical properties [5–7]. However, the defects (such as impurities and micro-pores) during the breakdown of pure PI will lead to a theoretical reduction in the actual breakdown strength of the material, which limits its demand in some special applications [8]. The traditional strategy of adding inorganic nanofillers (such as Al_2O_3 [9,10] and $BaTiO_3$ [11]) to the polymer matrix can improve the comprehensive properties of the composites and suppress the distorted electric field. However, the small particle size and large specific surface area of conventional nanoparticles lead to their poor dispersion in polymers. Recently, it was reported that two-dimensional nanofillers [12–19], such as boron nitride, graphene, mica and titanium dioxide nanosheets, play an important role in improving the performance, including the breakdown field strength, of polymer composites when they are evenly dispersed in the matrix.

Two-dimensional graphitic carbon nitride (g-C_3N_4) is a promising material commonly used in the fields of photocatalysis and heterogeneous catalysis due to its simple synthesis and low cost [20–22]. Recently, Zhu et al. [23] reported that the frictional properties of the composites were improved by introducing g-C_3N_4 as a filler in the polyimide matrix. This work provides an example of the application of g-C_3N_4 as an excellent nanofiller in polymer-based composites. Wang et al. [24] explored the potential application of carbon nitride nanosheets in improving the thermal conductivity of PI through experiments and simulations. The improvement in thermal conductivity benefited from the self-orientation and strong interaction of the fillers, along with the PI film. Although g-C_3N_4 has been used as a filler to improve polymers' properties, the research on its application as an electrical insulation filler in the field of high-voltage insulation is relatively rare [25]. Generally, the poor interfacial compatibility between the polymer matrix and the filler affects the dielectric breakdown strength of composites [26–28]. To solve this problem, an economical, efficient and easy way is to modify the inorganic fillers through surface treatment, modulating the interfacial properties and increasing the breakdown field strength of the composite dielectric. In our previous work [25], silicone rubber (SR)/g-C_3N_4 composites were prepared by in situ modification with vinyl tri-methoxysilane (VTMS), and the incorporation of VTMS reduced the defects in the prepared composites and improved their breakdown strength and mechanical properties.

The polydopamine coating is formed by spontaneous oxidative copolymerization of the dopamine on the surface of the substrate. Compared with traditional chemical modification methods, polydopamine can adhere to the surface of most materials without destroying the matrix structure. In addition, strong adhesion is formed between the polydopamine-encapsulated nanofiller and the polymer [13]. The covalent bonds are formed between the dicarboxylic anhydride of PI and the amine groups of polydopamine, further improving the compatibility of carbon nitride and the PI matrix.

In this work, the g-C_3N_4 composites were prepared from melamine through thermal condensation, and then carbon nitride nanosheets (CNNS) were prepared using the ultrasonic stripping method. Carbon nitride nanosheets modified by polydopamine (DCNNS) were prepared by self-polymerization of dopamine in a weak alkali environment, and then PI composite films with various CNNS and DCNNS were prepared by solution casting. The microstructure, dielectric, and breakdown properties of the composite films were investigated.

2. Materials and Methods

2.1. Materials

Melamine was purchased from Anhui Jinhe Co., Ltd. (Chuzhou, China). 4,4-oxydianiline (ODA) was purchased from Tianjin Guangfu Fine Chemical Research Institute (Tianjin, China). Pyromellitic dianhydride (PMDA) was purchased from Beijing Chemical Factory (Beijing, China). Dimethylacetamide (DMAc) was purchased from Beijing Innochem Science & Technology Co., Ltd. (Beijing, China). Tris(hydroxymethyl)aminomethane hydrochloride (Tris-HCl) and dopamine hydrochloride were purchased from Alfa Aesar Co., Ltd. (Shanghai, China). All reagents were used as received.

2.2. Sample Preparation

Preparation of the CNNS: Melamine powder covered with thin aluminum foil was heated to 500 °C with a heating rate of 3 °C·min^{-1} in a muffle furnace, and the temperature was maintained at 550 °C for 4 h. After cooling to room temperature, the bulk g-C_3N_4 was obtained. The ground bulk g-C_3N_4 (10 g) was added into 1 L of deionized water and then stirred with a high-speed mixer at a speed of 13,000 rpm for 2 h, followed by ultrasonically treating for 48 h. After standing for 5 h, the exfoliated CNNS was left in the upper white suspension. After evaporating most of the water at 80 °C, the upper suspension was concentrated from 1 L to 50 mL, and then the CNNS powder was obtained through freeze-drying.

Preparation of the DCNNS: The CNNS powder (4 g) was ultrasonically dispersed in 100 mL of deionized water for 30 min, and then 0.315 g of Tri-HCl and 0.379 g of dopamine hydrochloride was added to the CNNS aqueous suspension. After adjusting the pH to 8.5 by NaOH, the suspension was stirred for 4 h to complete the modification. Finally, the mixture was centrifuged and rinsed repeatedly 5 times followed by freeze-drying to obtain the DCNNS powder.

Preparation of the PI composite films: The PI/CNNS and PI/DCNNS composite films were prepared through a viscous prepolymer cast on the glass slide and the thermal imidization method. A certain amount of CNNS or DCNNS was ultrasonically dispersed in 30 mL of DMAC for 2 h, and then equimolar proportions of 3.064 g of ODA and 3.34 g of PMDA were added and stirred until the ODA was dissolved completely. Subsequently, the mixture was stirred under a nitrogen flow for 8 h, followed by degassing under vacuum for 2 h. The solid content of the PAA/CNNS and PAA/DCNNS suspension was 18%. After casting on a flat glass plate, the mixture was converted into films by scraping them with a glass rod, and then thermally imidized at 60 °C for 10 h, 120 °C, 200 °C, 250 °C and 320 °C for 1 h each. After the sample was naturally cooled to room temperature, the yellow and transparent PI composite films were obtained and denoted as PI/CNNS-X and PI/DCNNS-Y, where X% and Y% stand for the weight percentage of the CNNS and DCNNS. The thickness of the PI composites was approximately 30 μm.

2.3. Characterization and Measurement

The fractured surface of the PI/CNNS and PI/DCNNS composite films were observed by scanning electron microscopy (SEM, SU8010, Hitachi, Tokyo, Japan) with an accelerating voltage of 10 kV. The surface element composition was determined by the X-ray photoelectron spectrometer (XPS, Thermo Scientific K-Alpha, Waltham, MA, USA) equipped with an Al Ka X-ray source under an operating pressure below 8×10^{-10} Pa. The dielectric properties were tested by precision dielectric test apparatus (Novocontrol Concept 80, Montabaur, Germany). The dielectric breakdown strength (E_b) was tested using a voltage-withstand testing device (HCDJC; Beijing Huace Testing Instrument Co. Ltd., Beijing, China) at ambient temperature, with an increasing alternating voltage of 0.5 kV/s. The specimens were sandwiched between two copper rod electrodes with diameters of 25 mm and immersed in pure silicone oil to prevent surface flashover.

3. Results and Discussions

3.1. Structure of the CNNS and DCNNS

The photographs of the CNNS and DCNNS are shown in Figure 1a. It can be seen that the color of the CNNS powder is dark-yellow, while it changes to gray-brown when coated with polydopamine. As shown in Figure 1b, the DCNNS can form a very stable suspension in the DMAc, and no noticeable precipitate was observed even after 24 h. In contrast, the CNNS were very unstable in the DMAc, as all the CNNS settled to the bottom after 24 h. Since polydopamine with many hydroxyl groups is coated on the surfaces of the nanosheets, the DCNNS can form hydrogen bonds with the DMAc, which effectively improves the dispersion and stability of the nanosheets in the solvent.

XPS measurement was used to investigate the surface chemical composition of the elements in the CNNS and DCNNS. As compared between Figure 1c,f, both the CNNS and DCNNS contained the elements C, O, and N, while the DCNNS had more O element as compared to the CNNS. Owing to the numerous oxygen-containing functional groups, the content of the O element increased from 3.28% to 7.44%, and the C/O ratio decreased from 12.4 to 6.3 after the polydopamine was coated on the nanosheets. In Figure 1d, the C1s spectra of both the CNNS and DCNNS contained two peaks at 284.8 eV and 288.0 eV, which were assigned to C-C and N-C=N, respectively. The dominant peak of N-C=N originated from the aromatic nitrogen heterocyclic structure of the g-C_3N_4 [29]. However, one more peak at 286.1 eV related to the C-O peak was found in the spectrum of the DCNNS, which should be attributed to the amidogen and phenolic hydroxyl from the polydopamine. In Figure 1e, the N1s spectrum of the CNNS presented four component peaks, C=N-C at 398.6 eV, N-$(C)_3$ at 400.1 eV, N-H at 401.2 eV, and a heterocyclic charging effect of 404.5 eV [30]. In contrast to the CNNS, the DCNNS had the same peak, and the peak intensity of the C=N-C was relatively low, which can be attributed to the adhesion of the polydopamine on the surface of the CNNS, which reduced the N element content detected on the surface of the CNNS. The results mentioned above illustrate the successful attachment of the dopamine to the surface of the CNNS.

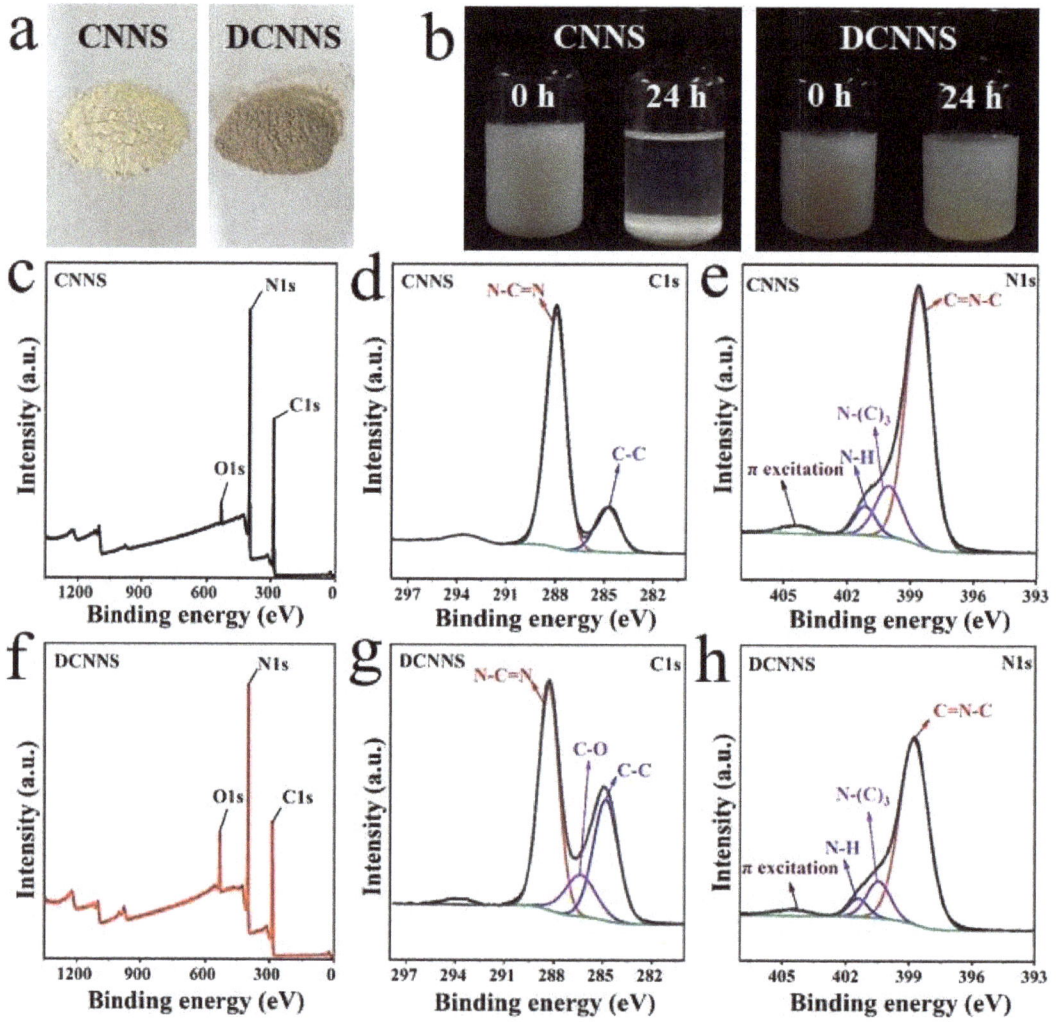

Figure 1. The photographs of (**a**) CNNS and DCNNS powder; (**b**) dispersing stability of CNNS and DCNNS in DMAC; XPS (**c**) wide-scan, (**d**) C1s and (**e**) N1s core-level spectra of CNNS; XPS (**f**) wide-scan, (**g**) C1s and (**h**) N1s core-level spectra of DCNNS.

3.2. Fractured Surface Morphologies

The SEM images of the fracture surface for the PI composite films are shown in Figure 2. For the PI/CNNS composite films, some voids and defects can be seen in the composites, indicating the poor interfacial compatibility between the CNNS and the PI that becomes worse with higher filler content. As for the PI/DCNNS composites, no voids or defects can be seen in the composites with 0.25 wt% and 0.5 wt% filler contents. Moreover, some voids and defects were observed in the composites with 0.75 wt% and 1.0 wt%, but still much fewer than those in the PI/CNNS composites with the same filler content. Only when the specific gravity of the DCNNS reached 1.0 wt% did a small number of protrusions appear, while the PI film with the CNNS added did cause filler agglomeration at 0.75 wt%. This can be attributed to the fact that the hydroxyl groups on the surface of the DCNNS can form hydrogen bonds with the PI matrix or covalent bonds with the carboxyl groups of polyamic acid (the medium product used during the synthesis of the PI). This provides a morphological basis for the better breakdown performance of the PI/DCNNS composites, as discussed below. Under the action of the high-voltage electric field, such structural defects in the composites might induce the distortion of the electric field, which would directly affect the breakdown characteristics of the composite materials.

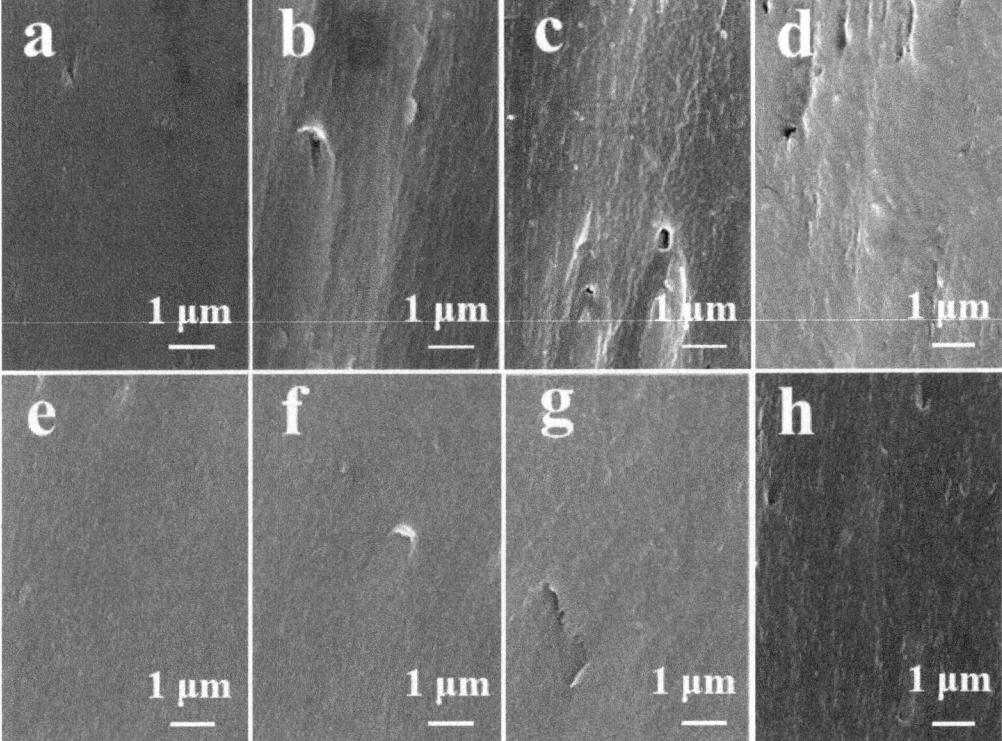

Figure 2. SEM images of the fractured surface for PI composites with (**a**) 0.25 wt%, (**b**) 0.5 wt%, (**c**) 0.75 wt% and (**d**) 1.0 wt% CNNS and (**e**) 0.25 wt%, (**f**) 0.5 wt%, (**g**) 0.75 wt% and (**h**) 1.0 wt% DCNNS.

3.3. Dielectric Performance

Breakdown strength is an important parameter used to evaluate dielectric materials.

Due to differences in structure and process, the scatter of breakdown times and breakdown strength may be large even for identical specimens. The researchers found that the breakdown voltage obeyed statistical probability distribution under the test standard [31].

The Weibull distribution is a statistical distribution proposed by the Swedish physicist Weibull in the fatigue test of materials, the main feature of which is to find the weakest independent small unit in the material and to calculate it analytically [32]. It is generally believed that the breakdown of polymer-insulating material occurs at the weakest point of the polymer. Therefore, in dielectric physics, the Weibull distribution is commonly used for statistical analysis of the breakdown field strength of polymer-insulating materials under DC and AC voltages [33]. The statistical model of the Weibull distribution can be used to analyze the magnitude and dispersion of the breakdown field strength of dielectric materials and to obtain the stability analysis results of the breakdown of dielectric materials.

The breakdown strength of the CNNS and DCNNS composites were analyzed by two parameter Weibull distribution approaches, which reflected the probability of the material being broken down under a certain electric field and the probability of failure after a certain electric field action time, as depicted in this formula:

$$P(E) = 1 - exp\left[-\left(\frac{E}{E_b}\right)^\beta\right]$$

where $P(E)$ is the cumulative probability of failure for electrical faults, E is the breakdown field strength for each data point, E_b is the characteristic breakdown field strength where the cumulative failure probability is 63% for the polymer, and β is the Weibull shape parameter used to assess the dispersion of the experimental data. In this experiment, for each sample, 11 data points were taken for testing to estimate the corresponding Weibull breakdown strengths.

Figure 3a shows the Weibull distribution DC breakdown strength of the pure PI and PI composite films. As shown in Figure 3a, the breakdown strength of all the PI/CNNS composite films was a little higher than that of the pure PI (~179 kV mm^{-1}). Nevertheless, all the PI/DCNNS composite films exhibited much higher breakdown strength than the PI/CNNS composite films with the same filler content, certainly significantly higher than the pure PI. In addition, as the content of DCNNS increased, the breakdown strength of the PI/DCNNS composite films first increased and then decreased. The PI/DCNNS composite film containing 0.5 wt% DCNNS showed the highest breakdown strength of ~300 kV/mm, which is 67.6% and 14.5% higher than the pure PI and the PI/CNNS with the same filler content, respectively.

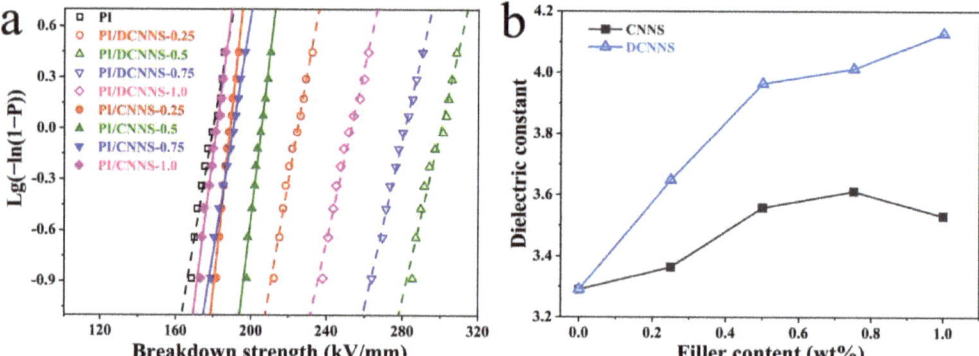

Figure 3. (a) Weibull distribution of electric strength and (b) dielectric constant change law of PI composites.

Owing to the high specific surface area of the two-dimensional nanomaterials, the interface between two-dimensional nanofillers and polymers occupies an important position even at low-filler content [34]. The interface layer surrounding the nanofiller is believed to dissipate the charge and improve the internal electric field distribution, while the dispersed nanofiller acts as a scattering center to help reduce the charge transport. The deep interface traps, introduced by incorporating the nanosheets, can capture carriers, leading to a decrease in mobility and thus an increase in the electrical strength at the working frequency. Compared with the PI/CNNS composite films, the stronger interfacial interaction results in much fewer defects in the PI/DCNNS composite films, and the tight combination of the PI and nanosheets also reduced the fluidity of the PI macromolecular chain and the carrier transport. Furthermore, more polar groups on the surface of the DCNNS enhanced the electron scattering. Therefore, a higher breakdown strength for the PI/DCNNS composite films was observed. The decrease in breakdown strength for filler content above 0.5 wt% should be due to the formation of the filler agglomerations in the PI matrix.

Moreover, the dielectric constant of PI composite films with various filler contents at 1 kHz is shown in Figure 3b. All the PI/DCNNS composite films exhibited a higher dielectric constant than the PI/CNNS composite films with the same filler content, which should also be attributed to the stronger filler–matrix interactions and better filler dispersion.

4. Conclusions

In this work, the DCNNS were prepared using ultrasonic stripping of the bulk g-C_3N_4 followed by polydopamine modification, which was confirmed by XPS. PI composite films with various contents of CNNS and DCNNS were prepared via the in-situ polymerization method. The modification of the g-C_3N_4 by polydopamine was revealed to strengthen the interfacial compatibility between the PI matrix and the DCNNS, and reduce the defects in the composites, resulting in an improvement in breakdown strength. Compared with the pure PI film, the PI/DCNNS composite films showed increased dielectric constant and a much higher breakdown strength. The PI/DCNNS composite film with 0.5 wt% DCNNS showed an increase in breakdown strength by 67.6% and 14.5% as compared to the pure PI and the PI/CNNS composite film with the same filler content, respectively. Therefore, we provided a simple and convenient method to prepare high-voltage insulating materials with excellent breakdown strength and good dielectric properties by adding only a small amount of nanosheets.

Author Contributions: Conceptualization, S.H. (Shaojian He); Data curation, Y.D., Z.W. and S.H. (Shouchao Huo); Formal analysis, Y.D. and Z.W.; Funding acquisition, J.L. and S.H. (Shaojian He); Investigation, Y.D. and S.H. (Shouchao Huo); Resources, J.L.; Supervision, J.L. and S.H. (Shaojian He); Writing—original draft, Y.D.; Writing—review and editing, S.H. (Shaojian He). All authors have read and agreed to the published version of the manuscript.

Funding: This work was supported by the National Natural Science Foundation of China [grant numbers 51973057, 51773058].

Institutional Review Board Statement: Not applicable.

Data Availability Statement: Not applicable.

Acknowledgments: The authors would like to thank the National Natural Science Foundation of China [grant numbers: 51973057, 51773058] for the financial support in completing this work.

Conflicts of Interest: The authors declare no conflict of interest.

References

1. Wan, B.Q.; Li, H.Y.; Xiao, Y.H.; Yue, S.S.; Liu, Y.Y.; Zhang, Q.W. Enhanced dielectric and energy storage properties of BaTiO$_3$ nanofiber/polyimide composites by controlling surface defects of BaTiO$_3$ nanofibers. *Appl. Surf. Sci.* **2020**, *501*, 144243. [CrossRef]
2. Chen, J.; Zhang, X.; Yang, X.; Li, C.; Wang, Y.; Chen, W. High Breakdown Strength and Energy Storage Density in Aligned SrTiO$_3$@SiO$_2$ Core–Shell Platelets Incorporated Polymer Composites. *Membranes* **2021**, *11*, 756. [CrossRef] [PubMed]
3. Cai, Z.M.; Wang, X.H.; Luo, B.C.; Hong, W.; Wu, L.W.; Li, L.T. Nanocomposites with enhanced dielectric permittivity and breakdown strength by microstructure design of nanofillers. *Compos. Sci. Technol.* **2017**, *151*, 109–114. [CrossRef]
4. Li, L.L.; Zhou, B.; Ye, J.F.; Wu, W.; Wen, F.; Xie, Y.C.; Bass, P.; Xu, Z.; Wang, L.W.; Wang, G.F.; et al. Enhanced dielectric and energy-storage performance of nanocomposites using interface-modified anti-ferroelectric fillers. *J. Alloys Compd.* **2020**, *831*, 154770. [CrossRef]

5. Zhao, H.; Yang, C.; Li, N.; Yin, J.H.; Feng, Y.; Liu, Y.Y.; Li, J.L.; Li, Y.P.; Yue, D.; Zhu, C.C.; et al. Electrical and mechanical properties of polyimide composite films reinforced by ultralong titanate nanotubes. *Surf. Coat. Technol.* **2019**, *360*, 13–19. [CrossRef]
6. Yin, P.; Shi, Z.; Sun, L.; Xie, P.; Dastan, D.; Sun, K.; Fan, R. Improved breakdown strengths and energy storage properties of polyimide composites: The effect of internal interfaces of C/SiO$_2$ hybrid nanoparticles. *Polym. Compos.* **2021**, *42*, 3000–3010. [CrossRef]
7. Ai, D.; Li, H.; Zhou, Y.; Ren, L.; Han, Z.; Yao, B.; Zhou, W.; Zhao, L.; Xu, J.; Wang, Q. Tuning Nanofillers in In Situ Prepared Polyimide Nanocomposites for High-Temperature Capacitive Energy Storage. *Adv. Energy Mater.* **2020**, *10*, 1903881. [CrossRef]
8. Li, F.R.; Zhao, J.Y.; Guo, H.Q.; Gao, L.X. Enhanced Energy Storage Performance of Polyimide-based Nanocomposites by Introducing Two-dimensional Nanosheets. *Acta Polym. Sin.* **2020**, *51*, 295–302. [CrossRef]
9. Duan, G.Y.; Cao, Y.T.; Quan, J.Y.; Hu, Z.M.; Wang, Y.; Yu, J.R.; Zhu, J. Bioinspired construction of BN@polydopamine@Al$_2$O$_3$ fillers for preparation of a polyimide dielectric composite with enhanced thermal conductivity and breakdown strength. *J. Mater. Sci.* **2020**, *55*, 8170–8184. [CrossRef]
10. He, S.J.; Hu, J.B.; Zhang, C.; Wang, J.Q.; Chen, L.; Bian, X.M.; Lin, J.; Du, X.Z. Performance improvement in nano-alumina filled silicone rubber composites by using vinyl tri-methoxysilane. *Polym. Test.* **2018**, *67*, 295–301. [CrossRef]
11. Sundar, U.; Lao, Z.; Cook-Chennault, K. Enhanced Dielectric Permittivity of Optimized Surface Modified of Barium Titanate Nanocomposites. *Polymers* **2020**, *12*, 827. [CrossRef] [PubMed]
12. He, S.J.; He, T.F.; Wang, J.Q.; Wu, X.H.; Xue, Y.; Zhang, L.Q.; Lin, J. A novel method to prepare acrylonitrile-butadiene rubber/clay nanocomposites by compounding with clay gel. *Compos. Part B Eng.* **2019**, *167*, 356–361. [CrossRef]
13. Qin, S.L.; Qiu, S.H.; Cui, M.J.; Dai, Z.D.; Zhao, H.C.; Wang, L.P. Synthesis and properties of polyimide nanocomposite containing dopamine-modified graphene oxide. *High Perform. Polym.* **2019**, *31*, 331–340. [CrossRef]
14. Chen, X.; Wang, J.Q.; Zhang, C.; Yang, W.; Lin, J.; Bian, X.M.; He, S.J. Performance of silicone rubber composites using boron nitride to replace alumina tri-hydrate. *High Volt.* **2021**, *6*, 480–486. [CrossRef]
15. Zhu, Y.; Yao, H.; Jiang, P.; Wu, J.; Zhu, X.; Huang, X. Two-Dimensional High-k Nanosheets for Dielectric Polymer Nanocomposites with Ultrahigh Discharged Energy Density. *J. Phys. Chem. C* **2018**, *122*, 18282–18293. [CrossRef]
16. Zhang, Y.-H.; Dang, Z.-M.; Xin, J.H.; Daoud, W.A.; Ji, J.-H.; Liu, Y.; Fei, B.; Li, Y.; Wu, J.; Yang, S.; et al. Dielectric Properties of Polyimide-Mica Hybrid Films. *Macromol. Rapid Comm.* **2005**, *26*, 1473–1477. [CrossRef]
17. He, S.J.; Luo, C.M.; Zheng, Y.Z.; Xue, Y.; Song, X.P.; Lin, J. Improvement in the charge dissipation performance of epoxy resin composites by incorporating amino-modified boron nitride nanosheets. *Mater. Lett.* **2021**, *298*, 130009. [CrossRef]
18. Xue, Y.; Wang, H.; Li, X.; Chen, Y. Synergy boost thermal conductivity through the design of vertically aligned 3D boron nitride and graphene hybrids in silicone rubber under low loading. *Mater. Lett.* **2020**, *281*, 128596. [CrossRef]
19. Liao, Y.F.; Weng, Y.X.; Wang, J.Q.; Zhou, H.F.; Lin, J.; He, S.J. Silicone rubber composites with high breakdown strength and low dielectric loss based on polydopamine coated mica. *Polymers* **2019**, *11*, 2030. [CrossRef]
20. Gaddam, S.K.; Pothu, R.; Boddula, R. Graphitic carbon nitride (g-C3N4) reinforced polymer nanocomposite systems—A review. *Polym. Compos.* **2020**, *41*, 430–442. [CrossRef]
21. Xu, S.; Wang, J.; Lin, L.; Valério, A.; He, D. Synthesis of carbon nitride nanosheets with tunable size by hydrothermal method for tetracycline degradation. *Mater. Lett.* **2020**, *264*, 127005. [CrossRef]
22. Kang, S.; He, M.; Chen, M.; Wang, J.; Zheng, L.; Chang, X.; Duan, H.; Sun, D.; Dong, M.; Cui, L. Ultrafast plasma immersion strategy for rational modulation of oxygen-containing and amino groups in graphitic carbon nitride. *Carbon* **2020**, *159*, 51–64. [CrossRef]
23. Zhu, L.; You, L.J.; Shi, Z.X.; Song, H.J.; Li, S.J. An investigation on the graphitic carbon nitride reinforced polyimide composite and evaluation of its tribological properties. *J. Appl. Polym. Sci.* **2017**, *134*, 45403. [CrossRef]
24. Wang, Y.Y.; Zhang, X.; Ding, X.; Zhang, P.; Shu, M.T.; Zhang, Q.; Gong, Y.; Zheng, K.; Tian, X.Y. Imidization-induced carbon nitride nanosheets orientation towards highly thermally conductive polyimide film with superior flexibility and electrical insulation. *Compos. Part B Eng.* **2020**, *199*, 108267. [CrossRef]
25. He, S.J.; Wang, J.Q.; Hu, J.B.; Zhou, H.F.; Nguyen, H.; Luo, C.M.; Lin, J. Silicone rubber composites incorporating graphitic carbon nitride and modified by vinyl tri-methoxysilane. *Polym. Test.* **2019**, *79*, 106005–106010. [CrossRef]
26. Song, H.J.; Li, L.Y.; Yang, J.; Jia, X.H. Fabrication of Polydopamine-Modified Carbon Fabric/Polyimide Composites With Enhanced Mechanical and Tribological Properties. *Polym. Compos.* **2019**, *40*, 1911–1918. [CrossRef]
27. Su, C.; Xue, F.; Xu, F.L.; Li, T.S.; Xin, Y.S.; Wang, M.M. Tribological Properties of Surface-Modified Graphene Filled Carbon Fabric/Polyimide Composites. *J. Macromol. Sci. Part B* **2019**, *58*, 603–618. [CrossRef]
28. McCaffrey, M.; Hones, H.; Cook, J.; Krchnavek, R.; Xue, W. Geometric analysis of dielectric failures in polyimide/silicon dioxide nanocomposites. *Polym. Eng. Sci.* **2019**, *59*, 1897–1904. [CrossRef]
29. Yu, Z.X.; Li, F.; Yang, Q.B.; Shi, H.; Chen, Q.; Xu, M. Nature-Mimic Method To Fabricate Polydopamine/Graphitic Carbon Nitride for Enhancing Photocatalytic Degradation Performance. *ACS Sustain. Chem. Eng.* **2017**, *5*, 7840–7850. [CrossRef]
30. Xia, P.; Mingjin, L.; Cheng, B.; Yu, J.; Zhang, L. Dopamine Modified g-C3N4 and Its Enhanced Visible-Light Photocatalytic H2-Production Activity. *ACS Sustain. Chem. Eng.* **2018**, *6*, 8945–8953. [CrossRef]
31. Kim, C.; Jiang, P.K.; Liu, F.; Hyon, S.; Ri, M.G.; Yu, Y.; Ho, M. Investigation on dielectric breakdown behavior of thermally aged cross-linked polyethylene cable insulation. *Polym. Test.* **2019**, *80*, 106045. [CrossRef]

32. Zhao, Y.K.; Zhang, G.Q.; Guo, R.R.; Yang, F.Y. The Breakdown Characteristics of Thermostable Insulation Materials under High-Frequency Square Waveform. *IEEE Trans. Dielectr. Electr. Insul.* **2019**, *26*, 1073–1080. [CrossRef]
33. Silau, H.; Stabell, N.B.; Petersen, F.R.; Pham, M.; Yu, L.Y.; Skov, A.L. Weibull Analysis of Electrical Breakdown Strength as an Effective Means of Evaluating Elastomer Thin Film Quality. *Adv. Eng. Mater.* **2018**, *20*, 1800241. [CrossRef]
34. Beier, C.W.; Sanders, J.M.; Brutchey, R.L. Improved Breakdown Strength and Energy Density in Thin-Film Polyimide Nanocomposites with Small Barium Strontium Titanate Nanocrystal Fillers. *J. Phys. Chem. C* **2013**, *117*, 6958–6965. [CrossRef]

Article

Molecular Dynamics Simulation of Cracking Process of Bisphenol F Epoxy Resin under High-Energy Particle Impact

Yunqi Xing [1,2], Yuanyuan Chen [1,2], Jiakai Chi [1,2], Jingquan Zheng [1,2], Wenbo Zhu [3] and Xiaoxue Wang [1,2,*]

1. State Key Laboratory of Reliability and Intelligence of Electrical Equipment, Hebei University of Technology, Tianjin 300401, China; yqxing@hebut.edu.cn (Y.X.); yychebut@163.com (Y.C.); jiakaichi@gmail.com (J.C.); jqzhebut@163.com (J.Z.)
2. Key Laboratory of Electromagnetic Field and Electrical Apparatus Reliability of Hebei Province, Hebei University of Technology, Tianjin 300401, China
3. China Southern Power Grid Research Institute Co., Ltd., Guangzhou 510080, China; zhuwb@csg.cn
* Correspondence: xxwang@hebut.edu.cn; Tel.: +86-1382-192-9690

Abstract: The current lead insulation of high-temperature superconductivity equipment is under the combined action of large temperature gradient field and strong electric field. Compared with a uniform temperature field, its electric field distortion is more serious, and it is easy to induce surface discharge to generate high-energy particles, destroy the insulation surface structure and accelerate insulation degradation. In this paper, the degradation reaction process of bisphenol F epoxy resin under the impact of high-energy particles, such as O_3^-, HO^-, H_3O^+ and NO^+, is calculated based on ReaxFF simulation. According to the different types of high-energy particles under different voltage polarities, the micro-degradation mechanism, pyrolysis degree and pyrolysis products of epoxy resin are analyzed. The results show that in addition to the chemical reaction of high-energy particles with epoxy resin, their kinetic energy will also destroy the molecular structure of the material, causing the cross-linked epoxy resin to pyrolyze, and the impact of positive particles has a more obvious impact on the pyrolysis of epoxy resin.

Keywords: epoxy resin; partial discharge; active product; electro-thermal dissociation; reactive molecular dynamics

1. Introduction

Epoxy resin is currently the polymer thermosetting insulating material with the highest share in industrial applications [1–3]. Among them, the bisphenol F epoxy resin is widely used in the insulation of extreme environment equipment, such as aerospace, superconductivity, medical and military industries due to its good low temperature performance [4,5]. For example, in the terminal of superconducting power equipment, the current lead is insulated with epoxy resin under the coupling action of a strong electric field and a large temperature gradient field, which easily induces partial discharge and accelerates insulation aging. Previous studies have shown that partial discharge (PD) is the main cause of insulation deterioration breakdown. The impact, corrosion and thermal effects of high-energy particles and active products produced by PD on the surface of insulating material will greatly influence the transient dielectric strength and aging process of insulating material [6–8].

At present, the research on insulation properties of polymer materials such as epoxy resins is mostly based on experiments. For example, there have been many studies on the extraction of characteristic parameters of partial discharge signal and the characterization methods [9–11] between partial discharge signal and material [12–14] insulation state and degradation process, and the aging cracking mechanism is analyzed based on the experimental data. However, it is difficult to reveal the aging micro-mechanism and degradation cracking process of insulation materials under partial discharge (PD) by experimental research, physical modeling and other methods. Furthermore, the research on simulation analysis at atomic level is mostly limited to the thermal decomposition process of macromolecule materials [15,16]. Few studies have been conducted on the influence mechanism of high-energy particle impact on aging and cracking of epoxy resin materials at atomic level.

ReaxFF force field uses quantum chemistry theory to judge the formation and breakage of chemical bonds and uses classical molecular dynamics force field to simulate the molecular conformation of the system. Combining the two effectively, ReaxFF force field can be widely used to simulate pyrolysis, detonation, particle impact, chemical reaction and other phenomena of various materials. By simulating the heating and cracking process of epoxy resin under microwave heating and conventional heating, Zhang Yiming reveals the cause of thermal runaway of macromolecular polymer system caused by microwave heating from a micro level [17]. Farzin Rahmani studied the mass loss, surface damage and penetration depth of AO in polyimide materials with different grafting methods bombarded by Atomic Oxygen (AO) based on ReaxFF force field [18]. Morrissey verified the feasibility of using ReaxFF force field to study the corrosion resistance of spacecraft metal materials under the impact of high-energy particles [19]. The above research results show that ReaxFF force field can be used to simulate the aging process of epoxy resin materials under the impact of partial discharge particles.

Under the action of strong electric field, air (main effective components are N_2, O_2, H_2O) collides with electrons emitted by electrodes and positive and negative ions generated by ionization during partial discharge, resulting in particles with high energy and active products [7]. The high-energy particles generated during the discharge process have high kinetic energy under the action of strong electric field, and they continuously impact the epoxy resin surface. When impact occurs, the kinetic energy will be converted into internal energy, which will make the local temperature of insulation material rise sharply. At the same time, high-energy particles themselves will also have complex chemical reaction with insulation material, destroying the cross-linking structure of epoxy resin and accelerating the aging process of insulation. Chen X and Pavlik M carried out corona discharge studies on air under different humidity conditions and analyzed the species and activities of its products. The results showed that the highest reactivity and content was O_3^-, followed by HO^- free radicals [20,21]. Wayne Sieck studied the corona discharge products of humid air and found that more than ten ion products with strong reaction activity were produced at the positive electrode, among which H_3O^+ and O_2^+ had the strongest reaction activity [22]. Y. Ehara et al. studied the degradation process of epoxy resin surface under partial discharge in different humidity air environments. The experiments proved that partial discharge can produce ozone, active oxygen atoms, nitrogen oxide, nitric acid and other active products [23]. Therefore, taking the above ion products as the object, it is of great significance to study the effect of high-energy particle impact produced by partial discharge on the cracking process of epoxy resin materials, so as to reveal its deterioration mechanism from the micro level.

In this paper, the interfacial model of cross-linked bisphenol F type epoxy resin and the models of hydration hydrogen ion (H_3O^+), nitric oxide ion (NO^+), hydroxide ion (HO^-) and ozone ion (O_3^-) were established. ReaxFF force field was used to inject the above four particles into the epoxy resin interface. The mechanism of high-energy particle impact caused by partial discharge on the aging and cracking process of epoxy resin and its reaction products were analyzed. The micro-aging mechanism of epoxy insulation material under partial discharge was revealed at atomic level.

2. Molecular Simulation Design
2.1. Model Building

In this paper, bisphenol F epoxy resin matrix and diethyltoluene diamine (DETDA) curing agent are used as the simulation research object to build the model. Figure 1 is the flow chart for establishing the cross-linked epoxy resin interface model. According to the steps in Figure 1, the periodic cross-linked epoxy resin interface model on the XOY plane is built, and the impact of high-energy particles on the surface model is simulated. According to the characteristics of ReaxFF force field, when the total number of atoms in the model reaches about 4000, stable and accurate simulation results can be obtained while guaranteeing calculation efficiency. Therefore, an amorphous mixture model with 60 DGEBF and 30 DETDA molecules and a density of 0.6 g/cm^3 is established in this paper. Before the establishment of the cross-linked epoxy resin model, the periodicity of the model in the Z-axis direction is blocked by adding a vacuum layer of 15 Å, in which the argon molecular layer is filled. The cross-linking process of the model in the Z-direction is isolated to establish the periodic interface model on the XOY plane and then cross-link the amorphous model [24]. After the cross-linking procedure is completed, the argon molecular layer on the surface is deleted, and a surface model of cross-linked epoxy resin with 95% degree of cross-linking is obtained. The obtained model is imported into AMS software and a density of 1.19 g/cm^3 model is obtained by 100 ps relaxation at 298 K and 0.1013 MPa using NPT system [24]. The addition of a vacuum layer in the Z-axis of the cell provides space for the motion of energetic particles, and the final cell volume is 30.195793 Å × 30.195793 Å × 100.00 Å. The position of high-energy particles is set on a plane at the top of the Z-axis of the cell. The horizontal position of each high-energy particle is randomly distributed, and the incident direction is vertical to the XOY plane.

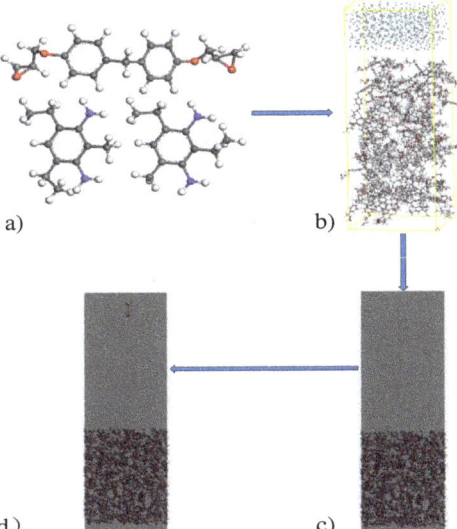

Figure 1. Construction process of cross-linked epoxy resin simulation model. (**a**) Molecular structure of bisphenol F epoxy resin monomer and curing agent monomer; (**b**) Constructing epoxy resin cross-linking interface model; (**c**) Import ReaxFF software; (**d**) Arrange the position and angle of incidence of high-energy particles.

2.2. Simulation Settings

In order to simulate the dynamic and thermodynamic effects of high-energy particles bombarding the insulation interface, a micro-canonical ensemble (NVE) is selected for the simulation system, in which the atomic number N, volume V and energy E remain unchanged. When high-energy particles with certain kinetic energy are incident, the energy E of the system rises, while the form and parameters of potential in NVE ensemble are fixed, and the position of the particles is determined by the motion of the particles themselves. The energy of the system will change the temperature of the system. Therefore, the use of

NVE ensemble can simultaneously reflect the kinetic and thermodynamic effects caused by high-energy electric particle bombardment.

The initial temperature of the simulation system is 298 K, and the air pressure is set to 0.1013 MPa. Due to the high kinetic energy of high-energy particles in the simulation process, it is necessary to set a short simulation step in order to make the simulation converge. The simulation step size is 0.05 fs, the total number of simulation steps is 2.0×10^6 steps, and the total time is 100 ps. The four energetic particles H_3O^+, NO^+, OH^- and O_3^- carry energy of 1, 4, 7 and 10 eV in different combinations and are incident along the epoxy resin interface in the negative direction of Z-axis at intervals of 1 ps. It takes 100 ps to simulate the accumulation effect of high-energy particle impact under pulsed partial discharge.

3. Simulation Results and Analysis

3.1. Analysis of Small Molecular Products of Epoxy Resin Impacted by High-Energy Particles

In the simulation of high-energy particle impact, all atoms in the epoxy resin surface model are cross-linked and can be regarded as one molecule. Therefore, the total number of molecules in a cell can represent the volume of debris and damage caused by impact of energetic particles. The volume of debris produced by four high-energy particles impacting the surface of insulating material with different energy tends to change, as shown in Figure 2.

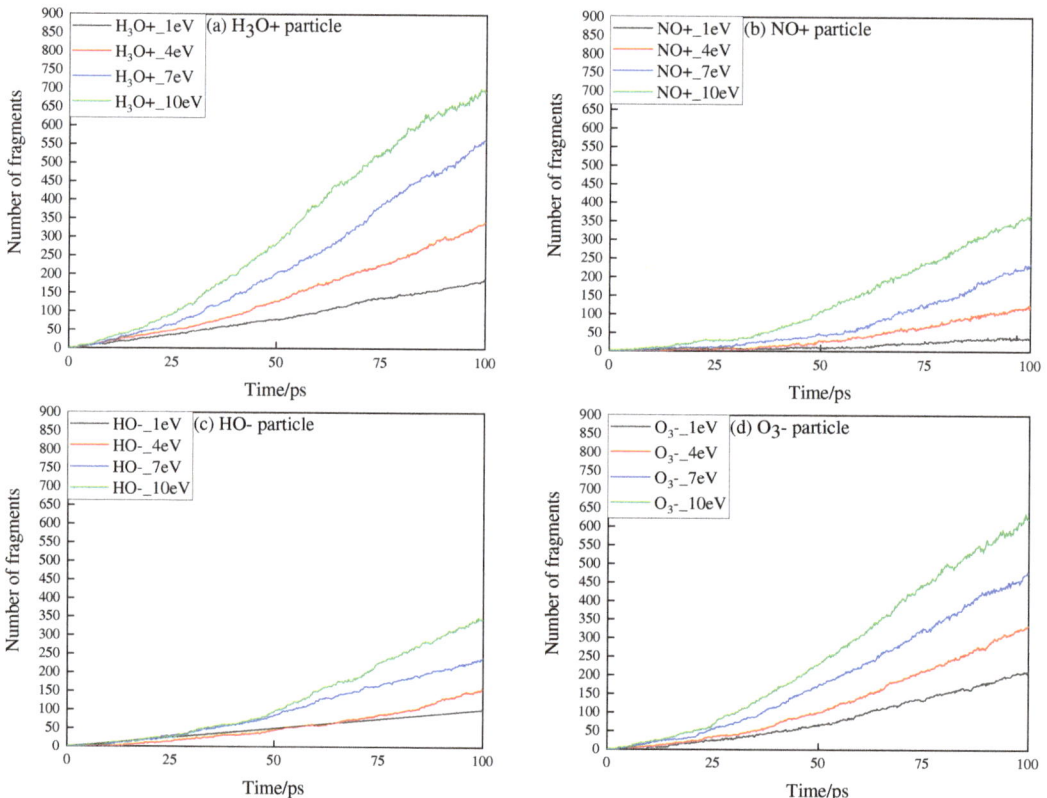

Figure 2. Variation trend of fragment number caused by impact of four high-energy particles on epoxy resin interface with different energy. (**a**) Changes in the number of H_3O^+ particle fragments with different energies; (**b**) Changes in the number of NO^+ ion fragments with different energies; (**c**) Changes in the number of HO^- ion fragments with different energies; (**d**) Changes in the number of O_3^- ion fragments with different energies.

First, as the energy carried increases, the volume of debris produced by the impact of various high-energy particles increases. Among them, H_3O^+ and O_3^- are the strongest ones to destroy the surface of epoxy resin. At the same time, the increase in the volume of debris produced is not linear with the increase in the energy of high-energy particles. When all types of high-energy particles carry more or less energy, there is little difference in the volume of debris generated in the first half of the simulation process, while the speed of debris generated by the impact of high-energy particles with higher energy in the second half is greatly increased. This is mainly due to the fact that the kinetic energy carried by high-energy particles when impacted on the surface of insulating material changes into thermal energy, which makes the cell temperature rise sharply.

The change trend of local temperature at the interface of epoxy resin impacted by high-energy particles with different energy is shown in Figure 3. During the impact process of H_3O^+ carrying different energies, the temperature in the cell will rise to over 4000 K, which will cause the temperature in the local area of the epoxy material surface to rise sharply. When the energy of high-energy particles is low, most of the incident particles cannot directly destroy the surface of the epoxy resin, usually only because the high-energy particles have their own chemical reaction activity to attack the epoxy resin. The collision between high-energy particles and the surface of epoxy resin results in heat energy which causes local temperature rise, while the thermal conductivity of epoxy resin is poor, at only 0.2~2.2 W/mK. Local temperature rise caused by particle impact cannot diffuse in time, which leads to rapid accumulation of local heat and temperature rise, making epoxy resin subject to more severe erosion.

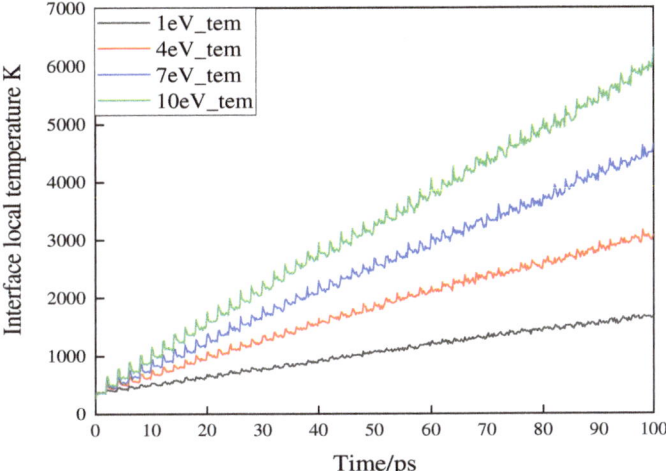

Figure 3. Local temperature change trend of epoxy resin.

The simulation results of main small molecular products generated by the impact of high-energy particles with various parameters are analyzed to study the aging process of small molecular products and insulating materials analyzed in partial discharge experiments. Epoxy resin is an organic insulating material, and the damage of carbon skeleton structure is directly related to the formation of carbon-containing, small molecular products. It can be found that when NO^+ and HO^- particles interact, the amount of C2 and C3 products in the small molecule products produced is very low, so the mass loss under the corresponding conditions is smaller. In the impact simulation of H_3O^+, the small molecular products of C1 and C2 produced by H_3O^+ are significantly higher than those of other particles. After 40 ps, as the temperature of the simulation system increases, the epoxy resin molecules are more likely to be damaged by particle impact, and the rate of small molecule products production in each system increases significantly. In addition to carbon-containing, small molecular products, the main small molecular products under the action of particles are H_2O, H_2, CO_2, etc. If there are few H_3O^+ and O_3^- particles in the environment where partial discharge occurs, it is difficult to cause significant damage to the surface of epoxy resin when there are only a few NO^+ and HO^- particles. H_3O^+ will be released, resulting in a large amount of hydrogen gas. Because of its strong oxidation, O_3^- particles will further corrode the epoxy resin while breaking the molecular structure of the epoxy resin under impact.

3.2. Surface Damage and Mass Loss Characteristics of Epoxy Resin

The interface structure change of epoxy resin under the action of high-energy particles is shown in Figure 4. The analysis of model structure changes during simulation shows that with the development of simulation process, the small molecular debris generated on the surface of epoxy resin diffuses in the empty area along the Z-axis when the energy of high-energy particles is low. However, most of the small molecular debris are still concentrated in the lower half, and the average molecular mass of the small molecular debris is higher than that of high-energy particles. At the same time, when the energy carried by the high-energy particles is low, during the impact process of the high-energy particles, most of the high-energy particles rebound after striking the surface of the epoxy resin without reacting with the epoxy resin or generating small molecular debris.

When the energy of high-energy particles is high, obvious damage occurs on the epoxy resin surface at an earlier stage. The high-energy particles with 7 eV already caused relatively long and deeper void damage at an earlier stage compared to the low-energy particles. In addition, most of the high-energy particles can impact on the surface of epoxy resin to produce small molecular debris, and the average incident depth is significantly increased.

Normalized mass statistics of epoxy resin under high-energy particle impact during simulation are shown in Figure 5. The trend of epoxy resin variation can indicate the degree of erosion of epoxy resin under impact of high-energy particles. According to GBT 20112-2006 (the evaluation and identification of electrical insulation structure), when the quality loss of insulation material reaches 5%, the insulation material can be considered to be ineffective [25]. It can be seen that the mass loss of epoxy resin material is small when the energy of NO^+ and HO^- particles is low, and the rate of mass loss is small during the simulation process of the first 50 ps or so. In combination with the side-cut snapshot of the simulation model in Figure 4, it can be found that although the impact action of high-energy particles will cause the cross-linkage structure of epoxy resin to react or break with high-energy particles, the amount of small molecular products that completely break off the surface of epoxy resin is small. The mass loss of epoxy resin under the action of H_3O^+ and O_3^- particles is significantly higher than that under the other two conditions. At the end of the simulation process, the mass loss of the model under H_3O^+ is higher than that under the same condition of O_3^- particles. However, O_3^- is easy to produce during impact, and its strong oxidation will also cause corrosion to the epoxy resin, which leads to the mass loss of epoxy resin under the action of O_3^- particles reaching 5% earlier, making the insulating material invalid earlier.

3.3. The Main Small Molecule Products of Epoxy Resin Surface Damage

It is very difficult to analyze the small molecular products produced by the impact of high-energy particles of partial discharge in experiments, and the generation of these small molecular products is directly related to the aging of insulating materials. Therefore, the generation and mechanism of small molecular products caused by the impact of various high-energy particles on epoxy resin are analyzed.

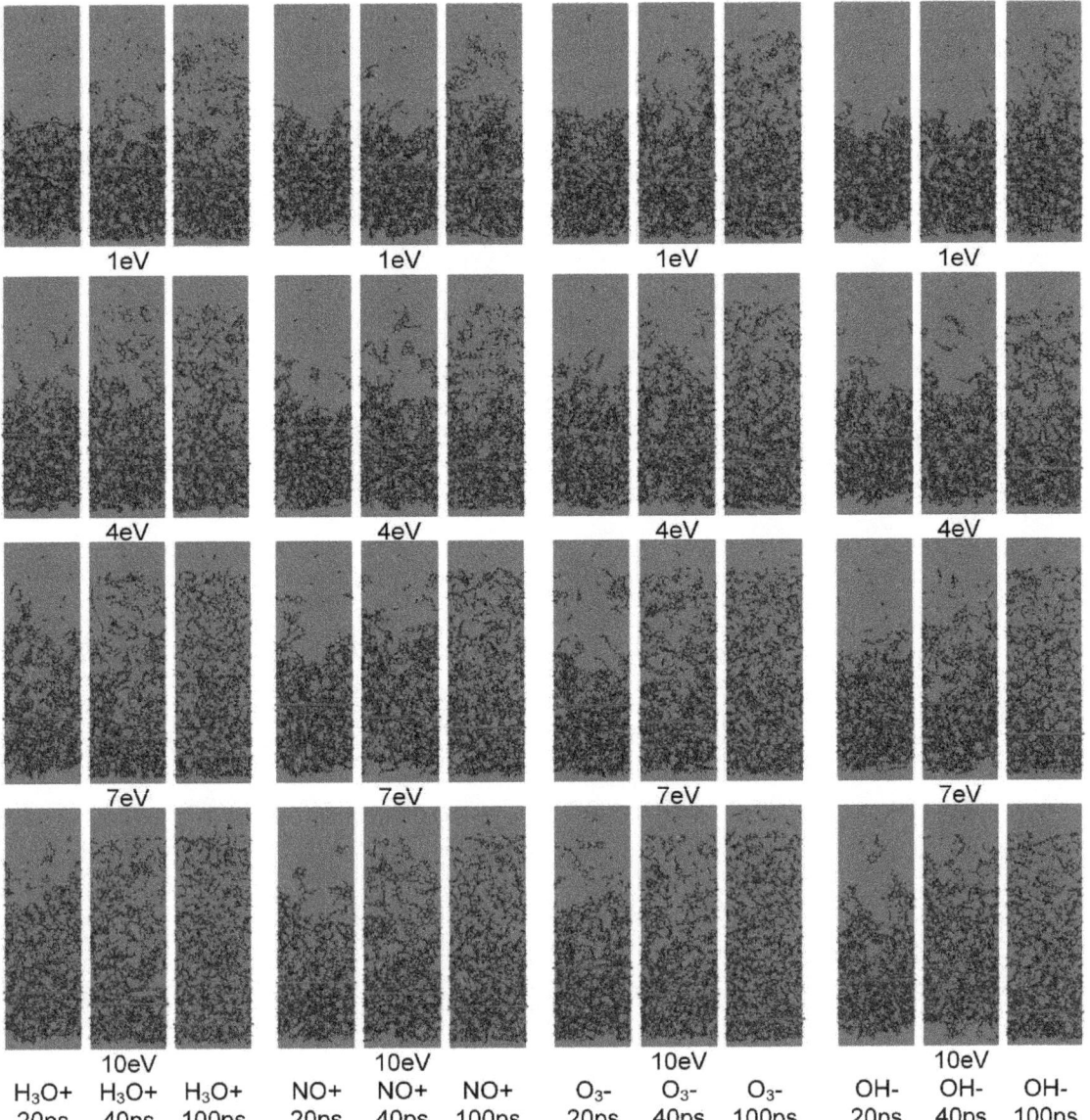

Figure 4. Structural change of interface model under the action of high-energy particles.

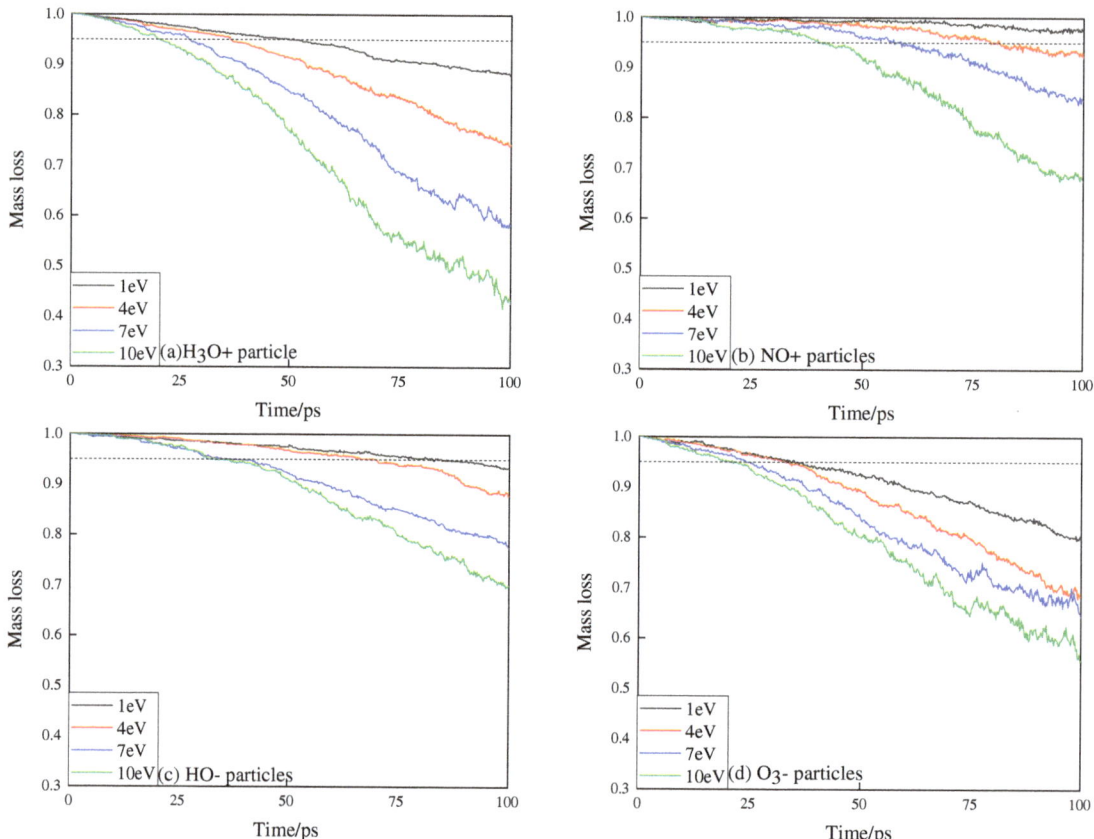

Figure 5. Normalized mass loss curve of epoxy resin under the action of high-energy particles. (**a**) Normalized mass loss curve of epoxy resin under the action of H_3O^+ particles; (**b**) Normalized mass loss curve of epoxy resin under the action of NO^+ particles; (**c**) Normalized mass loss curve of epoxy resin under the action of HO^- particles; (**d**) Normalized mass loss curve of epoxy resin under the action of O_3^- particles.

Epoxy resin is an organic insulating material, and the damage of carbon skeleton structure is directly related to the formation of carbon-containing, small molecular products. It can be found that when NO^+ and HO^- particles interact, the amount of C2 and C3 products in the small molecule products produced is very low, so the mass loss under the corresponding conditions is smaller. In the impact simulation of H_3O^+, the amount of small molecular products of C1 and C2 produced by H_3O^+ is significantly higher than that of other high-energy particles. After 40 ps, with the increase in the temperature of the simulation system, epoxy resin molecules are more likely to be damaged by high-energy particle impact, and the production rate of small molecular products in each system increases significantly. In addition to carbon-containing, small molecular products, the main small molecular products under the action of high-energy particles are H_2O, H_2, CO_2, etc.

Combined with the mass loss information in Figure 5, it can be found that if the H_3O^+ and O_3^- particles produced in the environment where partial discharge occurs are small, it is difficult to cause significant damage to the surface of epoxy resin when there are only a few NO^+ and HO^- particles. H_3O^+ will release hydrogen ions after impact contact with epoxy resin, resulting in a large amount of hydrogen gas. Because of its strong oxidation,

O_3^- particles will further corrode the epoxy resin while breaking the molecular structure of the epoxy resin under impact.

3.4. Polar Effect of Surface Damage of Epoxy Resin

In DC discharge experiments, the positive and negative discharge polarities often have effects on the starting discharge voltage, flashover voltage, damage to insulating materials, etc. Therefore, this paper presents a comparative study on the change trend of the number and mass loss of small molecular fragment products of epoxy resin under the combined impact of positive and negative high-energy particles, as shown in Figures 6 and 7.

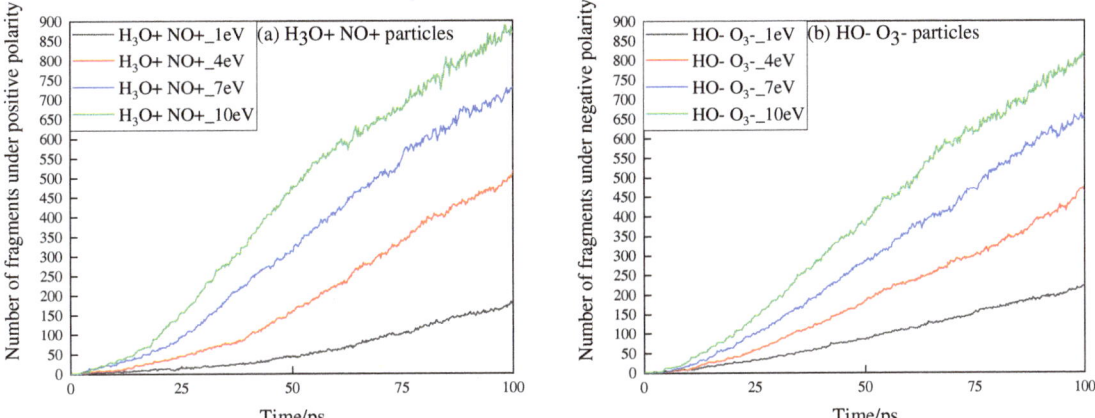

Figure 6. Variation trend of small molecular fragment products of epoxy resin under the action of positive and negative high-energy particles. (**a**) Variation trend of small molecular fragment products of epoxy resin under the action of positive high-energy particles; (**b**) Variation trend of small molecular fragment products of epoxy resin under the action of negative high-energy particles.

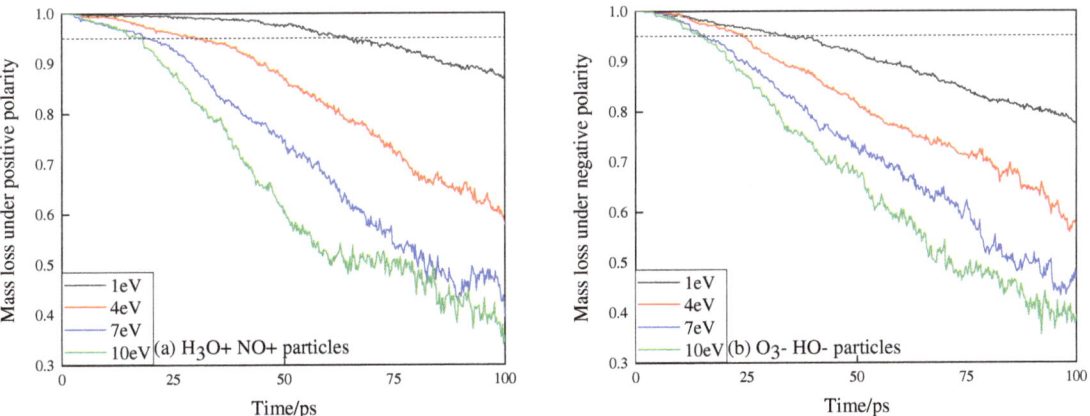

Figure 7. Normalized mass loss curve of epoxy resin under the action of positive and negative high-energy particles. (**a**) Normalized mass loss curve of epoxy resin under the action of positive high-energy particles; (**b**) Normalized mass loss curve of epoxy resin under the action of positive and negative high-energy particles.

In Figure 6, under various energy conditions carried by high-energy particles, the mass loss of epoxy resin under the action of negative high-energy particles increases faster than that under the action of positive high-energy particles. At 14.4 ps, the mass loss of the epoxy resin model subjected to the impact of 10 eV high-energy particles reaches 5%, which is higher than that of the epoxy resin model subjected to the simultaneous action of single positive-polar high-energy particles. Under the same energy condition, the mass loss of epoxy resin reaches 5% at 18 ps under the action of positive energetic particles. When the energy carried by the high-energy particles is low, the difference between the positive and negative high-energy particles on the erosion of epoxy resin is greater. When the energy carried by high-energy particles is 1 eV, the time for the mass loss to reach 5% under the action of positive and negative high-energy particles is 64.4 ps and 34 ps, respectively. When the energy of charged particles is 4 eV, it is 29.8 ps and 24 ps, respectively. This shows that the working life of epoxy resin is longer under the action of positive energetic particles.

With the development of impact, the temperature of the simulation system gradually increases. The local temperature rise of epoxy resin interface under the impact of positive and negative high-energy particles is limited, as shown in Figure 8, and the temperature under the action of high-energy particles with negative polarity is always higher than that under positive polarity. This leads to the impact of positive energetic particles at the same time, which will locally raise the temperature of the epoxy resin surface to a higher temperature and aggravate the erosion. At the same time, the unstable O_3^- and HO^- in the negative particles have strong oxidation after collision with the epoxy resin, which makes the cross-linking structure of the epoxy resin more fragile and the surface of the epoxy resin more vulnerable to erosion.

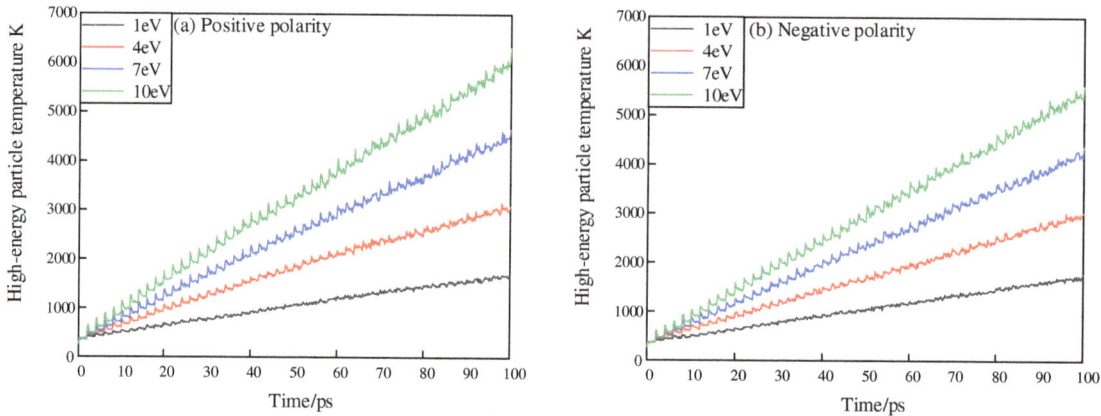

Figure 8. Temperature variation curve of simulation system under the action of positive and negative high-energy particles. (**a**) Temperature variation curve of simulation system under the action of positive high-energy particles; (**b**) Temperature variation curve of simulation system under the action of negative high-energy particles.

4. Conclusions

The effect of high-energy particle impact on the insulation degradation of epoxy resin is analyzed by ReaxFF simulation force field. The decomposition process, mechanism and characteristics of insulating materials were analyzed at the micro atomic level, and the following conclusions are drawn:

(1) The high-energy particles produced by partial discharge will cause irreversible erosion on the surface of epoxy resin material. The simulation results show that O_3^- particles have the strongest erosion effect on the surface of epoxy resin material under the action of high-energy particles alone. High-energy particles not only impact the surface of epoxy resin and interrupt its chemical structure, but also convert its kinetic energy into potential energy, resulting in a sharp rise in the system temperature and the cracking of epoxy resin. This also shows that when the insulating material is subjected to partial discharge, the local temperature will rise sharply, resulting in insulation deterioration.

(2) In the simulation of various high-energy particles carrying different sizes of energy, the erosion effect on epoxy resin is characterized by mass loss. It can be found that compared with charged particles with higher energy, the erosion effect of charged particles with lower energy on epoxy resin materials is less in the initial stage, but it is very obvious in the later stage of impact development and accumulation.

(3) In the study of the positive and negative polarity of high-energy particles, it is found that the corrosion effect of negative charged particles on epoxy resin is more obvious. This is because the temperature rise of the system increases under the impact of negative charged particles, and the negative particles O_3^- and HO^- are unstable and have strong oxidation, which makes the cross-linking structure of epoxy resin more fragile.

Author Contributions: Conceptualization, W.Z. and X.W.; methodology, J.Z.; software, J.C.; investigation, J.C.; resources, Y.X.; data curation, Y.C. and X.W.; writing—original draft preparation, Y.C. and Y.X.; writing—review and editing, Y.X., Y.C. and X.W.; funding acquisition, Y.X. All authors have read and agreed to the published version of the manuscript.

Funding: This research was funded by a project supported by the Young Scientists Fund of the National Natural Science Foundation of China, grant number 51907047. This research was funded by a project supported by the Young Scientists Fund of the Natural Science Foundation of Hebei Province, China, grant number E2020202159. This research was funded by a project supported by the Young Program of the Natural Science Foundation of Tianjin, China, grant number 20JCQNJC00690. This research was funded by a project supported by State Key Laboratory of Reliability and Intelligence of Electrical Equipment, grant number EERI_PI2020002, Hebei University of Technology.

Data Availability Statement: All data are available in the main text.

Conflicts of Interest: The authors declare no conflict of interest.

References

1. Li, Q.M.; Huang, X.; Liu, T.; Yan, J.; Wang, Z.; Zhang, Y.; Lu, X. Application Progresses of Molecular Simulation Methodology in the Area of High Voltage Insulation. *Trans. China Electrotech. Soc.* **2016**, *31*, 1–13.
2. Kolesov, S.N. The influence of morphology on the electric strength of polymer insulation. *IEEE Trans. Electr. Insul.* **1980**, *15*, 382–388. [CrossRef]
3. Gubanski, S.M. Modern outdoor insulation-concerns and challenges. *IEEE Electr. Insul. Mag.* **2005**, *21*, 5–11. [CrossRef]
4. He, F.; Su, S.; Fu, H.; Zhou, L. Review on Modification of Epoxy Resin Doped Micro-Nano Oxide Particles. *High Volt. Appar.* **2020**, *56*, 94–103.
5. Shan, X.R. Low Temperature Thermal Properties of Epoxy Resin Matrix Composites as Insulation Materials for Large Scale Superconducting Magnet. Ph.D. Thesis, Tianjin University, Tianjin, China, 2017.
6. Katz, M.; Theis, R.J. New high temperature polyimide insulation for partial discharge resistance in harsh environments. *IEEE Electr. Insul. Mag.* **1997**, *13*, 24–30. [CrossRef]
7. Cao, X.L.; Zhong, L.S. *Principles of Electrical Insulation Technology*; China Machine Press: Beijing, China, 2009; pp. 115–120.
8. Hillborg, H.; Gedde, U.W. Hydrophobicity recovery of polydimethylsiloxane after repeated exposure to corona discharges. Influence of crosslink density. In Proceedings of the 1999 Annual Report Conference on Electrical Insulation and Dielectric Phenomena, Austin, TX, USA, 17–20 October 1999; Volume 2, pp. 751–755.
9. Zhou, X.; Zhou, C.; Kemp, I.J. An improved methodology for application of wavelet transform to partial discharge measurement denoising. *IEEE Trans. Dielectr. Electr. Insul.* **2005**, *12*, 586–594. [CrossRef]
10. Pan, C.; Chen, G.; Tang, J.; Wu, K. Numerical modeling of partial discharges in a solid dielectric-bounded cavity: A review. *IEEE Trans. Dielectr. Electr. Insul.* **2019**, *26*, 981–1000. [CrossRef]
11. Liao, Y.; Weng, Y.; Wang, J.; Zhou, H.; Lin, J.; He, S. Silicone Rubber Composites with High Breakdown Strength and Low Dielectric Loss Based on Polydopamine Coated Mica. *Polymers* **2019**, *11*, 2030. [CrossRef]
12. Fabina, J.H.; Hartmann, S.; Hamidi, A. Partial discharge failure analysis of AlN substrates for IGBT modules. *Microelectron. Reliab.* **2004**, *44*, 1425–1430.
13. Li, J.H.; Han, X.T.; Liu, Z.; Li, Y. Review on Partial Discharge Measurement Technology of Electrical Equipment. *High Volt. Eng.* **2015**, *41*, 2583–2601.
14. He, S.; Wang, J.; Yu, M.; Xue, Y.; Hu, J.; Lin, J. Structure and Mechanical Performance of Poly(vinyl Alcohol) Nanocomposite by Incorporating Graphitic Carbon Nitride Nanosheets. *Polymers* **2019**, *11*, 610. [CrossRef]
15. Diao, Z.J.; Zhao, Y.M.; Chen, B.; Duan, C.L. Thermal Decomposition of Epoxy Resin Contained in Printed Circuit Boards from Reactive Dynamics Using the ReaxFF Reactive Force Field. *Acta Chim. Sin.* **2012**, *70*, 2037–2044. [CrossRef]

16. Diao, Z.; Zhao, Y.; Chen, B.; Duan, C.; Song, S. ReaxFF reactive forcefield for molecular dynamics simulations of epoxy resin thermal decomposition with model compound. *J. Anal. Appl. Pyrolysis* **2013**, *104*, 618–624. [CrossRef]
17. Zhang, Y.M.; Li, J.L.; Wang, J.P.; Yang, X.S.; Shao, W.; Xiao, S.Q.; Wang, B.Z. Research on epoxy resin decomposition under microwave heating by using ReaxFF molecular dynamics simulations. *RSC Adv.* **2014**, *4*, 17083–17090. [CrossRef]
18. Rahmani, F.; Nouranian, S.; Li, X.; Al-Ostaz, A. Reactive Molecular Simulation of the Damage Mitigation Efficacy of POSS-, Graphene-, and Carbon Nanotube-Loaded Polyimide Coatings Exposed to Atomic Oxygen Bombardment. *ACS Appl. Mater. Interfaces* **2017**, *9*, 12802–12811. [CrossRef] [PubMed]
19. Morrissey, L.S.; Handrigan, S.M.; Nakhla, S.; Rahnamoun, A. Erosion of Spacecraft Metals due to Atomic Oxygen: A Molecular Dynamics Simulation. *J. Spacecr. Rocket.* **2019**, *56*, 1231–1236. [CrossRef]
20. Pavlik, M.; Skalny, J.D. Generation of $[H_3O]^+.(H_2O)_n$ clusters by positive corona discharge in air. *Rapid Commun. Mass Spectrom.* **1997**, *11*, 1757–1766. [CrossRef]
21. Chen, X.; Lan, L.; Lu, H.; Wang, Y.; Wen, X.; Du, X.; He, W. Numerical simulation of Trichel pulses of negative DC corona discharge based on a plasma chemical model. *J. Phys. D Appl. Phys.* **2017**, *50*, 395202. [CrossRef]
22. Sieck, L.W.; Buckley, T.J.; Herron, J.T.; Green, D.S. Pulsed electron-beam ionization of humid air and humid air/toluene mixtures: Time-resolved cationic kinetics and comparisons with predictive models. *Plasma Chem. Plasma Process.* **2001**, *21*, 441–457. [CrossRef]
23. Ehara, Y.; Aono, K. Degradation analysis of epoxy resin surfaces exposed to partial discharge. In Proceedings of the 2016 IEEE Conference on Electrical Insulation and Dielectric Phenomena (CEIDP), Toronto, ON, Canada, 16–19 October 2016; pp. 881–884.
24. Xing, Y.Q.; Chi, J.K.; Xiao, M. Reactive molecular dynamics simulation on the pyrolysis characteristics of epoxy resin under the effect of partial discharge active products. *High Perform. Polym.* **2021**, *33*, 635–645. [CrossRef]
25. Yang, Y.Z.; You, W.M.; Gao, Y. A lifetime prediction method for insulated spindle motors at level B. *Coal Mine Mach.* **2016**, *37*, 201–202.

Article

Terahertz-Based Method for Accurate Characterization of Early Water Absorption Properties of Epoxy Resins and Rapid Detection of Water Absorption

Hongchuan Dong [1], Yunfan Liu [2,*], Yanming Cao [1], Juzhen Wu [1], Sida Zhang [2], Xinlong Zhang [2] and Li Cheng [2]

[1] State Grid Economic and Technological Research Institute Co., Ltd., Beijing 102200, China; fw9866@126.com (H.D.); litao199811@outlook.com (Y.C.); 18535965807@163.com (J.W.)
[2] State Key Laboratory of Power Transmission Equipment & System Security and New Technology, Chongqing 400044, China; zsdstar@126.com (S.Z.); zxl18789198200@163.com (X.Z.); chengl9877@gmail.com (L.C.)
* Correspondence: cqulyf@cqu.edu.cn; Tel.: +86-156-6570-2589

Abstract: Moisture is detrimental to the performance of epoxy resin material for electrical equipment in long-term operation and insulation. Therefore, moisture absorption is one of the critical indicators for insulation of the material. However, some relevant test methods, e.g., the direct weighing method, are time-consuming, and it usually takes months to complete a test. For this, it is necessary to have some modification to save the test time. Firstly, the study analyzes the present prediction method (according to ISO 62:2008). Under the same accuracy, the time required is reduced from 104 days to 71 days. Subsequently, the Langmuir curve-fitting method for water absorption of epoxy resin is analyzed, and the initial values of diffusion coefficient, bonding coefficient, and de-bonding coefficient are determined based on the results of molecular simulation, relevant experiment, and literature review. With the optimized prediction model, it takes only 1.5 days (reduced by 98% as compared with the standard prediction method) to determine the moisture absorbability. Then, the factors influencing the prediction accuracy are discussed. The results have shown that the fluctuation of balance at the initial stage will affect the test precision significantly. Accordingly, this study proposes a quantitative characterization method for initial trace moisture based on the terahertz method, by which the trace moisture in epoxy resin is represented precisely through the established terahertz time-domain spectroscopy system. When this method is used to predict the moisture absorbability, the experimental time may be further shortened by 33% to 1 day. For the whole water absorption cycle curve, the error is less than 5%.

Keywords: epoxy resins; Langmuir; terahertz; molecular simulation; prediction

1. Introduction

Epoxy resin has many advantages, e.g., high insulation strength, good chemical properties, and excellent environmental adaptability [1–7]. It is widely used in electrical equipment as a packaging and pouring material. Existing studies have shown that moisture will accelerate the aging of epoxy resin material for power equipment and reduce its insulation performance [8,9]. After the intrusion of water molecules, physical and chemical changes such as plasticization and hydrolysis will occur in epoxy resin, resulting in irreversible crack, structural damage, and performance degradation of epoxy resin [10–12]. Chemical aging of epoxy resin mainly involves its changes in chemical structure, including chemical processes such as bond breaking and cross-linking degradation caused by hydrolysis. Chemical changes will have an irreversible effect on the material itself. Physical changes refer to morphological changes such as stiffness degradation and crack, as well as damage on epoxy matrix. After swelling for water absorption, the crack will be enlarged with the diffusion of water molecules, and thus the moisture absorption will be accelerated, making the water molecules fully contact with the epoxy resin and further damage its matrix [13–15]. Therefore, moisture absorption is an important indicator for material aging, and it is necessary to characterize the moisture absorption of epoxy resin material so that epoxy resin of excessively high water absorption will not be used for the power grid.

Presently, the main method for studying water absorption of composite materials is weighing method, which may reflect the moisture absorbability of the material simply and intuitively. However, for organic polymer materials such as epoxy resin, usually, it takes more than 50 days to reach the saturation point for water absorption, which has seriously restricted the application of water absorption method in engineering practice [6,16–19]. In order to reduce the experimental time by the direct method for determining the moisture absorption and improve the practical utility of the method. ISO 62:2008 [20] put forward a water absorption prediction formula, combined with Fick's law and the curve method or calculation tool, that can estimate the saturated water absorption of the material. However, it is still time consuming, the error is large, and the prediction effect is not obvious. Based on Fick's Law, Simon Heid-Jørgensen [21] established the analytical homogenization model to predict the water absorption of epoxy-glass composites, which predicted the first stage of water absorption of the material successfully. However, for the second and third stages of water absorption, the prediction effect was unsatisfactory. Fu Yingqiang [22] predicted the water absorption of modified polyvinyl alcohol with the BP neural network method. The water absorption of the material was predicted successfully based on the measured water absorption values with a precision of 85%. Hui Li [23] established a water absorption prediction model for the composite in combination with Arrhenius's and Fick's Laws, which accurately predicted the moisture absorption of the composite any time as soaked in water at a constant temperature of 95 °C.

The existing moisture absorption prediction methods are based solely on experimental data, which has the following shortcomings: (1) for the vast majority of composite materials, relevant theory has proved that the moisture absorption should satisfy the Langmuir model [24]. However, the existing methods fail to consider the Langmuir model, and thus the prediction efficiency is low; that is, the experimental time is not shortened obviously. (2) For a certain material, a lot of tests need to be carried out first to determine its moisture absorption at different time points, thus causing heavy workload and poor universality.

In order to reduce the time required for moisture absorption test, this study conducted an epoxy resin moisture absorption prediction and proposed a shrinkage–expansion prediction algorithm based on the Langmuir diffusion model, which was conducive to engineering moisture absorption evaluation, because it shortened the experimental time greatly as guaranteeing the precision. This study first applied the water absorption prediction model of material proposed in ISO 62:2008 to the experimental samples and carried out a relevant error analysis. Then, we proposed a nonlinear fitting method based on shrinkage–expansion theory according to the characteristics of the Langmuir diffusion equation for polymers, explored the diffusion process for moisture absorption of epoxy resin through molecular simulation, and determined the search range of shrinkage–expansion algorithm. Additionally, through a comparison with the moisture absorbability of epoxy resin material measured experimentally, the study analyzed the relationship between the experimental time and the final prediction error, put forward the shortest experimental time, and obtained a preliminary moisture absorption behavior prediction model for epoxy resin. At the same time, in order to further reduce the influence for fluctuation of balance in the experiment and represent the initial trace water absorption in epoxy resin was represented precisely. The study established a terahertz time-domain spectroscopy system. As a result, the early water absorption of epoxy resin was represented precisely by the Terahertz time-domain spectroscopy test method, a theoretical verification was carried out through molecular simulation. Therefore, the prediction model was deeply optimized. The experimental time was further reduced to 24 h by terahertz spectroscopy of epoxy resin, and the precision of the model was up to 99%.

2. Materials and Methods

2.1. Materials and Experiments

In this experiment, diglycidyl ether of bisphenol-A (DGEBA) (Changzhou Runxiang Chemical Co., Ltd., Changzhou, China) was adopted, with methylhexahydrophthalic anhydride (MeHHPA) (Shanghai Aladdin Biotechnology Co., Ltd., Shanghai, China) as the curing agent and DMP30 (tertiary amine) (Jiangsu Dandelion Biotechnology Co., Ltd., Jiangsu, China) as the accelerant. The preparation process was as shown in Figure 1. Mix all the reagents according to the ratio of epoxy resin: curing agent: toughening agent: accelerant = 100:80:10:1, and stir at a constant temperature of 50 °C. Then, pour the mixture into a mold, and after vacuum defoamation, solidify at 90 °C and 110 °C for 2 h, respectively, in a thermostat (Shanghai qixin scientific instrument co., LTD, Shanghai, China). Cool and demold until flake samples of diameter 100 mm and thickness 1 mm are obtained.

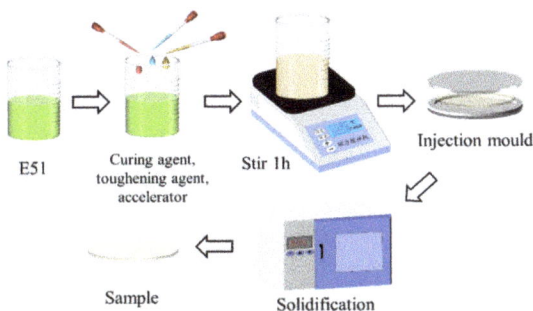

Figure 1. Sample preparation process.

The experimental process is shown in Figure 2, in which experimental samples are placed on the hollow iron frame in the constant temperature and humidity box. Before the experiment, the epoxy resin samples were dried at 60 °C in a drying oven and weighed regularly until the sample mass did not change. The average weight of the six samples after drying was recorded as m_0. Then, the constant temperature and humidity box was tested. The temperature and humidity meter was placed in the box, and the parameters were set as temperature = 25 °C and relative humidity = 95 %. When the display for the constant temperature and humidity chamber as well as the reading of built-in hygrothermograph remained unchanged, it was determined that the experimental environment was good, and the experimental test was carried out [25]. In the process of the experiment, all the dried samples were put into the constant temperature and humidity box, keeping the relative humidity unchanged; a high-precision balance (precision 1 mg) was regularly used to measure the weight, and the samples were put back immediately after each measurement. Six samples were repeated five times, and the average $m(t)$ of five experiments was taken; then, the water absorption percentage, $w(t)$, can be expressed by Formula (1).

$$w(t) = \frac{m(t) - m_0}{m_0} \times 100\%, \quad (1)$$

where $m(t)$ represents the mass of sample after experimental time t, and m_0, the mass of dried sample.

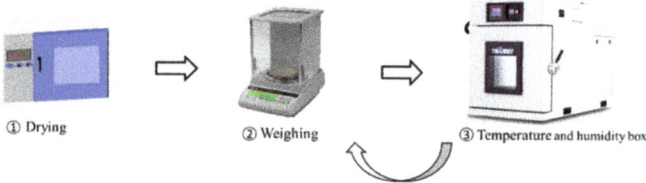

Figure 2. Experimental process.

2.2. Molecular Dynamics Simulation

In the early stage of moisture absorption process for the material, the water absorption rate is fast, and the initial moisture absorption curve is close to a straight line. Therefore, in this study, an epoxy resin containing water molecules/anhydride curing agent system was established, and the search range of diffusion coefficient D was determined through molecular dynamics simulation.

The system was universal DGEBA, with MeHHPA as the curing agent and DMP30 as accelerant. The model construction was as shown in Figure 3. Based on the flow for epoxy resin processing, the crosslinking temperature was determined as 580 K, the crosslinking pressure as 0.005 GPa, and the COMPASS field was selected as the crosslinking force field. The standard Forcite/COMPASS force field was for several times of epoxy resin molecule optimization. The system was performed a 500 ps of molecular dynamics equilibrium simulation, in which the equilibrium was determined based on the fluctuation ranges of temperature and energy. The epoxy resin containing water molecules/anhydride curing agent system contains a total of 2397 atoms, i.e., 274 O, 975 C, and 1148 H.

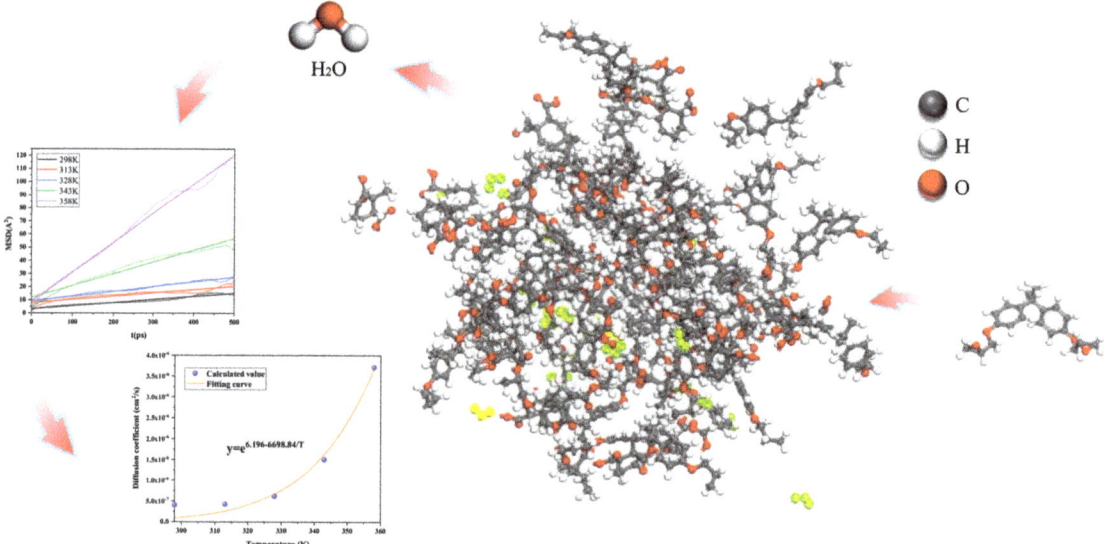

Figure 3. DGEBA Molecular Dynamics Model: the MSD curves at different temperatures and the relationship curves between diffusion coefficient and temperature were obtained by simulation.

The range for initial value of diffusion coefficient D was determined by simulating the diffusion of water molecules at different temperatures in the epoxy resin system.

The chain movement of water molecules in epoxy resin may be represented by mean square displacement:

$$MSD = \left\langle [r_i(t) - r_0(t)]^2 \right\rangle, \quad (2)$$

where $r_i(t)$ and $r_0(t)$ represent the position vectors of atom i at time t and time 0, respectively.

The diffusion coefficient of water molecules in epoxy resin may be solved by the Einstein Formula as follows:

$$D_a = \frac{1}{6N_a} \lim_{t \to \infty} \frac{d}{dt} \sum_{i=1}^{N_a} \left\langle [r_i(t) - r_0(t)]^2 \right\rangle, \tag{3}$$

where $\vec{r}_i(t)$ and $\vec{r}_0(t)$ represent the position vectors of atom i at time t and time 0, respectively. In Formula (3), when t is large enough, the diffusion coefficient D may be calculated with the mean square displacement:

$$D = \frac{\left|\vec{r}(t) - \vec{r}(0)\right|^2}{6t} = \frac{a}{6} \tag{4}$$

In Formula (4), a represents the slope of the fitting curve.

As shown in Figure 3, the MSD at 358 K was at maximum, and at 298 K was at minimum.

2.3. Terahertz Time-Domain Spectroscopy System

Considering that the requirement for initial experimental data was high for the model, this study represented water absorbability of epoxy resin precisely by terahertz time-domain spectroscopy to reduce the influence for fluctuation of balance. Due to the strong absorption effect of water molecules on terahertz waves, the nondestructive and sensitivity of terahertz technology to water molecules have attracted much attention and research from all walks of life. In recent years, a large number of articles [26–29] have studied the quantitative identification of moisture in various media by terahertz technology. Considering the stability and accuracy of THz measurement of water, we first combined it with the prediction method and proposed a faster water detection method.

The established terahertz platform was as shown in Figure 4. First, the femtosecond laser emitted a pulse less than 80 Fs at 1560 ± 20 nm, which was split into two beams perpendicular to each other by a beam splitter: pump light and the probe light. Then, the pump light was focused on the base surface of the photoconductive antenna through the reflector and time delay device, thus generating the THz pulse, which was focused on the tested sample after the parabolic mirror collimation. The THz pulse carrying sample information was transmitted through the samples, and then collimated and focused through another pair of parabolic mirrors, passing through a detector in alignment with the probe light. At last, the detector sent the signal to a computer for further data analysis.

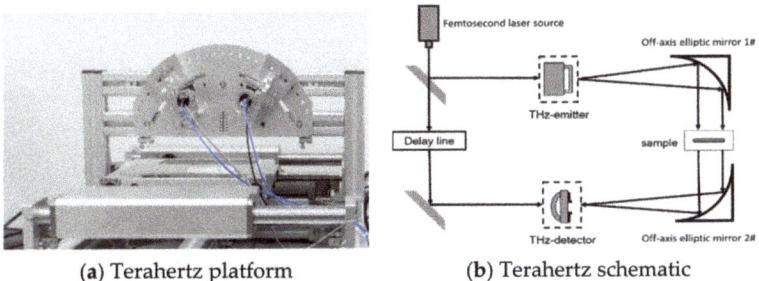

(a) Terahertz platform (b) Terahertz schematic

Figure 4. The transmission terahertz time-domain spectroscopy system.

2.4. DFT Methods

In order to explore the absorption peak of water-bearing epoxy resin in the terahertz band and verify the experimental data, a spectrum vibration analysis on water-bearing epoxy resin system was carried out with the Gaussian software (Gaussian, Inc, Gaussian 09W) to identify the characteristic absorption frequency of water molecules in the epoxy resin theoretically. DGEBA molecules of epoxy resin were selected for the model, with water molecules added for simulation, as shown in Figure 5. The DFT method was used for relevant calculation and simulation. The adopted method was mainly B3LYP commutative correlation functional from generalized gradient approximation, with def2-SVP defined as the base group of calculation, and the diffusion and polarization functions were added for all atoms.

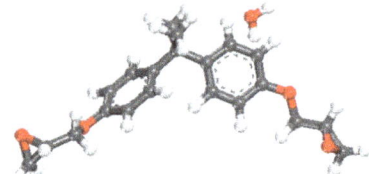

Figure 5. Molecular simulation of water-bearing epoxy resin system.

3. Study Methods

3.1. Physical Model of Water Absorption

Carter described a diffusion phenomenon by the Langmuir diffusion model [30]. Water molecules bind together temporarily (physically) or permanently (chemically) in the material, thus retarding the diffusion process and reducing the speed of water absorption, especially for a long-term water absorption process. Carter represented relevant physical and chemical interactions with bonding coefficient α and de-bonding coefficient β. In this case, $\alpha, \beta \ll \pi^2 Ds^{-2}$, and the water absorption can be expressed by the following formula:

$$\frac{w(t)}{w_s} = \left[\frac{\alpha}{\alpha+\beta}\exp(-\alpha t)(y(t)-1) + \exp(-\beta t)\left(\frac{\beta}{\alpha+\beta}-1\right)+1\right]$$
$$y(t) = 1 - \frac{8}{\pi^2}\sum_{j=0}^{\infty}\left\{\frac{1}{(2j+1)^2}\exp(-(\frac{2j+1}{2})^2)\cdot \pi^2 \cdot 4Dts^{-2}\right\} \quad (5)$$

3.2. Error Thershold Determination

In the water absorption test of epoxy resin sample, water loss is inevitable when the samples are removed from the constant temperature and humidity box. Therefore, the weight gain rate for water absorption fluctuates greatly. In the experiment, the applicable range for precision of model was determined through measuring the error, and thus, the error caused by the test itself was avoided.

The 2500 h of moisture absorption of epoxy resin samples were analyzed using the experimental apparatus as shown in Figure 6. For each sample, the analysis was repeated five times under the same environmental conditions, and the five sets of measurements were averaged at last, with the mean value as the baseline. As shown in Figure 6, the weight of the six samples after 2500 h of water absorption ranged from 9.692 g to 9.700 g, mean value 9.696 g, experimental error ±0.004 g, and the corresponding error of weight gain rate for water absorption about 5%.

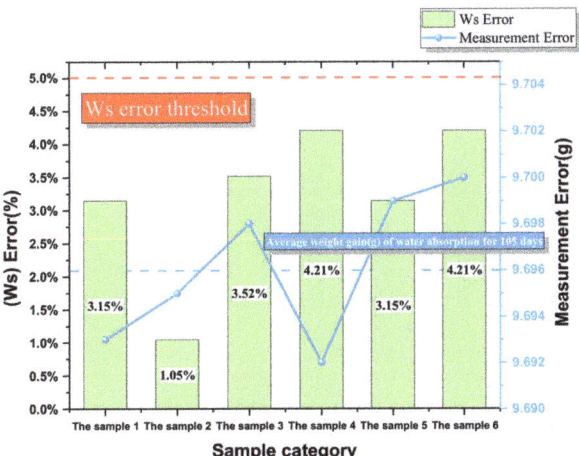

Figure 6. Test error of the testing apparatus. The right ordinate is the weight range of six samples after 2500 h of water absorption, and the left ordinate is the corresponding saturated water absorption error.

3.3. ISO 62:2008 Prediction Method

According to ISO 62:2008 [20], the measured D and C when constant mass is not achieved are as shown in Formula 6.

$$\sqrt{D} \approx \frac{1}{C_S} \times \frac{d}{0.52\pi} \times \frac{c(t)}{\sqrt{t}}, \qquad (6)$$

where C_S saturated water absorption; d sample thickness; t moisture absorption time; $C(t)$ water absorption rate measured at t.

As applied to the epoxy resin samples, its prediction precision was as shown in Figure 7. The prediction precision increased with the experimental days. Within the 5% range of threshold, the required experimental days was about 71 days, and for the samples, relevant fluctuation was large, and the prediction effect was not obvious.

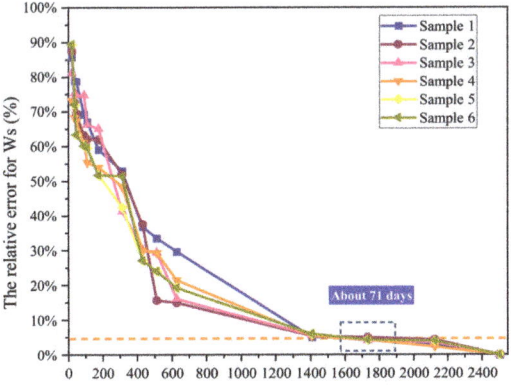

Figure 7. Relationship between experimental days and prediction accuracy of ISO 62:2008 model for six samples.

3.4. Fitting Method Based on Langmuir Formula

Current studies have shown that the water absorption of organic polymer materials such as epoxy resin conforms to the Langmuir's Law. Langmuir Formula is characterized by multiple parameters, cascade equation, high latitude, and high nonlinearity. Therefore, parameters D, α, β, and Ws were estimated by optimization fitting to determine the moisture absorbability of the material.

In this study, the shrinkage–expansion theory was applied to the optimization fitting of initial water absorption data of epoxy resin so as to establish the water absorption prediction model of epoxy resin. In the shrinkage–expansion theory, there are three stages, as follows: first, gradually reduce the step size in the initial search space; the second is to expand the step size for search; the third is to calculate the mean value and standard deviation for these degree points that meet relevant requirements, which are used to determine the step size of the next process and carry out iteration. The objective function can be searched in and out of multi-dimensional starting space, the search center and step size can be adjusted by the feedback information during the search process, and thus the optimal parameters for the given Langmuir Formula [31], i.e., D, α, β and Ws, may be approached by merely a few shrinkage-expansion cycles. This algorithm does not have to give the derivative or partial derivative of the formula, thus reducing the complexity of calculation, and therefore, this study introduced the shrinkage-expansion algorithm for relevant fitting. Since the initial values for the parameters were greater than 0, they were first all set to any value greater than 0.

As shown in Figure 8, for direct fitting based on all the 104 days of data, the direct fitting based on all 104 days of data had a good effect. However, water absorption prediction based on 36 h and 108 h initial experimental data had a poor effect, and the error relative to the true value was large. In order to further explore the relationship between the time required fir experiment and the final prediction error, the repeated five times of experimental data for six samples were averaged, and the fitting test was conducted with the experimental data at 18 h, 36 h, 45 h, 89 h, 108 h, 174 h, 311 h, 430 h, 511 h, 625 h, 1406 h, 1731 h, 2122 h, and 2500 h. The fitting results of the experimental data at 104 days were taken as the benchmark for observing and analyzing the error.

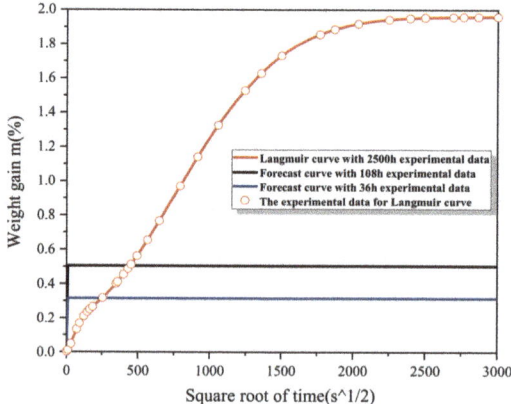

Figure 8. The experimental data of 36 h, 108 h, and 2500 h were used to directly apply the shrinkage-expansion algorithm to fit the obtained curves.

As shown in Figure 9, the shrinkage-expansion algorithm was directly used for optimization fitting of the Langmuir formula. Based on less than 94 days of experimental data, the error for the six samples was greater than 5.1%, and so, the prediction effect was poor, and the fitting error was large, that is, the order of magnitude for the results obtained for the parameters differed greatly from that of the true values. The reason for this was that, for the shrinkage—expansion algorithm, when there are multiple parameters and the nonlinear relationship is complex, the efficiency for optimization search is low, especially when the initial value is not so appropriate, the search process from the initial point to the optimal point is relatively slow, while the Langmuir formula contains four unknowns, with D, α, β, Ws as regression parameters to be estimated. Ws is merely parameter solution > 0, and it is impossible to estimate its initial value. Therefore, it is necessary to narrow the shrinkage range and determine the initial values of D, α and β.

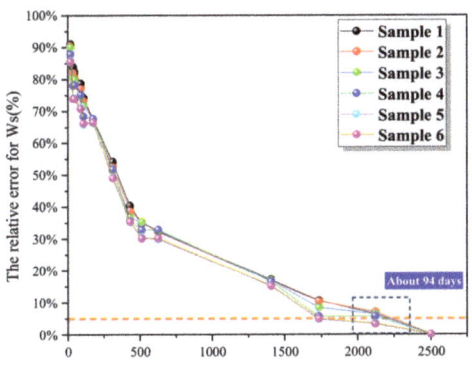

Figure 9. The relationship curve between experimental time and water absorption prediction error of the initial shrinkage–expansion prediction method directly applied to six samples.

4. Determination of the Initial Values of the Parameters

4.1. Search Range of Diffusion Coefficient D

Table 1 showed the diffusion coefficient of water molecules at 298 K, 313 K, 328 K, 343 K, and 358 K, with R^2 as the goodness of fit, which was greater than 0.98 for all, indicating that the simulation fitting results are credible. The diffusion coefficient of water molecules in this study was close to that in relevant literature [32], which indicated that the simulation method used in this study was effective and reasonable. According to Table 1, the diffusion coefficient of water molecules in epoxy resin increased with temperature.

Table 1. Diffusion coefficient of water molecules at different temperatures in epoxy resin.

Temperature	a	R^2	D/cm^2/s
298 K	0.024	0.98	3.96×10^{-7}
313 K	0.025	0.98	4.19×10^{-7}
328 K	0.037	0.98	6.11×10^{-7}
343 K	0.090	0.99	1.50×10^{-6}
358 K	0.222	0.99	3.71×10^{-6}

There is no unit for Slope a or correlation coefficient R^2, where D is diffusion coefficient.

According to the classical diffusion theory, the migration rate as well as the diffusion coefficient of water molecules increases with temperature. Microscopically rising temperature makes the model expand and the free volume of water molecules increase. At the same time, with the increase in temperature, the kinetic energy of water molecules increases, and the binding effect of epoxy resin on water molecules decreases. Therefore, the diffusion coefficient of water molecules increases with temperature.

According to the thermodynamic formula, the thermodynamic process of gas molecular diffusion follows the Arrhenius Formula, that is, the diffusion coefficient of water molecules has an exponential relationship with temperature. A temperature fitting was carried out for the diffusion coefficient of water molecules, and the fitting results were as shown in Figure 3. As a result, the diffusion coefficient functions of water molecules at different temperatures were obtained, which proved the accuracy of molecular simulation results.

The theoretical value for the magnitude order of diffusion coefficient of water molecules in pure epoxy resin was determined as about 10^{-7} through the establishment of molecular model. However, in practice, the penetration rate of water molecules through the epoxy resin may be hindered by the internal forces of the material. For this, the initial magnitude order for the diffusion coefficient may be selected as $10^{-7} \sim 10^{-8}$ for selecting the initial value of the algorithm. However, it should be noted that this is only a range for the initial value of the algorithm; that is, it is not the range for final value due to the influence of other environmental factors.

4.2. Search Range of α and β

In the initial moisture absorption stage of epoxy resin for rapid water absorption, water molecules interacting intensively diffuse into the polymer, while free water quickly fills the free volume of the epoxy resin. After a certain moisture absorption time, the diffusion is saturated, and the free water and the bound water are in equilibrium. This process is time consuming, while for the molecular simulation method, the time is relatively short, with ps as the unit, and thus, it cannot be simulated by molecular simulation. Therefore, the range of initial value for α and β was determined by an experiment and relevant literature in this study.

As shown in Table 2, the magnitude order of α and β for water absorption of epoxy resin measured in this experiment was $10^{-6} \sim 10^{-7}$ s^{-1}. Based on an enormous amount of literature, the magnitude order of α and β for organic polymer materials such as epoxy resin is roughly $10^{-6} \sim 10^{-8}$ s^{-1}. Therefore, the initial value was set as 10^{-7} s^{-1} for α and β of epoxy resin.

Table 2. Ranges of α and β of different materials.

	Water Absorption Test Value of Epoxy Resin/s^{-1}	Silicone Rubber [16]/s^{-1}	Epoxy Resin Adhesive (Type EC 2216 from 3M) [33]/s^{-1}
α	7.432×10^{-6}	1.5×10^{-6}	4×10^{-8}
β	9.071×10^{-7}	2.5×10^{-6}	8.1×10^{-8}

5. Prediction Results for Water Absorption of Epoxy Resin

In the experiment, the moisture absorption of epoxy resin material was measured, the relationship between the required experimental time for prediction and the final prediction error was analyzed, and the shortest time required for experiment was proposed. Thus, the moisture absorption behavior prediction model of epoxy resin was obtained.

The 36 h and 108 h experimental results were applied through the prediction model after the initial values were substituted. The results were as shown in Figure 10 and Table 3. Obviously, the prediction curve and saturated water absorption were more precise than before optimization of the model (Figure 8).

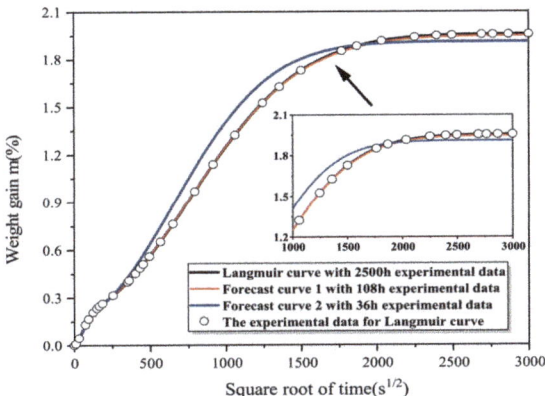

Figure 10. The experimental data of 36 h, 108 h, and 2500 h were used to compare the predicted curves of the optimized shrinkage-expansion algorithm.

Table 3. Parameter Comparison of 36 h, 108 h, and True Value.

The Experimental Time	D	The Relative Error	α/β	The Relative Error	W_s	The Relative Error
The real value	1.098×10^{-7}	0	8.248	0	1.958	0
108 h	1.082×10^{-7}	1.457%	8.003	2.970%	1.95	0.409%
36 h	1.052×10^{-7}	4.189%	9.113	10.487%	1.912	2.349%

The relationship between the required experimental time for prediction and the final prediction error was as shown in Figure 11. Based on the results for experimental error, the precision of the established water absorption prediction model of epoxy resin was further verified. Based on the precision of the testing apparatus mentioned above, the error threshold was determined as 5%, and thus, the shortest time required for experiment was 36 h. The prediction precision may be increased with experimental time. The optimized prediction model shortened the experimental time greatly as guaranteeing the precision as compared with the model before optimization and the prediction method proposed in ISO 62:2008.

However, the stability of prediction models for different samples needed to be further improved. As shown in Figure 11, the final error corresponding to the early experimental time of different samples fluctuated greatly, which was caused by the instability of balance. Therefore, it is necessary to further optimize the stability of the model.

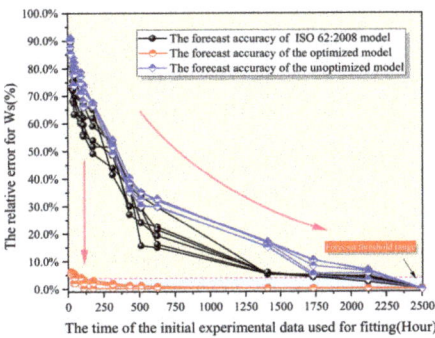

Figure 11. ISO 62:2008, Unoptimized Model and Optimized Model Prediction Error Comparison.

6. Precise Representation of Early Water Absorbability

The influence of water content on terahertz absorption was analyzed through calculating the absorption spectra of the samples. The THz-TDS test system was used to obtain the reference and sample time domain signals, which were $u_{ref}(t)$ and $u_{sample}(t)$, respectively. The $u_{ref}(t)$ and $u_{sample}(t)$ were converted to frequency domain through Fourier transform, and thus, frequency functions $E_{ref}(\omega)$ and $E_{sample}(\omega)$ were obtained. According to Bill Lambert's Law, the absorbance is directly proportional to the concentration of light absorbing material and the thickness of absorbing layer but inversely related to the transmittance. Therefore, Formula (7) may be used to calculate the absorption coefficient (α) of the samples.

$$\alpha = \frac{1}{d} ln(\frac{A_{ref}}{A_{sample}}), \tag{7}$$

where d is the sample thickness, and A_{ref} and A_{sample} are the amplitudes of the reference signal and sample signal frequency functions, respectively.

After testing, analyzing, and median filtering of epoxy resin, the absorption coefficient of the samples with different water content was obtained, as shown in Figure 12. Obviously, there was an intensive absorption peak near 1.9 THz, and the peak value increased with water content.

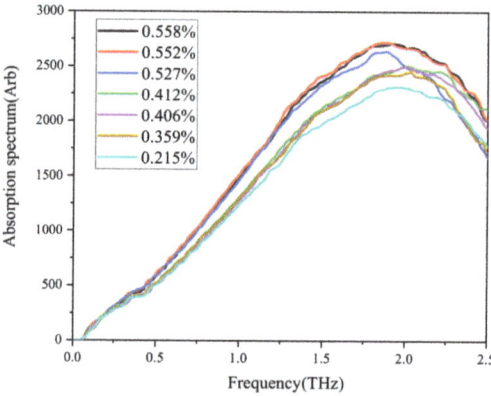

Figure 12. Absorption coefficient of samples with different water content.

The vibration spectra of the mixed system at 0–3 THz were as shown in Figure 13.

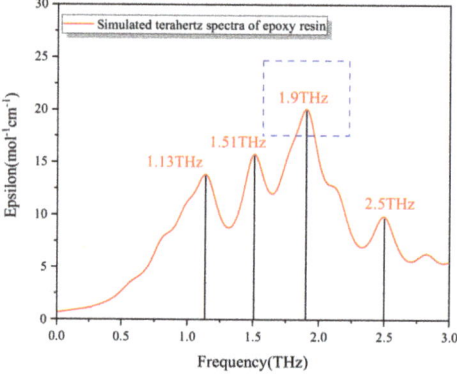

Figure 13. Terahertz spectrogram for molecular simulation.

According to Figure 13, the vibration simulation results of water-bearing epoxy resin system showed obvious absorption peaks at 1.13, 1.51, 1.9, and 2.5 THz, and the intensity of absorption peak at 1.9 THz was the highest, which was consistent with the test results.

For this, the peak value at 1.9 THz was taken as the reference for fitting with water absorption as shown in Figure 14. For absorption spectra of epoxy resin samples with different water content, the peak height at 1.9 THz fitted well with water content, with goodness of fit greater than 0.98. Therefore, the device may be used for preliminary testing.

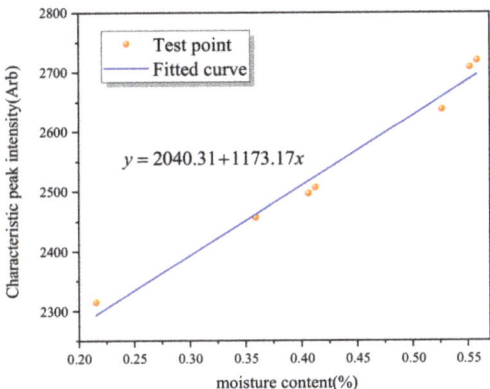

Figure 14. Peak height at 1.9 THz and relevant water content curve.

The error comparison of Terahertz with balance for testing was as shown in Figure 15, where there were water absorption data indirectly represented by terahertz and obtained by the balance. According to the obtained prediction curve, under an error threshold of 5%, the terahertz method may shorten the experimental time to 24 h and enhance relevant stability as compared with the testing of balance, thus further shortening the experimental time.

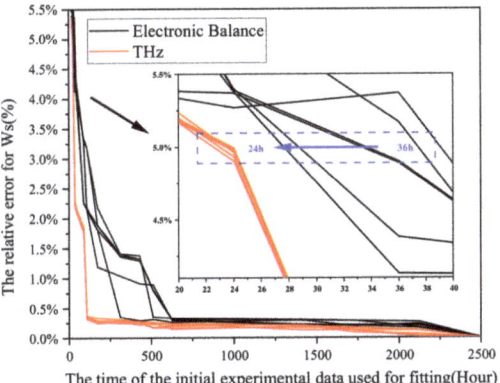

Figure 15. Comparison of model prediction precision between THz and electronic balance.

7. Conclusions

For the conventional experimental method, it takes a long time to test the water absorption of epoxy resin for electrical equipment. The core idea of this study is "to obtain long-term water absorption over time with short experimental time". With the prediction model proposed in this study, the long-term water absorption may be predicted, and the water absorption curve may be drawn simply by substituting 24 h of initial experimental data into the model.

(1) Based on the nonlinearity and high latitude of the Langmuir diffusion equation, this study has proposed a shrinkage-expansion algorithm-based fitting method, which may realize the fitting of the water absorption process, but the algorithm is still to be improved, for the experimental time is merely reduced from 104 days to 94 days.
(2) The diffusion coefficient of epoxy resin at different temperatures has been simulated and analyzed with a molecular simulation method, and the diffusion coefficient and the temperature satisfy the Arrhenius formula.
(3) Based on the results of molecular simulation, experiment, and the relevant literature, the initial values of diffusion coefficient, bonding coefficient, and de-bonding coefficient have been determined, and the prediction model has been optimized. Thus, the experimental time is further reduced from 108 days to 1.5 days, and the prediction error is no greater than the experimental error of moisture absorption test (5%), and so the engineering requirements may be set generally.
(4) In order to reduce the influence for fluctuation of balance and further shorten the experimental time required by the model, this study has proposed a method to improve the precision of the model based on terahertz. Under an error threshold of 5%, the terahertz method may shorten the experimental time from 36 h to 24 h.

Author Contributions: Conceptualization, H.D. and Y.L.; methodology, Y.L.; software, J.W.; validation, Y.C.; formal analysis, S.Z.; investigation, L.C.; resources, L.C.; data curation, X.Z.; writing—original draft preparation, Y.L.; writing—review and editing, H.D.; visualization, J.W.; supervision, Y.L.; project administration, Y.C.; funding acquisition, L.C. All authors have read and agreed to the published version of the manuscript.

Funding: This research was funded by National Natural Science Foundation of China (No.52077013) and Science and technology project of State Grid Corporation of China (No.52090020002Q).

Data Availability Statement: Not Applicable.

Conflicts of Interest: The authors declare no conflict of interest.

References

1. Awais, M.; Chen, X.; Dai, C.; Meng, F.B.; Paramane, A.; Tanaka, Y. Tuning epoxy for medium frequency transformer application: Resin optimization and characterization of nanocomposites at high temperature. *IEEE Trans. Dielectr. Electr. Insul.* **2021**, *28*, 1751–1758. [CrossRef]
2. Yin, Z.; Sun, P.; Sima, W.; Li, L.; Du, W.; Sui, H. Thermal response properties and surface insulation failure mechanism of epoxy resin under arc ablation. *IEEE Trans. Dielectr. Electr. Insul.* **2019**, *26*, 1503–1511. [CrossRef]
3. Zhang, C.; Ma, Y.; Kong, F.; Wang, R.; Ren, C.; Shao, T. Surface charge decay of epoxy resin treated by AP-DBD deposition and direct fluorination. *IEEE Trans. Dielectr. Electr. Insul.* **2019**, *26*, 768–775. [CrossRef]
4. Zhang, X.; Wu, Y.; Chen, X.; Wen, H.; Xiao, S. Theoretical study on decomposition mechanism of insulating epoxy resin cured by anhydride. *Polymers* **2017**, *9*, 341. [CrossRef]
5. He, S.J.; Luo, C.M.; Zheng, Y.Z.; Xue, Y.; Song, X.P.; Lin, J. Improvement in the charge dissipation performance of epoxy resin composites by incorporating amino-modified boron nitride nanosheets. *Mater. Lett.* **2021**, *298*, 130009. [CrossRef]
6. Wang, Y.; Zeng, Z.; Gao, M.; Huang, Z. Hygrothermal aging characteristics of silicone-modified aging-resistant epoxy resin insulating material. *Polymers* **2021**, *13*, 2145. [CrossRef]
7. Liao, Y.F.; Weng, Y.X.; Wang, J.Q.; Zhou, H.F.; Lin, J.; He, S.J. Silicone rubber composites with high breakdown strength and low dielectric loss based on polydopamine coated mica. *Polymers* **2019**, *11*, 2030. [CrossRef]
8. Wang, Y.; Meng, Z.; Zhu, W. Hygrothermal aging behavior and aging mechanism of carbon nanofibers/epoxy composites. *Constr. Build. Mater.* **2021**, *294*, 123538. [CrossRef]
9. Wang, B.; Li, D.; Xian, G.; Li, C. Effect of Immersion in water or alkali solution on the structures and properties of epoxy resin. *Polymers* **2021**, *13*, 1902. [CrossRef]
10. Zou, C.; Fothergill, J.C.; Rowe, S.W. The effect of water absorption on the dielectric properties of epoxy nanocomposites. *IEEE Trans. Dielectr. Electr. Insul.* **2008**, *15*, 106–117. [CrossRef]
11. Nakamura, T.; Tabuchi, H.; Hirai, T. Effects of silane coupling agent hydrophobicity and loading method on water absorption and mechanical strength of silica particle-filled epoxy resin. *J. Appl. Polym. Sci.* **2020**, *137*, 48615. [CrossRef]
12. Cabanelas, J.C.; Prolongo, S.G.; Serrano, B.; Bravo, J.; Baselga, J. Water absorption in polyamino siloxane-epoxy thermosetting polymers. *J. Mater. Process. Tech.* **2003**, *143–144*, 311–315. [CrossRef]

13. Liang, X.; Bao, W.; Gao, Y. Decay-like fracture mechanism of silicone rubber composite insulator. *IEEE Trans. Dielectr. Electr. Insul.* **2018**, *25*, 110–119. [CrossRef]
14. Gao, Y.; Wang, J.; Liang, X.; Yan, Z.; Liu, Y.; Cai, Y. Investigation on permeation properties of liquids into HTV silicone rubber materials. *IEEE Trans. Dielectr. Electr. Insul.* **2014**, *21*, 2428–2437. [CrossRef]
15. Chen, X.; Wang, J.Q.; Zhang, C.; Yang, W.; Lin, J.; Bian, X.M. Performance of silicone rubber composites using boron nitride to replace alumina tri-hydrate. *High Volt.* **2021**, *6*, 480–486. [CrossRef]
16. Lutz, B. Water absorption and water vapor permeation characteristics of HTV silicone rubber material. In Proceedings of the 2012 IEEE International Symposium on Electrical Insulation, San Juan, PR, USA, 10–13 June 2012; pp. 478–482.
17. Wang, Z.; Zhao, L.H.; Jia, Z.D.; Guan, Z.C. Water and moisture permeability of high-temperature vulcanized silicone rubber. *IEEE Trans. Dielectr. Electr. Insul.* **2017**, *24*, 2440–2448. [CrossRef]
18. Wang, Z.; Jia, Z.D.; Fang, M.H.; Guan, Z.C. Absorption and permeation of water and aqueous solutions of high-temperature vulcanized silicone rubber. *IEEE Trans. Dielectr. Electr. Insul.* **2015**, *22*, 3357–3365. [CrossRef]
19. He, S.J.; Wang, J.Q.; Hu, J.B.; Zhou, H.F.; Nguyen, H.; Luo, C.M. Silicone rubber composites incorporating graphitic carbon nitride and modified by vinyl tri-methoxysilane. *Polym. Test* **2019**, *79*, 106005. [CrossRef]
20. Plastics—Determination of Water Absorption (ISO 62:2008). UNE-EN ISO 62-2008. Available online: https://www.iso.org/standard/41672.html (accessed on 1 February 2008).
21. Simon, H.; Claus, H.; Michal, K. Temperature dependence of moisture diffusion in woven epoxy-glass composites: A theoretical and experimental study. *Mater.Today Commun.* **2021**, *29*, 102844.
22. Fu, Y.Q.; Yao, L.; Jiang, C.Q. Prediction of water absorption of modification polyvinyl alcohol using BP Neural network. *J. Anhui Univ. Tech. Sci. (Nat. Sci.)* **2007**, *22*, 44–47.
23. Hui, L.; Wang, Y.G.; Xu, L.; Ma, S.H.; Zhang, X.; Fei, B.Q. Moisture absorption model of composites considering water temperature effect. *J. Mater. Eng.* **2016**, *44*, 83–87.
24. Wang, L.; Nie, Z.; Zhao, C.; Zhou, J.; Geng, W. Water permeation characteristic in cycloaliphatic epoxy resin composite insulator sheath. *High Volt. Eng.* **2019**, *45*, 173–180.
25. Gao, Y.; Wang, J.; Yan, Z.; Liang, X.; Liu, Y. Investigation on water diffusion property into HTV silicone rubber. *Proc. CSEE* **2015**, *35*, 231–239.
26. Hassan, A.M.; Hufnagle, D.C.; El-Shenawee, M. Terahertz imaging for margin assessment of breast cancer tumors. In *IEEE MTT-S international Microwave Symposium Digest*; IEEE: Montreal, QC, Canada, 2012; pp. 1–3.
27. Singh, A.K.; Pérez-López, A.V.; Simpson, J. Three-dimensional water mapping of succulent Agave victoriae-reginae leaves by terahertz imaging. *Sci. Rep.* **2020**, *10*, 1404. [CrossRef]
28. Li, B.; Zhang, X.; Wang, R. Leaf water status monitoring by scattering effects at terahertz frequencies. *Spectrochim. Acta. A* **2020**, *245*, 118932. [CrossRef]
29. Pagano, M.; Baldacci, L.; Ottomaniello, A. THz water transmittance and leaf surface area: An effective nondestructive method for determining leaf water content. *Sensors* **2019**, *19*, 4838. [CrossRef]
30. Carter, H.G.; Kibler, K.G. Langmuir-type model for anomalous moisture diffusion in composite resins. *J. Compos. Mater.* **1978**, *12*, 118–131. [CrossRef]
31. Gu, L.S.; Wan, L.S.; Huang, L.J.; Wang, W.P. Improved contraction-expansion algorithm for curve and surface fitting. *Acta Agron. Sin.* **2007**, *4*, 583–589.
32. Yin, L. Research on water absorption characteristic of cycloaliphatic epoxy resin insulator. In Proceedings of the 2020 IEEE 1st China International Youth Conference on Electrical Engineering (CIYCEE), Wuhan, China, 1–4 November 2020; pp. 1–5.
33. Popineau, S.; Rondeau-Mouro, C.; Sulpice-Gaillet, C. Free/bound water absorption in an epoxy adhesive. *Polymer* **2005**, *46*, 10733–10740.

Article

Deep Insight into the Influences of the Intrinsic Properties of Dielectric Elastomer on the Energy-Harvesting Performance of the Dielectric Elastomer Generator

Yingjie Jiang [1], Yujia Li [1], Haibo Yang [1], Nanying Ning [1,2,*], Ming Tian [1,2,*] and Liqun Zhang [1,2]

[1] Beijing Advanced Innovation Center for Soft Matter Science and Engineering, Beijing University of Chemical Technology, Beijing 100029, China; jyj_940719@163.com (Y.J.); yujiali10930@163.com (Y.L.); yanghb@mail.buct.edu.cn (H.Y.); zhanglq@mail.buct.edu.cn (L.Z.)
[2] Key Laboratory of Carbon Fiber and Functional Polymers, Ministry of Education, Beijing University of Chemical Technology, Beijing 100029, China
* Correspondence: ningny@mail.buct.edu.cn (N.N.); tianm@mail.buct.edu.cn (M.T.)

Abstract: The dielectric elastomer (DE) generator (DEG), which can convert mechanical energy to electrical energy, has attracted considerable attention in the last decade. Currently, the energy-harvesting performances of the DEG still require improvement. One major reason is that the mechanical and electrical properties of DE materials are not well coordinated. To provide guidance for producing high-performance DE materials for the DEG, the relationship between the intrinsic properties of DE materials and the energy-harvesting performances of the DEG must be revealed. In this study, a simplified but validated electromechanical model based on an actual circuit is developed to study the relationship between the intrinsic properties of DE materials and the energy-harvesting performance. Experimental verification of the model is performed, and the results indicate the validity of the proposed model, which can well predict the energy-harvesting performances. The influences of six intrinsic properties of DE materials on energy-harvesting performances is systematically studied. The results indicate that a high breakdown field strength, low conductivity and high elasticity of DE materials are the prerequisites for obtaining high energy density and conversion efficiency. DE materials with high elongation at break, high permittivity and moderate modulus can further improve the energy density and conversion efficiency of the DEG. The ratio of permittivity and the modulus of the DE should be tailored to be moderate to optimize conversion efficiency (η) of the DEG because using DE with high permittivity but extremely low modulus may lead to a reduction in η due to the occurrence of premature "loss of tension".

Keywords: dielectric elastomer; intrinsic property; energy harvesting

1. Introduction

The dielectric elastomer transducer (DET) has been a hot area of research in recent decades due to its high flexibility, light weight, large mechanical strain, simple structure and low cost [1–4]. A typical DET device consists of a dielectric elastomer (DE) film sandwiched by two compliant electrodes [5,6]. In generator mode, the DET is called the dielectric elastomer generator (DEG), and is able to convert mechanical energy into electrical energy during the stretch-release process due to its stretching variable capacitance property. As a new type of generator, the DEG provides a simple and feasible solution for harvesting energy from nature motion sources, such as waves, tides and human movements [7–10]. Therefore, the DEG has attracted much attention in recent years [11–15].

The working principle of the DEG is illustrated in Figure 1. The DE film is first electrically excited under low voltage at stretched state with high capacitance. Then, the film is released, and higher voltage across the film at the released state with lower capacitance can be obtained. The harvested electrical energy in this process is called the single-cycle generated energy (ΔU), which can be calculated as follows:

$$\Delta U = U_{out} - U_{in} = \frac{1}{2}C_2 V_2^2 - \frac{1}{2}C_1 V_1^2 \quad (1)$$

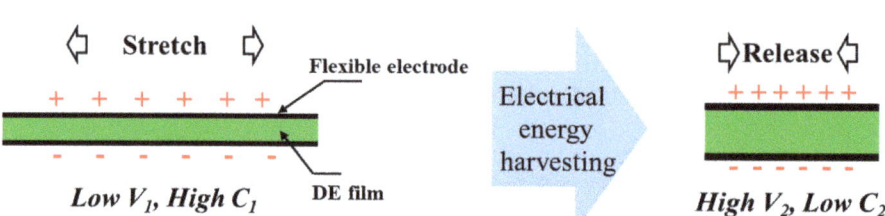

Figure 1. Schematic diagram of the working principle of the DEG. Mechanical energy is converted into electrical energy by releasing a stretched and charged DE film.

Among them, U_{out} and U_{in} represent the input electrical energy in the stretched state and the output electric energy in the released state, respectively. C, V and the subscripts 1 and 2 represent the capacitance, voltage, the stretched state and the released state, respectively. The calculation formula of capacitance is:

$$C = \frac{\varepsilon_0 \varepsilon_r S}{z} \quad (2)$$

where ε_0 represents the vacuum permittivity; and ε_r, S and z, represent the relative permittivity (hereinafter referred to as permittivity), the area and the thickness of DE film, respectively. C, as well as ΔU, will be affected by the film size. Therefore, the gravimetric energy density (w_m) and electromechanical conversion efficiency (η), as more important performances that avoid the influence of film shape [16], can be achieved by dividing ΔU by the mass of the DE film (m) and input mechanical work (W_{mech}), respectively.

$$w_m = \frac{\Delta U}{m} \quad (3)$$

$$\eta = \frac{\Delta U}{W_{mech}} \times 100\% \quad (4)$$

Previous studies on the DEG have mainly focused on the circuit or device structure design to improve the energy-harvesting performance of the DEG based on commercial elastomers trademarked as VHB4905/10 [17–23]. On the other hand, some efforts have been made to prepare DE materials with high permittivity to enhance the performances of the DEG [24–29]. Ellingford et al. introduced a polar functional group into a styrene-butadiene triblock copolymer (SBS) to enhance the permittivity, but the significant charge leakage reduced the generated energy, which may have been caused by the relatively high conductivity of the modified SBS [28]. Yang et al. employed nature rubber (NR) and barium titanate (BT) as a high-elasticity DE matrix and dielectric filler, respectively, while the achieved conversion efficiency was relatively low, which may have resulted from the high modulus of the composite [26]. In these studies, the mechanical and electrical properties of the prepared materials were not well coordinated. Therefore, the energy-harvesting performances of the as-prepared DE materials were not satisfied. Therefore, to provide guidance for the preparation of high-performance DE materials for the DEG, the relationship between the intrinsic properties of DE materials and the energy-harvesting performances must be studied and revealed.

To date, scarce studies have reported the influences of the intrinsic properties of DE materials on the energy-harvesting performances of the DEG. Koh et al. established an electromechanical model to calculate the theoretical maximum energy density based on several failure mechanisms, and explored the influences of material parameters, such as Young's modulus (Y), the product of permittivity and the square of electrical breakdown strength (E_b), on the energy density [30]. This study provides a preliminary guidance for the preparation of high-performance DE materials for the DEG, but the results on energy density calculated by this model cannot be obtained in an actual circuit. Moreover, this model does not take the mechanics/charge loss of the DE into consideration. Therefore, the simulation results differ from the experimental results [22].

Therefore, an electromechanical model based on an actual circuit is necessary to reveal the relationship between the intrinsic properties of DE materials and the energy-harvesting performance, thereby guiding the design and preparation of DE. An electromechanical model would also help to predict the theoretical energy-harvesting performances of the DEG. Based on the energy generation mechanism of the DEG, the main influencing factors involved in the energy conversion are as follows. During the electrical excitation and harvesting process, the permittivity and strain, electrical breakdown strength and the bulk conductivity influence ΔU by affecting the capacitance, bias voltage and charge loss, respectively [31,32]. Therefore, the main influencing factors involved in energy conversion include the intrinsic properties of DE and the external environment, as summarized in Figure 2 [16,30]. Among these intrinsic properties, the mechanical-related properties include the Young's modulus, elongation at break and mechanical loss, and the electrical related properties include the permittivity, electrical breakdown strength and bulk conductivity. The external environment factor can be divided into device variables, including the stretching mode and circuit design, and the operating variables, including stretch ratio and bias voltage. It is noted that some intrinsic properties of the material limit the maximum value of the operating variable. That is, the elongation at break and the electrical breakdown strength of the DE material limit the maximum stretch ratio and the maximum bias voltage that can be applied, respectively. Therefore, the energy-harvesting performance of the DE with different elongation at break and breakdown field strength can be equivalently investigated under different stretch ratios and bias voltages, respectively. Under the premise of fixed device variables, the relationship between the intrinsic properties of DE materials and the energy-harvesting performances can be studied.

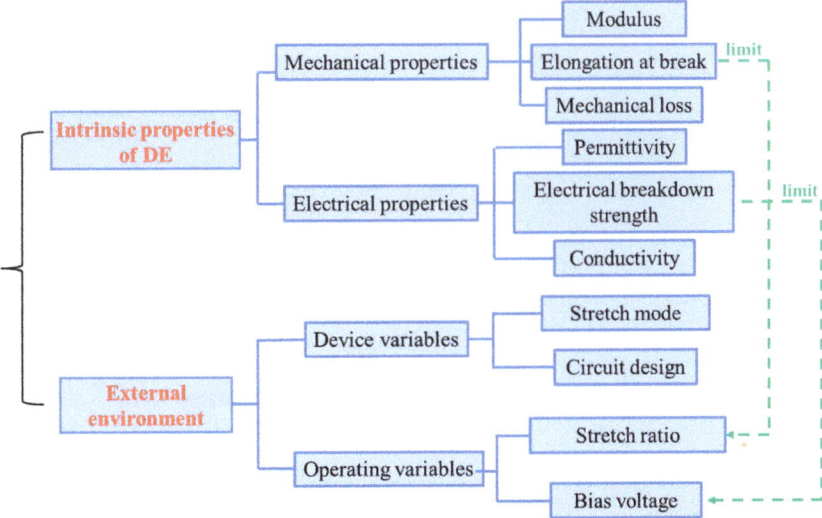

Figure 2. Influencing factors on the energy-generation performances of the DEG.

In this study, a simplified but validated electromechanical model based on an actual circuit was developed to describe the relationship between intrinsic properties of DE material and the energy-harvesting performance. Linear elastic proposition with strain relaxation parameter was employed, and charge leakage was considered while studying the energy conversion mechanism. Experimental verification of the model was performed. The influences of six intrinsic properties of DE materials (including the modulus, elongation at break, mechanical loss property, permittivity, electrical breakdown strength and conductivity) on the energy-harvesting performances (including energy density and conversion efficiency) were systematically studied. Furthermore, guidance for the preparation of DE materials with high energy-harvesting performance was proposed. In addition, the ratio of permittivity and modulus of the DE material on the energy conversion efficiency of the DEG was discussed.

2. Modeling of DEG

2.1. Setup of Device Variables

The device variables contain the stretch mode and circuit design. The stretch modes reported in the literature have mainly included cone stretch, equibiaxial stretch and diaphragm inflatable stretch [21]. Among these stretch modes, equibiaxial stretch, a kind of uniform stretch in a plane, is the most widely used. This is because the uniformity of the thickness of the DE film can be maintained during stretching, and the highest energy-harvesting performances can be obtained under equibiaxial stretch [22,31]. Figure 3a shows a schematic diagram of equibiaxial stretch. The radius of the circular DE film before and after stretching are r_0 and r, respectively. In this case, the equibiaxial stretch ratio λ is used to describe the degree of stretching, which is calculated using the formula $\lambda = r/r_0$. The relationship between the stretch ratio and strain ε is $\lambda = 1 + \varepsilon$.

The circuit shown in Figure 3d was adopted according to Samuel Shian's study [22]. The parallel transfer capacitors (C_P) with the capacitance of $C_P = 1.3\ C_1$ in this circuit can reduce the voltage rise caused by film releasing to prevent the film from electrical breakdown during the releasing process and ensure the completion of the cycle. The DEG energy-harvesting process was performed through the following four steps: (i) the stretching process, where the equibiaxial stretch was performed on the DE film and changed its radius from r_0 to r_1; (ii) the voltage boosting process, where S1 was closed, and the DC source with preset input voltage V_1 was made to fully charge the DE film and C_P; (iii) the releasing process, where S1 was disconnected, and the DE film was released with a charged state, during which a higher voltage V_2 can be obtained across the DE and C_P. Because of the existence of Maxwell stress, a "loss of tension" occurred before it releasing to r_0, and the radius of DE film at the released state was r_k ($r_k > r_0$); (iv) the harvesting process, where S2 was closed to release the charges across the DE film and C_P.

The input mechanical work (W_{mech}) is equal to the difference between the work done by the stretch device on the DE film (W_s) during the stretch process and the work done by DE film on the stretching device (W_r) during the release process, that is:

$$W_{mech} = W_s - W_r = \int_{r_0}^{r_1} F_s dr - \int_{r_k}^{r_1} F_r dr \tag{5}$$

where F_s, F_r and r_1 represent the stretching force, restoring force and the radius of stretched state, respectively.

Both the equibiaxial stretch force and Maxwell stress perform work on the DE film during the energy-harvesting process. Figure 4 shows the schematic diagram of the Maxwell stress and equibiaxial stretch force acting on the DE film. Performing the Maxwell stress and equibiaxial stretch force from the thickness direction and horizontal direction on the DE film with an initial thickness of z_0 and initial radius of r_0, respectively, to make the film produce a slight deformation, the thickness becomes z_0-dz, and the radius becomes $r_0 + dr$.

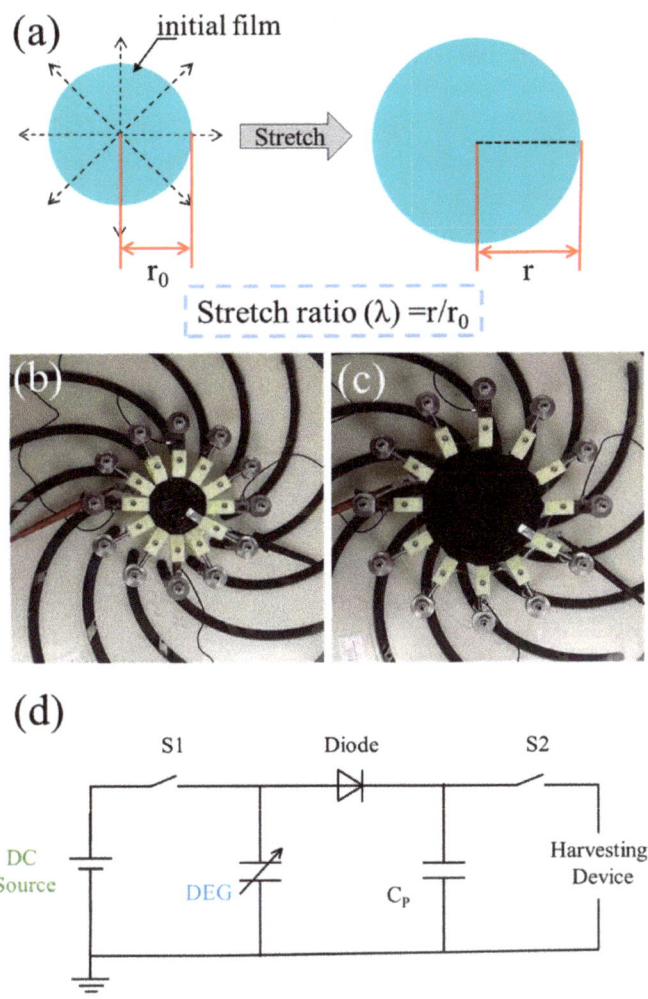

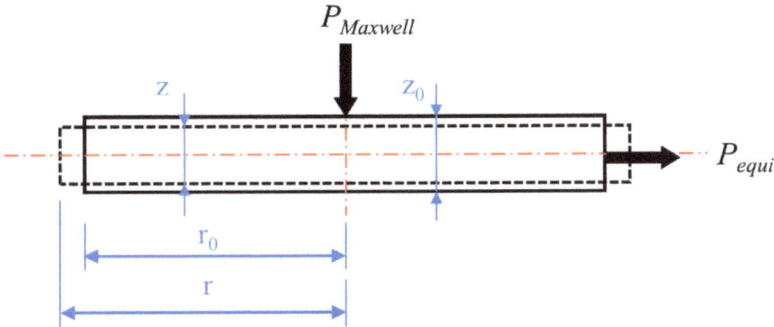

Figure 3. (**a**) Schematic diagram of equibiaxial stretch. (**b**) Released state and (**c**) stretched state of homemade equibiaxial stretch device. (**d**) Circuit principle adopted in this study.

Figure 4. Schematic diagram of the Maxwell stress and equibiaxial stretch force acting on the DE film.

Since the energy consumed by the two deformation methods is the same, the work performed by the Maxwell stress ($P_{Maxwell}$) in the vertical direction is equal to the work performed by the equibiaxial stretch force (P_{equi}) in the horizontal direction, that is:

$$(P_{Maxwell} \pi r_0^2) \cdot dz = (P_{equi} 2\pi r_0 \cdot z_0) \cdot dr \tag{6}$$

Assuming that the volume of the DE film remains unchanged, that is:

$$\pi r_0^2 dz = 2\pi r_0 \cdot z_0 dr \tag{7}$$

Combine the Equations (6) and (7):

$$P_{Maxwell} = P_{equi} \tag{8}$$

Therefore, the effect caused by P_{equi} in horizontal direction is equivalent to that caused by the same magnitude of $P_{Maxwell}$ in the vertical direction.

During the stretching process, the stretch force applied on the circumference of the film causes the film to stretch in the radial direction and shrink in the thickness direction. At this time, the recovery force of the DE film is equal to its stretch force. Since the Maxwell stress also tends to shrink the film in the thickness direction and expand in the radial direction, the generation of the Maxwell stress caused by exerting bias voltage reduces the restoring force of the film. The input mechanical work is calculated based on the force-displacement relationship in the thickness direction. From Equation (8), during the releasing process of charged film, the restoring force is equal to equivalent P_{equi} in the vertical direction minus $P_{Maxwell}$. The value of $P_{Maxwell}$ under the action of the electric field strength (E) is $\varepsilon_0 \varepsilon_r E^2$. Therefore, during the releasing process of charged film, the vertical restoring force (P_r) of the film under the action of the electric field is:

$$P_r = P_{equi} - P_{Maxwell} = P_{equi} - \varepsilon_0 \varepsilon_r E^2 \tag{9}$$

Expressing the relationship between force and deformation in terms of Hooke's law can simplify the model. Assuming that during the stretching process, the P_{equi} and the stretch ratio λ satisfy the following linear relationship:

$$P_{equi} = (\lambda - 1)M \tag{10}$$

The proportional coefficient M is called the elastic coefficient. Since the effective modulus of equibiaxial stretching is twice of Young's modulus [33], the relationship between M and Young's modulus is $M = 2Y(1 + \varepsilon) = 2\lambda Y$, where ε represents the strain. The higher Young's modulus of the material results in the greater elastic coefficient M, so M can also reflect the ability of a material to resist elastic deformation under the action of external force.

2.2. The Description of Mechanical Loss Behavior

Rubber is a viscoelastic material [34]. During the stretch-release process, the movement of the molecular chain of the viscoelastic material needs to overcome the internal resistance to do work, and it must convert a part of the energy into heat energy, thus causing mechanical loss. After being stretched to a certain strain, the stress of rubber gradually decreases with time, a stress relaxation phenomenon [35,36]. During the voltage boosting process, it takes time for the voltage to increase from zero to the bias voltage value. Therefore, the DE film undergoes a stretching-relaxation-release process in the energy-harvesting cycle. The stress relaxation property is used in this work to describe the mechanical loss in the conversion process. The difference in the recovery force between the end of the stretching process and the beginning of releasing process comes from two aspects: One is the decrease caused by the Maxwell stress, and the other is the decrease caused by stress relaxation. In order to simplify the model description, the stress relaxation ratio θ was introduced to describe the mechanical loss behavior of the DE (i.e., the ratio between the relaxed stress and maximum stress) [37]. A smaller θ value indicates the better elasticity. At the beginning of the release process, the reduction of the recovery force caused by stress relaxation is directly deducted from the recovery force. DE is still regarded as a linear elastic material in the stretch and release process. In this study, the viscoelasticity of DE was considered, as shown in Figure 5. In this case, the restoring force of the film during the releasing process is as follows:

$$P_r = P_{equi} - \theta \cdot P_{equi}(\lambda_1) - P_{Maxwell} \tag{11}$$

where $P_{equi}(\lambda)$ represents the P_{equi} under λ, and λ_1 represents the stretched state ratio. In addition, λ_k represents the released state ratio when the recovery force of the film drops to 0 and the film cannot continue to shrink. λ_k can be calculated as follows:

$$P_{equi}(\lambda_k) - \theta \cdot P_{equi}(\lambda_1) - P_{Maxwell} = 0 \tag{12}$$

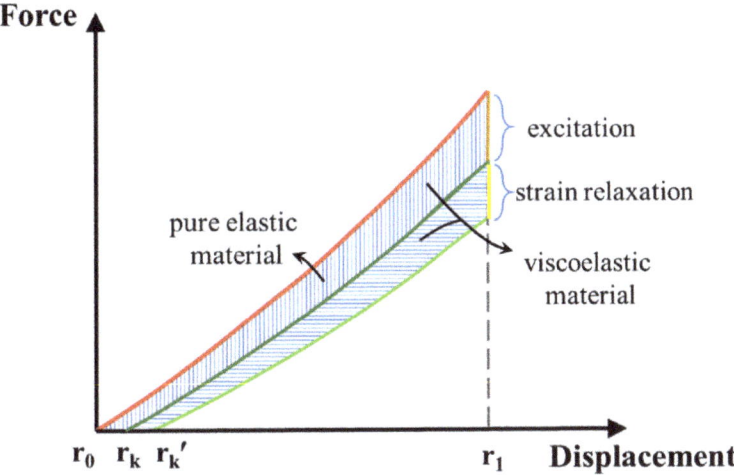

Figure 5. Force-displacement curve of the DEG under different elasticity conditions.

2.3. The Description of Electrical Loss Behavior

Equation (14) describes the single-cycle generated energy without charge loss. However, the DE film is not an ideal insulator, so the charge will be consumed due to the tiny leakage current inside the DE. The leakage of charge is a continuous process which occurs from the charging of the DE film to the release of the charge.

Figure 6 shows the force-displacement curve under three different charge leakage conditions: (a) leakage-free condition, (b) actual condition and (c) leakage-first condition. The differences of these conditions are mainly exhibited in the release process. At the beginning of the release process (λ_1, F_1), since no charge has been consumed yet, the restoring force in the actual condition is equal to that in the leakage-free condition. During the release process, the continuous leakage of charge causes the voltage in the actual condition to be relatively lower than that in the leakage-free condition, that is, the Maxwell stress across the film during the actual condition is relatively low. Therefore, the restoring force in the actual condition is higher than that in the leakage-free condition. Therefore, the r_k of the actual condition (marked as r_{k2}) is smaller than the that of leakage-free condition (marked as r_{k1}), that is, $r_{k2} < r_{k1}$. Since the voltage and area of the film are constantly changing, other parameters, such as the stretching rate and time must be introduced in order to accurately express the charge leakage, which greatly complicates the model. To simplify the model, the charge leakage ratio δ is defined as the percentage of the leakage charge in the input charge. The higher conductivity of the DE material results in the higher leakage current. Therefore, δ is positively related to the conductivity of DE material, which is used to describe the electrical loss behavior of the material. δ can be calculated by the following formula:

$$\delta = \frac{Q_{in} - Q_{out}}{Q_{in}} = \frac{(C_1 + C_P)V_1 - (C_2 + C_P)V_2}{(C_1 + C_P)V_1} \tag{13}$$

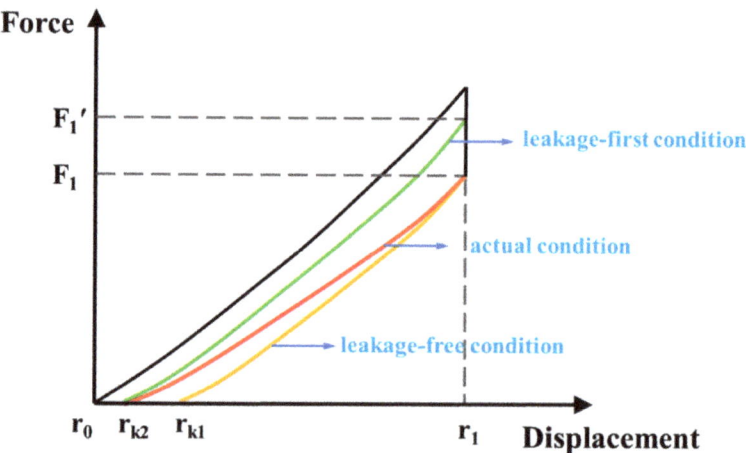

Figure 6. Force-displacement curve of the DEG under different charge leakage conditions.

At the end of the release process, the voltage across the DE film becomes:

$$V_2 = \frac{(C_1 + C_P)V_1(1 - \delta)}{(C_2 + C_P)} \tag{14}$$

Therefore, the actual single-cycle generated energy that takes the electrical loss into account can be rewritten from Equation (1) as:

$$\Delta U = \frac{1}{2}(C_1 + C_P)V_1^2 \left[\frac{(C_1 + C_P)}{(C_2 + C_P)}(1 - \delta)^2 - 1 \right] \tag{15}$$

Under the definition of δ, the "leakage-first" condition is constructed, which means that the DE loses δ of the charge at the beginning of the process, and then the release process is performed with a constant charge. At the beginning of the release process, since no charge has been consumed yet in the actual condition, the voltage and the Maxwell stress in the actual condition are higher than that in the leakage-first condition, that is, $F_1 < F_1'$. As the film releases, the continuous leakage of the charge causes a decrease in the difference in the restoring force between the actual and leakage-first condition. Finally, when the restoring force in both cases drops to 0, the charge leakage ratio of both cases is δ, so the actual condition curve and the leakage-first condition curve intersect at the point $(r_{k2}, 0)$. Clearly, the release curve in the actual process is between that in the leakage-first condition and the leakage-free condition, so the average value of W_r in the leakage-first condition and leakage-free condition can be used to approximate the W_r in actual condition, which is:

$$W_{r,actual} = \frac{W_{r,leakage-free} + W_{r,leakage-first}}{2} \tag{16}$$

2.4. The Model of Energy Harvesting Performances of DEG

By combining Equations (2) and (15), and Equations (5), (10), (12) and (16), the model on the energy-harvesting performances of the DEG can be obtained. ΔU and W_{mech} can be written as:

$$\Delta U = \frac{2.3\varepsilon_0\varepsilon_r \pi r_0^2 \lambda_1^4 V_1^2}{2z_0} \left[\frac{2.3\lambda_1^4}{1.3\lambda_1^4 + \lambda_{k2}^4} \times (1-\delta)^2 - 1 \right] \quad (17)$$

$$W_{mech} = 2\pi r_0^2 z_0 M(\lambda_1 - 1 - \ln\lambda_1)$$
$$- \pi r_0^2 z_0 \left\{ \begin{array}{l} \int_{\frac{1}{\lambda_{k1}^2}}^{\frac{1}{\lambda_1^2}} \left[(\lambda-1)M - (\lambda_1-1)M\theta - \frac{\varepsilon_0\varepsilon_r V_1^2}{z_0^2} \frac{(2.3\lambda_1^4)^2 \lambda^4}{(1.3\lambda_1^4+\lambda^4)^2} \right] \lambda^2 d\left(\frac{1}{\lambda^2}\right) \\ + \int_{\frac{1}{\lambda_1^2}}^{\frac{1}{\lambda_{k2}^2}} \left[(\lambda-1)M - (\lambda_1-1)M\theta - \frac{\varepsilon_0\varepsilon_r V_1^2(1-\delta)^2}{z_0^2} \frac{(2.3\lambda_1^4)^2 \lambda^4}{(1.3\lambda_1^4+\lambda^4)^2} \right] \lambda^2 d\left(\frac{1}{\lambda^2}\right) \end{array} \right\} \quad (18)$$

where r_0, z_0, M, θ, δ and ε_r represent the initial radius, initial thickness, elastic coefficient, stress relaxation ratio, charge leakage ratio and permittivity of DE film, respectively; and V_1, λ_1, λ, λ_{k1} and λ_{k2} represent bias voltage, stretch ratio, stretched state ratio and released state ratio without or with charge leakage, respectively. λ_{k1} and λ_{k2} can be obtained by:

$$(\lambda_{k1} - 1)M - (\lambda_1 - 1)M\theta = \frac{\varepsilon_0\varepsilon_r V_1^2}{z_0^2} \frac{(2.3\lambda_1^4)^2 \lambda_{k1}^4}{(1.3\lambda_1^4 + \lambda_{k1}^4)^2} \quad (19)$$

$$(\lambda_{k2} - 1)M - (\lambda_1 - 1)M\theta = \frac{\varepsilon_0\varepsilon_r V_1^2 (1-\delta)^2}{z_0^2} \frac{(2.3\lambda_1^4)^2 \lambda_{k2}^4}{(1.3\lambda_1^4 + \lambda_{k2}^4)^2} \quad (20)$$

3. Results and Discussion

3.1. Experimental Validation

Before further exploration, the actual energy-harvesting performance of the VHB4905 material was measured under a homemade test platform in Section S1.3 in the Supporting Information and then compared with simulation results to verify the accuracy of the proposed model. The materials parameters used in the simulation were obtained by the characterization of VHB4905 (see Sections S2.1–S2.5 and Table S1 in Supporting Information). The experimental and simulated energy density and electromechanical conversion efficiency of the VHB4905 material under $\lambda_1 = 2$ and different V_1 values are shown in Figure 7. The results show that the error between the experimental value and simulated value of the energy density and conversion efficiency was less than 15% and 20%, respectively. These favorable results verify the feasibility of the proposed model in describing the relationship between the intrinsic properties of DE material and the energy-harvesting performances of the DEG. Moreover, the results indicate that the proposed model can predict the energy-harvesting performances of the DE.

It should be mentioned that the focus of this study is to establish the relationship between the intrinsic properties of materials and energy-harvesting performances, so some of the assumptions and approximations used in this model sacrifice accuracy. First, the linear elastic model with stress relaxation parameter is obviously different from the actual stretch-release process, so there is a certain error in the calculation of input mechanical work. Second, some studies have shown that the dielectric constant of the material changes with tension [38,39]. The dielectric constant was set as a constant in this model, which led to a certain error in the calculation of generated energy. The above factors will lead to an error between the model and the experimental value.

3.2. Influences of Intrinsic Properties of DE Materials on Energy-Harvesting Performance

The parameters related to the material properties in the model include the elastic coefficient M, stress relaxation ratio θ, permittivity ε_r, bias voltage V_1, stretched state ratio λ_1 and charge leakage ratio δ. As explained in Section 2, the influences of the above six variables on the simulation results correspond to the influences of the modulus, viscoelasticity, permittivity, breakdown field strength, elongation at break and conductivity on the energy-harvesting performances, respectively. The six variables were divided into two groups: V_1, θ and δ as one group, and permittivity ε_r, elastic coefficient M and stretched state ratio λ_1 as the other group. The influences of each variable on the energy-harvesting performances were studied separately.

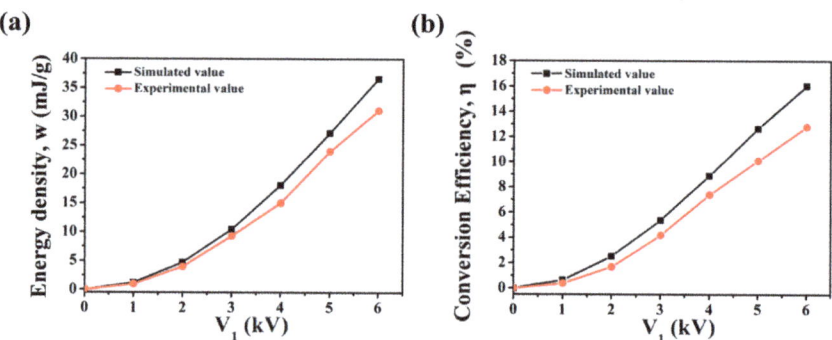

Figure 7. The experimental and simulated (**a**) energy density and (**b**) electromechanical conversion efficiency of the VHB4905 material under $\lambda_1 = 2$ and different V_1 values.

To provide better guidance for preparing high-performance DE materials for the DEG, the values of these variables for the simulation were chosen according to experimental values. The shape of the DE film and the variables used in the simulation of energy-harvesting performances are shown in Table 1. Since the VHB material has atypical and serious viscoelasticity loss, the setting of θ value in this work refers to silicone rubber with high elasticity. The stress relaxation rate of the silicone rubber is around 0.05 to 0.10 [40]. Referring to this value, we set 0, 0.05 and 0.1, respectively, indicating no mechanical loss, high elastic material and moderate elastic material.

Table 1. Summary of the shape of the DE film and variables in this study.

DE Film Shape and Variables	Value
thickness, z_0/mm	0.5
radius, r_0/mm	20
bias voltage, V_1/kV	0, 1, 2, 3, 4, 5
stress relaxation ratio, θ	0, 0.05, 0.1
charge leakage ratio, δ	0, 0.05
elastic coefficient, M/MPa	0.2, 0.1
permittivity, ε_r	4.2, 8.4
stretched state ratio, λ_1	2, 3

The energy-harvesting performances of DE materials under different bias voltage V_1, stress relaxation ratio θ and charge leakage ratio δ. The other three variables are fixed (elastic coefficient $M = 0.2$ MPa, permittivity $\varepsilon_r = 4.2$, stretched state ratio $\lambda_1 = 2$). Figure 8a shows the influences of the three variables above on the single-cycle generated energy ΔU. The increase in V_1 greatly enhanced ΔU. The increase in V_1 from 1 kV to 5 kV caused am increase in ΔU of 24 times. This is because the increase in V_1 increased the work done by the Maxwell stress during the releasing process by enhancing the Maxwell stress. Thus, more mechanical energy can be converted into electrical energy. This also means that under the same film thickness, DE with higher breakdown strength can withstand a higher V_1, which is able to obtain a higher ΔU. ΔU was significantly affected by charge leakage. Even a small charge leakage of 0.05 resulted in a significant decrease in ΔU of about 25%.

Therefore, DE with lower conductivity can increase ΔU in the form of reducing charge leakage. The increase in θ increased the reduction of restoring force during the stress relaxation process, which increased λ_k as well as C_2, and thus decreased ΔU. ΔU was not sensitive to stress relaxation, but the effect of θ on ΔU increased with the increase in V_1. In the case of $V_1 = 1$ kV, $\delta = 0.05$ and $\theta = 0.05$, ΔU decreased by 2.9%. In the case of $V_1 = 5$ kV, $\delta = 0.05$ and $\theta = 0.05$, ΔU decreased by 4.3%. Although ΔU was less affected by θ, DE with lower stress relaxation characteristics can obtain higher power generation, that is to say, high elasticity is needed for the ideal DE.

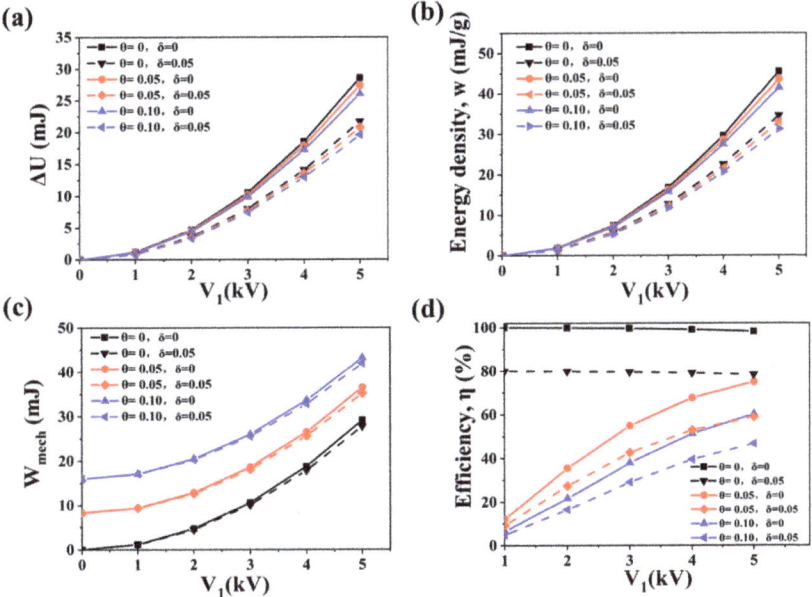

Figure 8. Influences of bias voltage V_1, stress relaxation ratio θ and charge leakage ratio δ on the energy-harvesting performances of DE materials: (**a**) generated energy, (**b**) energy density, (**c**) input mechanical work, (**d**) electromechanical conversion efficiency. The abscissa represents the bias voltage; the black, blue and red curves represent the $\theta = 0$, 0.05, 0.1, respectively; and the solid line and the dashed line represent the $\delta = 0$ and 0.05, respectively.

Figure 8b shows the influences of V_1, θ and δ on the energy density w. The energy density was obtained by dividing ΔU by the mass of the DE in the effective working area, which reflected the energy-harvesting performance of DE per unit mass. The mass of the effective working area of DE remained unchanged under different variables. Therefore, the influences of the variables on the energy density were very similar to that on ΔU, except the scale of the ordinate was different. Similarly, the increase in V_1, and the decrease of δ or θ all resulted in the increase in the energy density. Therefore, low conductivity, high breakdown strength and high elasticity are needed for DE materials with high energy density.

Figure 8c shows the variation of input mechanical work W_{mech} with V_1, θ and δ. The increase in V_1 increases the λ_k through enhancing the Maxwell stress, thus reduces the external work done by the film during the release process, which in turn leads to an increase in W_{mech}. The increase in θ greatly increases the W_{mech}, which reflects the energy loss caused by molecular chain rearrangement and slippage during the stretch-release process, and it is especially significant under low V_1. In the case of low V_1, the increase in W_{mech} caused by θ is much higher than that caused by the V_1. This means that most of W_{mech} is consumed due to the viscoelastic loss, and only a small amount of W_{mech} has been converted into electrical energy. The increase in δ slightly reduces the W_{mech}. This is because the charge leakage reduces the work done by Maxwell stress, as analyzed in Section 2.3.

Figure 8d shows the influences of V_1, θ and δ on the conversion efficiency η. The conversion efficiency was obtained by dividing ΔU by the input mechanical work W_{mech}. η decreased with the increase in δ. Compared with the condition of $\delta = 0$, $\delta = 0.05$ will result in a decrease in η of nearly 20%. This is because the increase in δ greatly reduces ΔU but less affects W_{mech}. Since the increase in θ had little effect on ΔU but largely increased W_{mech}, η also decreased with the increase in θ. For example, under $V_1 = 1$ kV and $\delta = 0.05$, more W_{mech} will be consumed in the viscoelastic loss by increasing θ, so the value of η drops from 79.97% ($\theta = 0$) to 9.27% ($\theta = 0.05$), and further reduces to 4.93% ($\theta = 0.1$). This shows that using low elastic materials at low bias voltages is an inefficient way of harvesting energy. The change of η with V_1 varied with the value of θ. In the case of $\theta = 0$, no energy was consumed in the mechanical loss, and η decreased slightly with the increase in V_1. In the case of $\theta \neq 0$, a large amount of the mechanical work was consumed in the mechanical loss, and the increase in V_1 enlarged the Maxwell stress, thus increasing the proportion of the work done by the Maxwell stress to W_{mech} during the conversion process, so the η was improved. The value of η gradually increased from 4.93% ($V_1 = 1$ kV, $\delta = 0.05$, $\theta = 0.1$) to 46.79% ($V_1 = 5$ kV, $\delta = 0.05$, $\theta = 0.1$). Therefore, the high breakdown strength, high elasticity and low conductivity of the material are of great significance for achieving high η.

Figure 9 shows the energy-harvesting performances of DE materials with different elastic coefficient M, permittivity ε_r and stretched state ratio λ_1 under different V_1 values. It is noticed that the black "Reference" curve represents the DE with the fitting conditions of $\varepsilon_r = 4.2$, $M = 0.2$ MPa, $\lambda_1 = 2$; the red "$M = 0.1$ MPa" curve represents the DE with the fitting conditions of $\varepsilon_r = 4.2$, $M = 0.1$ MPa, $\lambda_1 = 2$; the bule "$\varepsilon_r = 8.4$" curve represents the DE with the fitting conditions of $\varepsilon_r = 8.4$, $M = 0.2$ MPa, $\lambda_1 = 2$; amd the green "$\lambda_1 = 3$" curve represents the DE with the fitting conditions of $\varepsilon_r = 4.2$, $M = 0.2$ MPa, $\lambda_1 = 3$. The other two variables were fixed ($\theta = 0.05$, $\delta = 0.05$).

Figure 9a shows the influences of M, ε_r and λ_1 on ΔU under different V_1 values. Since the increase in both ε_r and λ_1 increased the Maxwell stress, and the increase in λ_1 also enlarged the stretch displacement, ΔU increased with the increase in ε_r, and greatly increased with the increase in λ_1. The simulation results show that under $V_1 = 5$ kV, the increase in ε_r to 2 times that of the reference sample caused the increase in ΔU from 20.77 mJ (Reference) to 35.84 mJ ($\varepsilon_r = 8.4$), an increase of 72.6%. The increase in λ_1 from 2 to 3 caused the increase in ΔU from 20.77 mJ (Reference) to 122.84 mJ ($\lambda_1 = 3$), an increase of 491%. On the other hand, the decrease in M led to a slight decrease in ΔU under low V_1, but when ΔU reached $V_1 = 5$ kV, the decrease in M by half caused the decrease in ΔU from 20.77 mJ (Reference) to 17.92 mJ ($M = 0.1$ MPa), a decrease of 13.8%. This is because the decrease in M had no effect on Maxwell stress, but it reduced the recovery force, which increased λ_k as well as the released state capacitance C_2, thus reducing ΔU.

Figure 9b shows the influences of M, ε_r and λ_1 on the energy density w under different V_1 values. Similar to Figure 8b, the influences of variables on w was very similar to that on ΔU, except that the scale of the ordinate was different. Therefore, the energy density increased with the increase in ε_r, λ_1 and M. The simulation results show that the material with the properties of $\theta = 0.05$, $\delta = 0.05$, $\varepsilon_r = 4.2$, $M = 0.2$ MPa, and $\lambda_1 = 3$ can obtain a theoretical energy density of up to 195.61 mJ/g under $V_1 = 5$ kV.

Figure 9c shows the influences of M, ε_r and λ_1 on W_{mech} under different V_1 values. The simulation results indicate that W_{mech} increased with the increase in M, ε_r and λ_1. The reduction in M reduced the useless work due to the viscoelastic loss, thus reducing W_{mech}. For example, in the case of $V_1 = 5$ kV, W_{mech} reduced from 35.20 mJ (Reference) to 30.04 mJ ($M = 0.1$ MPa), a decrease of 14.7%. The increase in ε_r indicatesan increase in Maxwell stress, which significantly increased W_{mech}. Except for the increase in the work done by the Maxwell stress, the increase in λ_1 also increased the mechanical loss, and thus largely increased W_{mech}. In the case of $V_1 = 5$ kV, the increase in ε_r to 2 times that of the reference sample caused the increase in W_{mech} from 35.20 mJ (Reference) to 60.08 mJ ($\varepsilon_r = 8.4$), an increase of 70.7%. The increase in λ_1 from 2 to 3 caused the increase in W_{mech} from 35.20 mJ (Reference) to 179.20 mJ ($\lambda_1 = 3$), an increase of 409%.

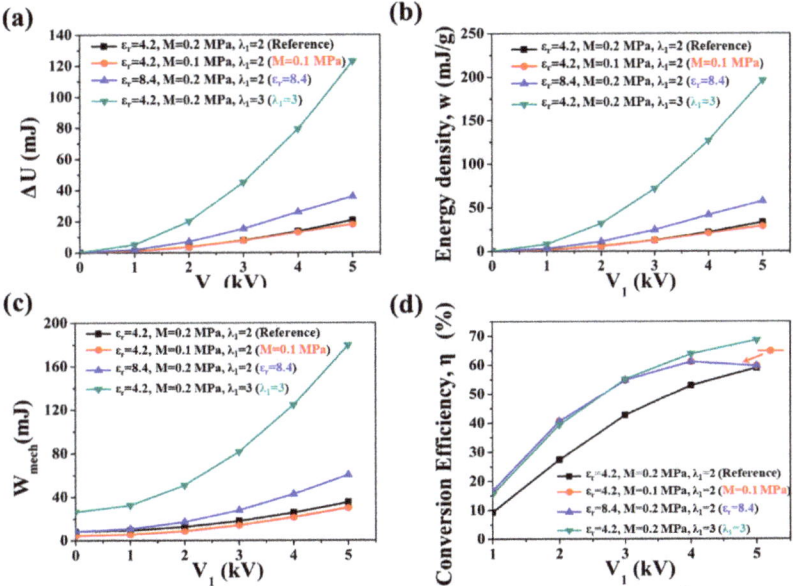

Figure 9. Influences of elastic coefficient M, permittivity ε_r and stretched state ratio λ_1 on the energy-harvesting performances of materials with under different V_1: (**a**) generated energy, (**b**) energy density, (**c**) input mechanical work, (**d**) electromechanical conversion efficiency. The black "Reference" curve represents the material with fitting conditions of $\varepsilon_r = 4.2$, M = 0.2 MPa, $\lambda_1 = 2$; the red curve represents the material with fitting conditions of $\varepsilon_r = 4.2$, M = 0.1 MPa, $\lambda_1 = 2$; the blue curve represents the material with fitting conditions of $\varepsilon_r = 8.4$, M = 0.2 MPa, $\lambda_1 = 2$; the green curve represents the material with fitting conditions of $\varepsilon_r = 4.2$, M = 0.2 MPa, $\lambda_1 = 3$.

Figure 9d shows the influences of M, ε_r and λ_1 on η under different V_1 values. Increasing λ_1 can effectively improve η, especially under low V_1. In the case of $V_1 = 1$ kV, η increases from 9.27% (Reference) to 15.48% ($\lambda_1 = 3$), an increase of 66.9%. In the case of $V_1 = 5$ kV, η increases from 59.00% (Reference) to 68.55% ($\lambda_1 = 3$), an increase of 16.2%. Interestingly, the blue and red curves almost overlapped, which shows that the influence of reducing M by half on η had the equivalent effect as that of increasing ε_r by two-fold. The M and ε_r values in the fitting condition of the blue curve (M = 0.2 MPa, $\varepsilon_r = 8.4$) were twice those in red curve (M = 0.1 MPa, $\varepsilon_r = 4.2$). Through extracting M or ε_r, Equations (19) and (20) can be rewritten to Equations (21) and (22), respectively, from which λ_{k1} and λ_{k2} are equal in the two conditions under any V_1 value.

$$M \cdot [(\lambda_{k1} - 1) - (\lambda_1 - 1)\theta] = \varepsilon_r \cdot \frac{\varepsilon_0 V_1^2}{z_0^2} \frac{(2.3\lambda_1^4)^2 \lambda_{k1}^4}{(1.3\lambda_1^4 + \lambda_{k1}^4)^2} \quad (21)$$

$$M \cdot [(\lambda_{k2} - 1) - (\lambda_1 - 1)\theta] = \varepsilon_r \cdot \frac{\varepsilon_0 V_1^2 (1-\delta)^2}{z_0^2} \frac{(2.3\lambda_1^4)^2 \lambda_{k2}^4}{(1.3\lambda_1^4 + \lambda_{k2}^4)^2} \quad (22)$$

Similarly, through extracting M or ε_r, Equations (17) and (18) can be rewritten as Equations (23) and (24), respectively.

$$\Delta U = \varepsilon_r \cdot \frac{2.3\varepsilon_0 \pi r_0^2 \lambda_1^4 V_1^2}{2z_0} \left[\frac{2.3\lambda_1^4}{1.3\lambda_1^4 + \lambda_{k2}^4} \times (1-\delta)^2 - 1 \right] \tag{23}$$

$$\begin{aligned} W_{mech} &= M\pi r_0^2 z_0 \cdot \left\{ 2(\lambda_1 - 1 - \ln\lambda_1) - \int_{\frac{1}{\lambda_{k1}^2}}^{\frac{1}{\lambda_1^2}} [(\lambda-1) - (\lambda_1-1)\theta]\lambda^2 d\left(\frac{1}{\lambda^2}\right) - \int_{\frac{1}{\lambda_{k2}^2}}^{\frac{1}{\lambda_1^2}} [(\lambda-1) - (\lambda_1-1)\theta]\lambda^2 d\left(\frac{1}{\lambda^2}\right) \right\} \\ &+ \varepsilon_r \pi r_0^2 z_0 \left\{ \int_{\frac{1}{\lambda_{k1}^2}}^{\frac{1}{\lambda_1^2}} \left[\frac{\varepsilon_0 V_1^2}{z_0^2} \frac{(2.3\lambda_1^4)^2 \lambda^4}{(1.3\lambda_1^4 + \lambda^4)^2} \right] \lambda^2 d\left(\frac{1}{\lambda^2}\right) + \int_{\frac{1}{\lambda_{k2}^2}}^{\frac{1}{\lambda_1^2}} \left[\frac{\varepsilon_0 V_1^2 (1-\delta)^2}{z_0^2} \frac{(2.3\lambda_1^4)^2 \lambda^4}{(1.3\lambda_1^4 + \lambda^4)^2} \right] \lambda^2 d\left(\frac{1}{\lambda^2}\right) \right\} \end{aligned} \tag{24}$$

Therefore, the ΔU and W_{mech} in the fitting condition of blue curve was twice of these in red curve under any V_1, respectively. Thus, the value of η was the same under the two conditions. In this case, the two curves in the η graph vs. the V_1 graph completely coincide. In addition, the η in these two fitting conditions first increased and then decreased with the increase in V_1, reaching a maximum value of 61.11% at $V_1 = 4$ kV. This indicates that DE with too high ε_r or too low M is not conducive to obtain high η under high working voltage since a premature "loss of tension" may occur [31]. Therefore, it is necessary to balance the relationship between E_b and ε_r or M when designing high-η materials: Under low E (lower than 32 kV/mm), it is suggested to enhance ε_r or reduce the modulus to improve η. Under high E (higher than 32 kV/mm), the ratio of ε_r and M of the DE should be tailored to optimize η, and the recommended ε_r/M value of the DE should be between 20/MPa and 40/MPa.

The results indicate that the ΔU and w were greatly affected by a small amount of charge leakage, but were not sensitive to mechanical loss property, so they could be significantly improved by increasing the insulation performance of the DE material. W_{mech} was less affected by charge leakage but was quite sensitive to mechanical loss. In addition, ΔU and w can be further enhanced with the increase in λ_1 and ε_r. Appropriately reducing M is beneficial to improve η, but excessively low M will cause the reduction in η under high V_1 since a premature "loss of tension" may occur. Since the reduction of the modulus and the increase in ε_r have equivalent effects on η, DE materials with high ε_r should have a moderate modulus. The recommended ε_r/M value of DE should be between 20/MPa and 40/MPa.

To sum up, high breakdown field strength, low conductivity and high elasticity of DE materials are the prerequisites for obtaining high energy density and conversion efficiency. DE materials with high elongation at break, high permittivity and moderate modulus can further improve the energy density and conversion efficiency of the DEG.

4. Conclusions

Herein, an electromechanical model of DEG was established to reveal the relationship between the intrinsic properties of DE materials and energy-harvesting performances. The good agreement between the simulation and experimental results was verified, indicating that this coupling model can well predict the energy-harvesting performance of the material under the preset conditions. By tailoring the fitting condition in the model, the relationship between the intrinsic properties of DE materials (including the modulus, elongation at break, mechanical loss property, permittivity, breakdown field strength and conductivity) and the energy-harvesting performances (including the energy density and conversion efficiency) of the DEG was revealed. The results indicate that DE materials with high breakdown field strength, low conductivity and high elasticity are the prerequisites for achieving high energy density and high conversion efficiency of the DEG. In addition, DE materials with high elongation at break, high permittivity and moderate modulus can further improve the energy density and conversion efficiency of the DEG.

Supplementary Materials: The following are available online at https://www.mdpi.com/article/10.3390/polym13234202/s1, Figure S1: Test platform contains equibiaxial stretching device, force/displacement sensor, digital electrometer, DC source and PC, Figure S2: Simulation of the elastic coefficient M of VHB4905 material. The slope of linear fitting curve of M = 0.2 MPa is adopted in simulation of the energy harvesting performances of VHB4905, Figure S3: The relationship between stress relaxation of VHB material and relaxation time under uniaxial strain of 100%. Stress relaxation ratio when time equal to 5s, that is θ = 0.35 is adopted in simulation of the energy harvesting performances of VHB4905, Figure S4: Permittivity vs. frequency graph of VHB4905. Permittivity under 100 Hz of ε_r = 4.2 is adopted in simulation of the energy harvesting performances of VHB4905, Figure S5: Calculated charge leakage ratio δ under different V_1. Average value of δ is adopted in simulation of the energy harvesting performances of VHB4905, Table S1: Summary of value of DE film shape and variables in simulation of the energy harvesting performances of VHB4905.

Author Contributions: Conceptualization, Y.J. and M.T.; methodology, Y.J. and H.Y.; formal analysis, Y.J. and Y.L.; investigation, Y.J., Y.L. and N.N.; data curation, Y.L.; writing—original draft preparation, Y.J.; writing—review and editing, N.N.; supervision, L.Z., N.N. and M.T. All authors have read and agreed to the published version of the manuscript.

Funding: The authors are grateful for the financial support from the National Natural Science Foundation of China (Grant No. 51525301) and the Talent Cultivation of State Key Laboratory of Organic-Inorganic Composites (Nos. OIC-D2021002).

Institutional Review Board Statement: Not applicable.

Informed Consent Statement: Not applicable.

Data Availability Statement: The data presented in this study are available on request from the corresponding author.

Acknowledgments: The authors are grateful for the financial support from the National Natural Science Foundation of China (Grant No. 51525301) and the Talent Cultivation of State Key Laboratory of Organic-Inorganic Composites (Nos. OIC-D2021002).

Conflicts of Interest: The authors declare no conflict of interest.

References

1. Osmani, B.; Seifi, S.; Park, H.S.; Leung, V.; Topper, T.; Muller, B. Nanomechanical probing of thin-film dielectric elastomer transducers. *Appl. Phys. Lett.* **2017**, *111*, 093104. [CrossRef]
2. Brochu, P.; Pei, Q. Advances in dielectric elastomers for actuators and artificial muscles. *Macromol. Rapid Commun.* **2010**, *31*, 10–36. [CrossRef] [PubMed]
3. Liu, X.; Sun, H.; Liu, S.; Jiang, Y.; Yu, B.; Ning, N.; Tian, M.; Zhang, L. Mechanical, dielectric and actuated properties of carboxyl grafted silicone elastomer composites containing epoxy-functionalized TiO_2 filler. *Chem. Eng. J.* **2020**, *393*, 124791. [CrossRef]
4. Chu, B.; Zhou, X.; Ren, K.; Neese, B.; Lin, M.; Wang, Q.; Bauer, F.; Zhang, Q. A dielectric polymer with high electric energy density and fast discharge speed. *Science* **2006**, *313*, 334–336. [CrossRef] [PubMed]
5. Pelrine, R.E.; Kornbluh, R.D.; Joseph, J.P. Electrostriction of polymer dielectrics with compliant electrodes as a means of actuation. *Sens. Actuators A Phys.* **1998**, *64*, 77–85. [CrossRef]
6. Carpi, F.; Anderson, I.; Bauer, S.; Frediani, G.; Gallone, G.; Gei, M.; Graaf, C.; Jean-Mistral, C.; Kaal, W.; Kofod, G.; et al. Standards for dielectric elastomer transducers. *Smart Mater. Struct.* **2015**, *24*, 105025. [CrossRef]
7. Pelrine, R.; Kornbluh, R.; Eckerle, J.; Jeuck, P.; Oh, S.J.; Pei, Q.B.; Stanford, S. Dielectric elastomers: Generator mode fundamentals and applications. In *Smart Structures and Materials 2001: Electroactive Polymer Actuators and Devices*; BarCohen, Y., Ed.; Spie-Int Soc Optical Engineering: Bellingham, WA, USA, 2001; Volume 4329, pp. 148–156.
8. Maas, J.; Graf, C. Dielectric elastomers for hydro power harvesting. *Smart Mater. Struct.* **2012**, *21*, 064006. [CrossRef]
9. Kornbluh, R.D.; Pelrine, R.; Prahlad, H.; Wong-Foy, A.; McCoy, B.; Kim, S.; Eckerle, J.; Low, T. Dielectric elastomers: Stretching the capabilities of energy harvesting. *MRS Bull.* **2012**, *37*, 246–253. [CrossRef]
10. Goudar, V.; Potkonjak, M. Dielectric Elastomer Generators for Foot Plantar Pressure Based Energy Scavenging. In Proceedings of the 11th IEEE Sensors Conference, Taipei, Taiwan, 28–31 October 2012; pp. 1001–1004.
11. Bortot, E.; Gei, M. Harvesting energy with load-driven dielectric elastomer annular membranes deforming out-of-plane. *Extrem. Mech. Lett.* **2015**, *5*, 62–73. [CrossRef]
12. Chen, S.E.; Deng, L.; He, Z.C.; Li, E.; Li, G.Y. Temperature effect on the performance of a dissipative dielectric elastomer generator with failure modes. *Smart Mater. Struct.* **2016**, *25*, 55017. [CrossRef]
13. Zhang, C.L.; Lai, Z.H.; Rao, X.X.; Zhang, J.W.; Yurchenko, D. Energy harvesting from a novel contact-type dielectric elastomer generator. *Energy Convers. Manag.* **2020**, *205*, 112351. [CrossRef]

14. Zhang, C.L.; Lai, Z.H.; Zhang, G.Q.; Yurchenko, D. Energy harvesting from a dynamic vibro-impact dielectric elastomer generator subjected to rotational excitations. *Nonlinear Dyn.* **2020**, *102*, 1271–1284. [CrossRef]
15. Moretti, G.; Righi, M.; Vertechy, R.; Fontana, M. Fabrication and Test of an Inflated Circular Diaphragm Dielectric Elastomer Generator Based on PDMS Rubber Composite. *Polymers* **2017**, *9*, 283. [CrossRef] [PubMed]
16. Jiang, Y.; Liu, S.; Zhong, M.; Zhang, L.; Ning, N.; Tian, M. Optimizing energy harvesting performance of cone dielectric elastomer generator based on VHB elastomer. *Nano Energy* **2020**, *71*, 104606. [CrossRef]
17. McKay, T.; O'Brien, B.; Calius, E.; Anderson, I. Self-priming dielectric elastomer generators. *Smart Mater. Struct.* **2010**, *19*, 055025. [CrossRef]
18. Wang, H.; Zhu, Y.; Wang, L.; Zhao, J. Experimental investigation on energy conversion for dielectric electroactive polymer generator. *J. Intell. Mater. Syst. Struct.* **2012**, *23*, 885–895.
19. Wang, H.; Wang, C.; Yuan, T. On the energy conversion and efficiency of a dielectric electroactive polymer generator. *Appl. Phys. Lett.* **2012**, *101*, 033904. [CrossRef]
20. McKay, T.; O'Brien, B.; Calius, E.; Anderson, I. An integrated, self-priming dielectric elastomer generator. *Appl. Phys. Lett.* **2010**, *97*, 062911. [CrossRef]
21. Moretti, G.; Rosset, S.; Vertechy, R.; Anderson, I.; Fontana, M. A Review of Dielectric Elastomer Generator Systems. *Adv. Intell. Syst.* **2020**, *2*, 2000125. [CrossRef]
22. Shian, S.; Huang, J.; Zhu, S.; Clarke, D.R. Optimizing the electrical energy conversion cycle of dielectric elastomer generators. *Adv. Mater.* **2014**, *26*, 6617–6621. [CrossRef]
23. Fan, P.; Chen, H. Optimizing the Energy Harvesting Cycle of a Dissipative Dielectric Elastomer Generator for Performance Improvement. *Polymers* **2018**, *10*, 1341. [CrossRef]
24. Yin, G.; Yang, Y.; Song, F.; Renard, C.; Dang, Z.; Shi, C.; Wang, D. Dielectric elastomer generator with improved energy density and conversion efficiency based on polyurethane composites. *ACS Appl. Mater. Interfaces* **2017**, *9*, 5237–5243. [CrossRef]
25. Pan, C.; Markvicka, E.J.; Malakooti, M.H.; Yan, J.; Hu, L.; Matyjaszewski, K.; Majidi, C. A Liquid-Metal-Elastomer Nanocomposite for Stretchable Dielectric Materials. *Adv. Mater.* **2019**, *31*, e1900663. [CrossRef] [PubMed]
26. Yang, D.; Xu, Y.; Ruan, M.; Xiao, Z.; Guo, W.; Wang, H.; Zhang, L. Improved electric energy density and conversion efficiency of natural rubber composites as dielectric elastomer generators. *AIP Adv.* **2019**, *9*, 025035. [CrossRef]
27. Yang, Y.; Gao, Z.; Yang, M.; Zheng, M.; Wang, D.; Zha, J.; Wen, Y.; Dang, Z. Enhanced energy conversion efficiency in the surface modified BaTiO3 nanoparticles/polyurethane nanocomposites for potential dielectric elastomer generators. *Nano Energy* **2019**, *59*, 363–371. [CrossRef]
28. Ellingford, C.; Zhang, R.; Wemyss, A.M.; Zhang, Y.; Brown, O.B.; Zhou, H.; Keogh, P.; Bowen, C.; Wan, C. Self-Healing Dielectric Elastomers for Damage-Tolerant Actuation and Energy Harvesting. *ACS Appl. Mater. Interfaces* **2020**, *12*, 7595–7604. [CrossRef] [PubMed]
29. Zhang, L.; Song, F.L.; Lin, X.; Wang, D.R. High-dielectric-permittivity silicone rubbers incorporated with polydopamine-modified ceramics and their potential application as dielectric elastomer generator. *Mater. Chem. Phys.* **2020**, *241*, 7. [CrossRef]
30. Koh, S.J.A.; Keplinger, C.; Li, T.; Bauer, S.; Suo, Z. Dielectric elastomer generators: How much energy can be converted? *IEEE/ASME Trans. Mechatron.* **2011**, *16*, 33–41. [CrossRef]
31. Koh, S.J.A.; Zhao, X.; Suo, Z. Maximal energy that can be converted by a dielectric elastomer generator. *Appl. Phys. Lett.* **2009**, *94*, 262902. [CrossRef]
32. Foo, C.C.; Koh, S.J.A.; Keplinger, C.; Kaltseis, R.; Bauer, S.; Suo, Z. Performance of dissipative dielectric elastomer generators. *J. Appl. Phys.* **2012**, *111*, 094107.
33. Eilaghi, A.; Flanagan, J.G.; Tertinegg, I.; Simmons, C.A.; Brodland, G.W.; Ethier, C.R. Biaxial mechanical testing of human sclera. *J. Biomech.* **2010**, *43*, 1696–1701. [PubMed]
34. Ehabe, E.; Bonfils, F.; Aymard, C.; Akinlabi, A.K.; Sainte Beuve, J. Modelling of Mooney viscosity relaxation in natural rubber. *Polym. Test.* **2005**, *24*, 620–627. [CrossRef]
35. Oman, S.; Nagode, M. Observation of the relation between uniaxial creep and stress relaxation of filled rubber. *Mater. Des.* **2014**, *60*, 451–457.
36. TuanDung, N.; Li, J.; Sun, L.; DanhQuang, T.; Xuan, F. Viscoelasticity Modeling of Dielectric Elastomers by Kelvin Voigt-Generalized Maxwell Model. *Polymers* **2021**, *13*, 2203.
37. Carniel, E.L.; Fontanella, C.G.; Stefanini, C.; Natali, A.N. A procedure for the computational investigation of stress-relaxation phenomena. *Mech. Time-Depend. Mater.* **2013**, *17*, 25–38. [CrossRef]
38. Kofod, G. The static actuation of dielectric elastomer actuators: How does pre-stretch improve actuation? *J. Phys. D Appl. Phys.* **2008**, *41*, 215405. [CrossRef]
39. Vu-Cong, T.; Jean-Mistral, C.; Sylvestre, A. Impact of the nature of the compliant electrodes on the dielectric constant of acrylic and silicone electroactive polymers. *Smart Mater. Struct.* **2012**, *21*, 105036. [CrossRef]
40. Cha, H.-S.; Yu, B.; Lee, Y.-K. Changes in stress relaxation property and softness of soft denture lining materials after cyclic loading. *Dent. Mater.* **2011**, *27*, 291–297. [CrossRef] [PubMed]

Article

Influence of Silicone Rubber Coating on the Characteristics of Surface Streamer Discharge

Xiaobo Meng [1], Liming Wang [2,*], Hongwei Mei [2,*] and Chuyan Zhang [3]

[1] School of Mechanical and Electrical Engineering, Guangzhou University, Guangzhou 510006, China; mengxb@gzhu.edu.cn
[2] Tsinghua Shenzhen International Graduate School, Tsinghua University, Shenzhen 518055, China
[3] School of Information Engineering, China University of Geosciences (Beijing), Beijing 100083, China; zcy@cugb.edu.cn
* Correspondence: wanglm@sz.tsinghua.edu.cn (L.W.); mei.hongwei@sz.tsinghua.edu.cn (H.M.); Tel.: +86-0755-26036695 (L.W.)

Abstract: A pollution flashover along an insulation surface—a catastrophic accident in electrical power system—threatens the safe and reliable operation of a power grid. Silicone rubber coatings are applied to the surfaces of other insulation materials in order to improve the pollution flashover voltage of the insulation structure. It is generally believed that the hydrophobicity of the silicone rubber coating is key to blocking the physical process of pollution flashover, which prevents the formation of continuously wet pollution areas. However, it is unclear whether silicone rubber coating can suppress the generation of pre-discharges such as corona discharge and streamer discharge. In this research, the influence of silicone rubber coating on the characteristics of surface streamer discharge was researched in-depth. The streamer 'stability' propagation fields of the polymer are lower than that of the polymer with silicone rubber coating. The velocities of the streamer propagation along the polymer are higher than those along the polymer with silicone rubber coating. This indicates that the surface properties of the polymer with the silicone rubber coating are less favorable for streamer propagation than those of the polymer.

Keywords: surface streamer discharge; silicone rubber coating; three-electrode arrangement; thermally stimulated current method; surface properties

Citation: Meng, X.; Wang, L.; Mei, H.; Zhang, C. Influence of Silicone Rubber Coating on the Characteristics of Surface Streamer Discharge. *Polymers* **2021**, *13*, 3784. https://doi.org/10.3390/polym13213784

Academic Editor: Shaojian He

Received: 16 October 2021
Accepted: 28 October 2021
Published: 31 October 2021

Publisher's Note: MDPI stays neutral with regard to jurisdictional claims in published maps and institutional affiliations.

Copyright: © 2021 by the authors. Licensee MDPI, Basel, Switzerland. This article is an open access article distributed under the terms and conditions of the Creative Commons Attribution (CC BY) license (https://creativecommons.org/licenses/by/4.0/).

1. Introduction

Pollution flashover along the insulation surface occurs widely in electrical power systems, which threatens the safe and reliable operation of the power grid. The hydrophobicity of silicone rubber coating can prevent the formation of continuously wet pollution areas, and thus it can block the physical process of pollution flashover along the insulation surface. Therefore, silicone rubber coatings have typically been applied to the surfaces of other insulation materials in order to increase the pollution flashover voltage of the insulation structure.

In the long-term operation of an electrical power system, a partial pre-discharge may occur on the surface of the silicone rubber coating and cause it to gradually lose hydrophobicity. At the same time, a partial arc can also develop more easily due to the existence of a partial discharge, and then the pollution flashover voltage will decrease [1]. However, it is unclear whether the silicone rubber coating suppresses or promotes the generation of pre-discharges such as corona discharges and streamer discharges. Therefore, the influence of silicone rubber coating on the characteristics of partial pre-discharges needs to be researched in depth. It is necessary to find ways to suppress the partial pre-discharge on the surface of silicone rubber coating.

There have been many studies on the engineering applications of silicone rubber coating in electrical power systems, which have provided many theoretical bases for the engineering applications of silicone rubber coatings [2–10]. However, research on the characteristics of the partial discharge on the surfaces of silicone rubber coatings is rare. Streamer discharge is the most complex physical process in partial discharge, which develops into leader discharge and surface flashover within high-enough electric fields [11–18]. The streamer 'stability' propagation fields in air were lower than those on the insulation surface [13]. When there was streamer propagation along the insulation surface, there were 'surface' and 'air' components of the streamer discharge [14–16]. We have previously obtained photographs of the streamer discharge, observing the 'surface' component of the streamer propagated along the insulation surface having a higher velocity, and the 'air' component of streamer propagated in the air having a lower velocity [17]. In [17], the influence of the dielectric materials on the characteristics of the streamer discharge was also researched; the conclusion was that both the permittivity and the surface properties of dielectric materials affected the streamer discharge, which affected the subsequent flashover processes. Therefore, research on the characteristics of streamer propagation along the surface of the silicone rubber coating is conducive to a deeper understanding of the mechanism of partial discharge. If the partial arc discharge can be suppressed during the streamer propagation stage, the external insulation performances of silicone rubber coating will be greatly improved.

The paper [18] designed an experiment to describe the quantitative influence of permittivity and surface properties on the characteristics of streamer propagation along insulation surfaces. In this paper, a test of the characteristics of the streamer propagation along the polymer and the polymer with a silicone rubber coating was designed, which measured them using three photomultipliers and an ultraviolet camera. Because the silicone rubber coating was very thin, the overall permittivity of the polymer with the silicone rubber coating hardly changed. The differences between the streamer propagation along the polymer and the polymer with the silicone rubber coating were determined by comparing the characteristics of the surface streamer discharge from those materials. Not only could the test results be used as a verification of the previous test results in [18], but also the influence of a silicone rubber coating on the surface streamer propagation process was analyzed. In addition, the characteristics of the streamer propagation along the silicone rubber coatings produced by different manufacturers were compared, which provided a feasible method for evaluating the insulation performances of silicone rubber coatings.

2. Experiment Arrangement and Measurement System

Figure 1 is a schematic diagram of the test equipment and measurement system used. Two flat electrodes and one needle electrode formed a three-electrode structure. The diameter of the parallel plates was 250 mm, and the distance between the upper and lower plates was 100 mm. The needle electrode was located at the circular hole (10 mm in diameter) in the center of the lower plates. The needle electrode was 2–3 mm above the plane of the lower plate, and was insulated from the lower plate. A negative DC voltage was applied to the upper plate, which was divided by a resistor divider, and then connected to a voltage measuring instrument via a coaxial cable. The lower plate was grounded. A square pulse voltage with adjustable amplitude and pulse width (1~6 kV, 100~250 ns) was applied to the needle electrode to trigger the positive polarity discharge. The square pulse voltage was divided by a high-voltage probe (Tektronix P6015A) that served as the trigger signal of a 4-channel 2 GHz oscilloscope (Agilent DSO7104A).

Three photomultipliers, each with a narrow slit (1 mm wide), were respectively directed at grazing incidence to the needle electrode, the middle position of the parallel plates, and the upper plate. The photomultiplier could monitor the development process of the streamer, because the head of the streamer would radiate photons into the space. The 'DayCor@Superb' UV imaging detector made by Ofil Corporation was used to take the photographs of the streamer discharge.

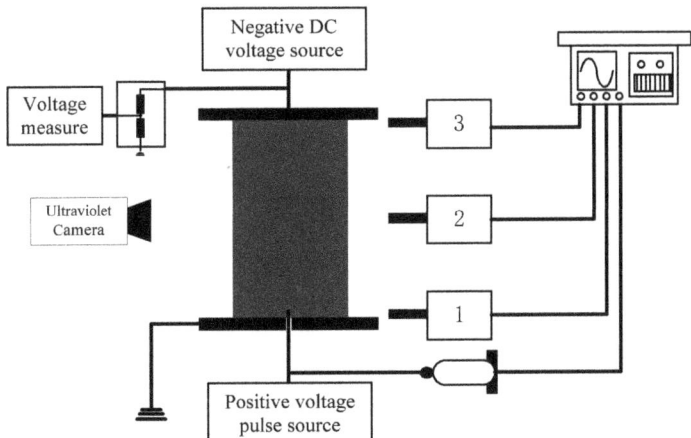

Figure 1. Schematic of the experiment arrangement and measurement equipment.

The polymer sheet used in the test was made of polyamide, and was placed vertically between two parallel plates. The polymer sheet was a square plate with a length of 100 mm, a width of 100 mm, and a thickness of 5 mm. The first polymer sheet was clean and had no silicone rubber coating, namely it was a polymer sheet. The second polymer sheet was coated with the first silicone rubber coating, namely Coating A. The last polymer sheet was coated with the second silicone rubber coating made by another manufacturer, namely Coating B. The permittivity of the polymer was 5, the permittivity of the first silicone rubber coating was 3.6, and the permittivity of the second silicone rubber coating was 3.8. The dielectric strength of Coating A was 22.2~22.8 kV/mm, and the dielectric strength of Coating B was 22.5~23.2 kV/mm. Their volume resistivity was 1.6×10^{14}~1.8×10^{14} Ω·m. For this study, the silicone rubber coatings were sprayed onto the surface of the polymer sheets, their thickness was 0.5 mm, and their surface drying time was 18–25 min. During the test, the indoor temperature was stable at about 25 °C, the relative humidity was maintained at about 65%, and the air pressure was the standard atmospheric pressure.

3. Experimental Results

3.1. Streamer Propagation Fields

Allen defined the applied electric field with a probability of 97.5% of the streamer propagating to the cathode plate as the streamer 'stability' propagation electric field E_{st} [13]. This definition was adopted in this study. The measurement method of the streamer stable propagation electric field was briefly as follows. The pulse voltage amplitude on the needle electrode was kept at a certain value U_{pulse}, and the DC voltage U_{app} applied between the plates was increased gradually. The DC voltage between the two parallel plates gradually increased from 450 kV/m to 750 kV/m. At each DC voltage value U_{app}, voltage pulses were applied to the needle electrode 20 times with a pulse interval of 20 s. That setting of the pulse interval was to ensure that the remaining ions from the previous streamer discharge were fully diffused. As the voltage between the two parallel plates gradually increased, the propagation probability of the streamer gradually increased from 0% to 100%. The streamer propagation probability and the applied electric field satisfied the Gaussian distribution function as shown in Figure 2. The Gaussian distribution function (1) was used to fit the statistical distribution curve of the streamer propagation probability with the electric field.

$$y = y_0 + \frac{A}{w\sqrt{\pi/2}} e^{-2\frac{(E-E_c)^2}{w^2}} \tag{1}$$

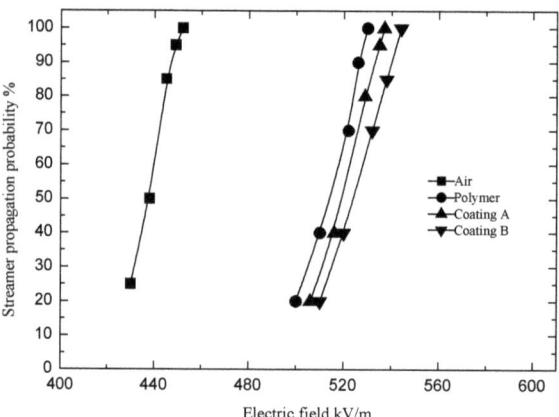

Figure 2. Relationship between the probability of streamer propagation and the guiding electric field.

In Formula (1), E is the electric field, E_c is the mean value of the electric field; w is the variance; y is the streamer propagation probability; A and y_0 are undetermined coefficients.

Then, the streamer 'stability' propagation fields E_{st} (streamer propagation probability of 97.5%) were obtained as shown in Figure 3. It was found that the streamer 'stability' propagation fields decreased linearly with the increase in the applied pulse amplitude. The reason is that the energy initially obtained by the streamer from the applied pulse increases with its amplitude, and the subsequent streamer propagation becomes easier. In addition, it can be seen that the streamer 'stability' propagation fields with the polymer with the silicone rubber coating are stronger than those with the polymer. Furthermore, the electric fields for the streamer stable propagation along the surface of the different silicone rubber coatings are quite different.

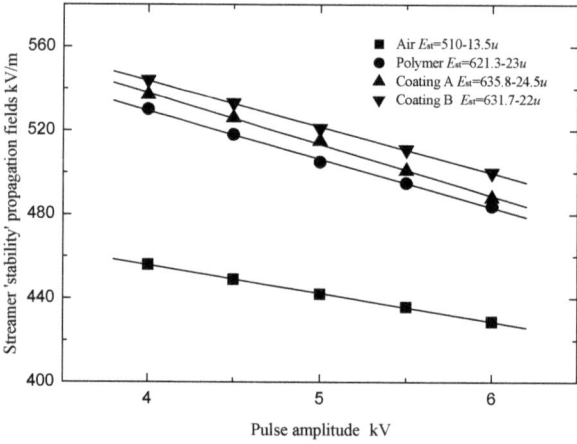

Figure 3. Relationship between the streamer 'stability' propagation fields and the pulse amplitude.

The fitting Formula (2) was used in Figure 3 to fit the curve. E_{st} is the streamer 'stability' propagation field, E_0 is the streamer stable propagation electric field when the pulse amplitude is 0 kV, u is the pulse amplitude, and α is an undetermined coefficient.

$$E_{st} = E_0 - \alpha u\,(\mathrm{kV/m}) \qquad (2)$$

3.2. Light Emission

A large number of photons are generated during the process of streamer discharge. Some of them participate in the photoionization in the discharge area, while others escape to the outside. The photoionization plays a vital role in the generation and development of streamers. The secondary electron avalanches generated by the photoionization in front of the streamer head supply the streamer discharge with positive and negative charges, and then the streamer channel moves forward.

Based on the physical phenomenon of photons being emitted outward during the streamer discharge, the photomultipliers and UV imaging detector were used to observe the process of the streamer discharge. In our previous articles [17,18], we found that there were 'surface' and 'air' components of the streamer discharge when streamer propagation occurred along the insulation surface. The 'surface' components of the streamers propagated along the insulation surfaces at a higher velocity, and the 'air' components of the streamers propagated in the air at a lower velocity. However, there were only 'air' components of the streamer discharges when the streamers propagated in air alone. The same conclusion was reached in this article. The photomultiplier detected two peaks of light at the cathode plate when the streamers propagated along either the polymer or the polymer with the silicone rubber coating as shown in Figure 4. Therefore, the 'surface' and 'air' components of the streamer discharges also occurred along both the polymer and the polymer with the silicone rubber coating. The 'surface' components of the streamers had higher velocities and their propagation paths lay along the insulation surfaces. The 'air' components of the streamers had lower velocities and their propagation paths were in the air.

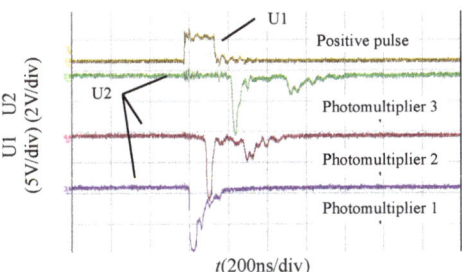

Figure 4. Typical signals from the photomultiplier monitoring the streamer propagation along the surface of the polymer.

The UV imaging detector was used to take the photographs of the streamer discharges. The light emitted from a single propagation process of a streamer was able to be recorded in a clear image. The white spots on each image are the signal displayed by the light emitted from a streamer discharge. Figures 5 and 6 show the streamer propagation photographs for the polymer and the polymer with the silicone rubber coating.

It can be observed that the 'surface' component of a streamer propagates along the insulation surface, while the 'air' component of a streamer propagated in the air and was away from the insulation surface. Within the same electric field, the luminous intensity of the streamer propagation along the polymer was greater than that along the polymer with the silicone rubber coating. Furthermore, the luminous intensity of the streamer propagation along the polymer with the different silicone rubber coatings was also different. It was determined that the luminous intensity of a streamer was closely related to the subsequent photoionization. The stronger the luminous intensity of a streamer was, the more intense the subsequent photoionization would be, and it would promote the development of the subsequent streamer. This also explains that the electric fields required for the streamer stable propagation along the polymer with the silicone rubber coating were greater than that along the polymer.

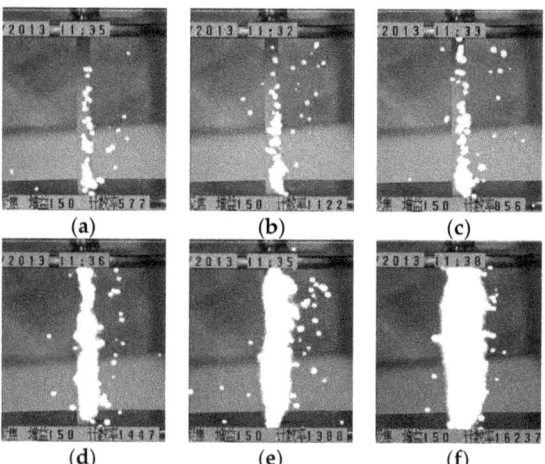

Figure 5. Streamer propagation photographs for the polymer. (**a**) 500 kV/m, (**b**) 530 kV/m, (**c**) 550 kV/m (**d**) 590 kV/m, (**e**) 620 kV/m, (**f**) 660 kV/m.

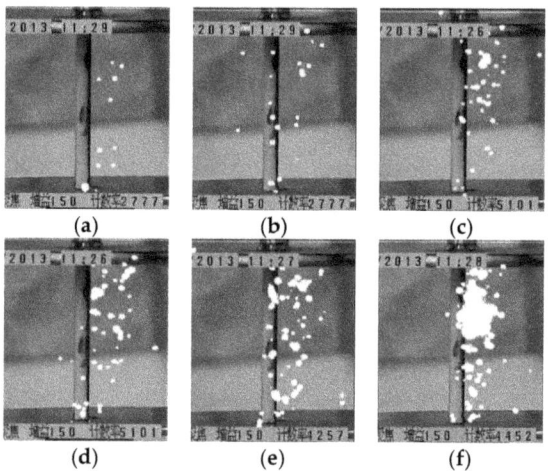

Figure 6. Streamer propagation photographs for the polymer with the silicone rubber coating. (**a**) 510 kV/m, (**b**) 540 kV/m, (**c**) 560 kV/m, (**d**) 590 kV/m, (**e**) 630 kV/m, (**f**) 660 kV/m.

3.3. Streamer Propagation Velocity

The propagation velocity of the streamers was calculated by the ratio of the vertical distance between the three photomultipliers and the time difference ΔT between the starting points of the rising edge of the light signals from the three photomultipliers (Figure 4). The streamer 'stability' propagation velocity V_{st} was defined as the streamer propagation velocity within the 'stability' electric field. Figure 7 shows the relationship between the streamer 'stability' propagation velocities and the pulse amplitude. The 'stability' velocities of the streamer propagation along the surface of the polymer with a coating were linearly related to the pulse amplitude. For Figure 7, Equation (3), which relates the streamer 'stability' propagation velocities to the pulse amplitude, was used to fit the curves. u is the pulse amplitude, kV; V_0 is the streamer stable propagation velocity when the pulse amplitude is 0 kV, 10^5 m/s; β is the undetermined coefficient.

$$V_{st} = V_0 + \beta u \qquad (3)$$

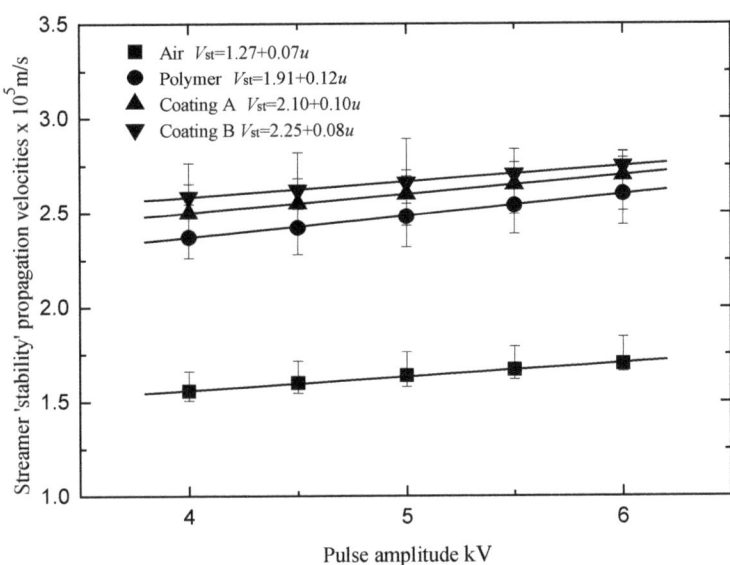

Figure 7. Relationship between the streamer 'stability' propagation velocities and the pulse amplitudes.

The 'surface' and 'air' components of the streamers occurred when the applied electric fields were larger than the streamer 'stability' propagation fields. The velocities of the 'surface' and 'slow' components under the varied electric fields are displayed in Figures 8 and 9, respectively. Equation (4) was used to draw the fitting curves in Figures 8 and 9. E_{st} and V_{st} come from Equations (2) and (3), and n and γ are the undetermined coefficients listed in Table 1.

$$V_s = V_{st}\left(\frac{E}{E_{st}}(1+\gamma)\right)^n \tag{4}$$

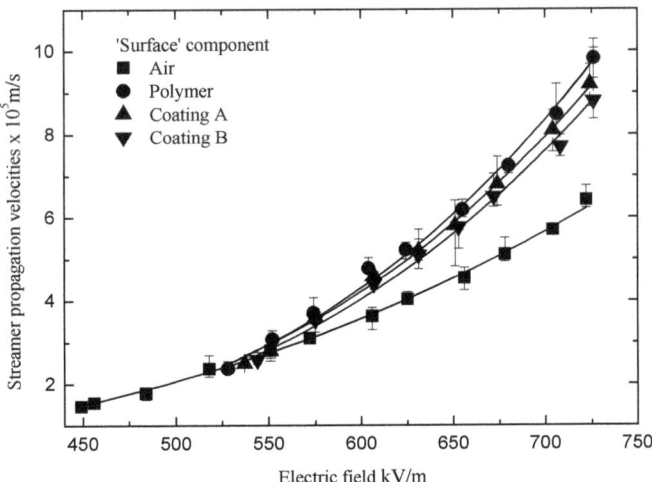

Figure 8. Velocities of the 'surface' components under the varied electric fields.

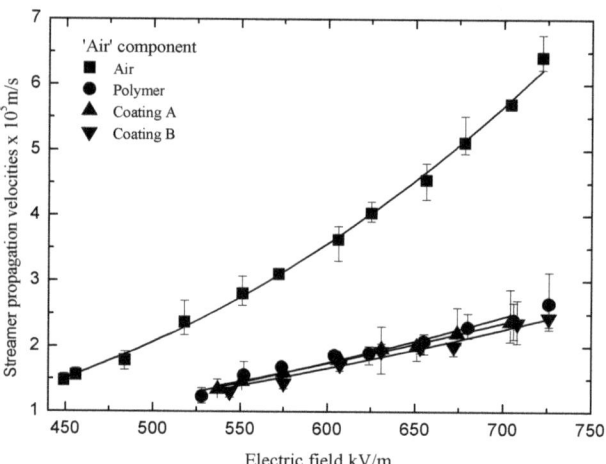

Figure 9. Velocities of the 'air' components under the varied electric fields.

Table 1. Corresponding parameters in Equation (3).

Material	E_{st}	'Surface' Component			'Air' Component		
		V_{st}	$\gamma \times 100$	n	V_{st}	$\gamma \times 100$	n
Air	456	1.56	0.22	3	1.56	0.23	3
Polymer sheet	528	2.37	0.15	4.3	1.23	3.24	2.2
Coating A	537	2.50	1.69	4.1	1.34	0.75	2.1
Coating B	544	2.58	1.16	4.1	1.3	2.48	2.2

E_{st} and V_{st} in Equation (4) were replaced by Equations (2) and (3) to become Equation (5). It describes the streamer propagation velocities under any pulse amplitude and applied electric field.

$$V_s = (V_0 + \beta u)\left(\frac{E(1+\gamma)}{E_0 - \alpha u}\right)^n \qquad (5)$$

In Figures 8 and 9, it can be seen that the velocities of the 'surface' components of the streamers were higher than those in the air alone, and they increased with the applied electric field significantly. In contrast, the velocities of the 'air' components of the streamers were lower than those in the air alone, and they increased with the applied electric field slowly. It can be explained that the electric field in the head of the 'air' component of a streamer is suppressed by the charge in the head of the 'surface' component. Furthermore, it can be seen that the velocities of the 'surface' components of the streamer decreased after the silicone rubber coating was applied to the polymer. In addition, the velocities of the 'surface' components of the streamers propagating along the different silicone rubber coatings were also different. However, the differences between the velocities of the 'air' components of the streamers propagating along the different insulation surfaces were relatively small.

4. Discussion

4.1. Permittivity

The main factors that affected the characteristics of the streamer propagation along the insulation surfaces were the permittivity and surface properties (the attachment of the charge to the surface, photoemission of secondary electrons from the surface, etc.) [18]. First, the influence of the silicone rubber coating on the permittivity of the polymer sheet was analyzed. Figure 10 shows the variation of the electric field from the needle electrode up to 1 mm along the insulation surface. The thickness of the silicone rubber coating was considered to be 0.5 mm. It can be seen that the electric fields from the needle electrode up to 1 mm along both the polymer and the polymer with the silicone rubber coating were basically the same, but the electric field along the polymer with the silicone rubber coating was slightly strengthened. The permittivity of the silicone rubber coating was smaller than that of the polymer. After the silicone rubber coating was applied to the polymer, the volume of the polymer sheet with the silicone rubber coating became larger than that of the polymer sheet. The overall capacitance (permittivity) increased, so the electric field along the polymer sheet with the silicone rubber coating increased. However, the silicone rubber coating only had a small increase in the electric field at the tip of the needle electrode, which indicates that the silicone rubber coating produced a small change in the overall permittivity (capacitance) of the polymer sheet. Therefore, the change in the overall permittivity caused by the silicone rubber coating had a tiny impact on the electric field distribution in the gap.

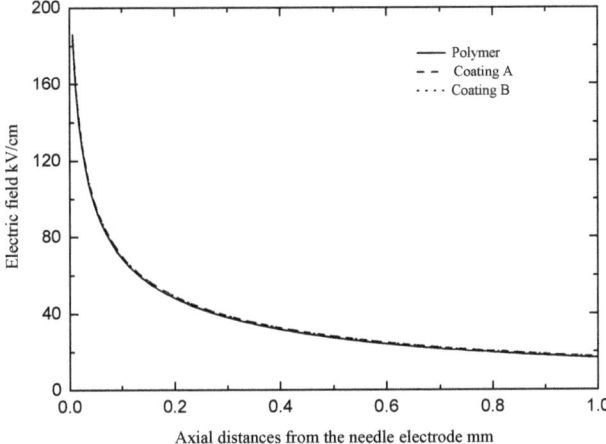

Figure 10. Variation of the electric field from the needle electrode up to 1 mm at axial distances.

With the increase in the permittivity, the charge (ions or electrons) accumulated on the insulation surface increased [19–21]. On the one hand, there were many negative charges accumulated on the insulation surface with the negative direct voltage applied to the cathode plane [22–24]. Those would have reduced the electric fields in the latter half of the sheet as shown in Figure 11. The negative charges on the surface weakened the electric fields in the latter half of the sheet. Hence, the streamer propagation along the sheet with the larger permittivity required higher electric fields [17]. On the other hand, the sheet with the larger permittivity would have attached more positive charges in the streamer. The ionization efficiency at the head of the streamer would have weakened, which made the streamer propagation difficult and required high electric fields. Because the silicone rubber coating made the overall permittivity (capacitance) of the polymer sheet slightly increase, the silicone rubber coating caused a slight increase in the positive charge in the streamers attached to the surface, which would have suppressed the development of those streamers to a certain extent. However, the stable streamer propagation fields of the polymer sheet and the polymer sheet with the silicone rubber coating displayed a large difference. The change in the overall permittivity caused by the silicone rubber coating should not have caused such a large difference. It must have been caused by the change in the surface properties of the polymer sheet after spraying the silicone rubber coating.

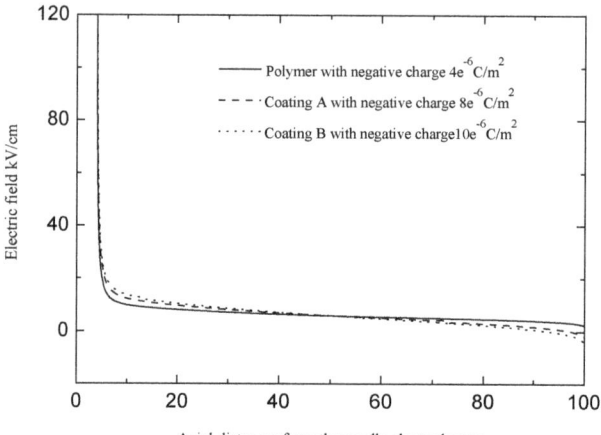

Figure 11. Variation of the electric field in the gap at axial distances.

4.2. Surface Properties

When the silicone rubber coating was applied to the polymer sheet, the surface condition changed greatly. The roughness of the materials was measured using the roughness gauge. The roughness of the three materials was as follows: Coating B was the largest (Ra of 0.93 μm), Coating A was second (Ra of 0.76 μm), and the polymer was the smallest (Ra of 0.65 μm). The larger the surface roughness of the material was, the more serious the accumulation of the surface charge was [21]. Therefore, the negative charges accumulated on the surface increased with the increases in the surface roughness, which had two influences on the streamer discharges. One is that the electric fields in the latter half of the sheet weakened due to the negative surface charges; the other is that the ionization efficiency at the head of the streamers weakened due to the attachment of the positive charges in the streamer to the surface. It was more difficult for the streamers to propagate along the sheet with the higher surface roughness. That is the reason why the streamer 'stability' propagation fields for the polymer with the silicone rubber coatings were larger than those for the polymer. In addition, the electric fields for the streamer stable propagations along Coating B were larger than those along Coating A at the different surface roughness levels.

Figure 12 shows the surface conditions of the three insulation materials measured by the scanning electron microscope. It can be seen that the surface of the polymer had more microporous defects than the polymer with the silicone rubber coatings. The trap charges (nC) and trap levels (eV) of the three insulation materials were tested using the method of the thermally stimulated current (TSC). The trap charges on the polymer surface were greater than those on the silicone rubber coatings. The microporous defects on the insulation surfaces could reflect the surface trap distributions [21]. The trap charges on the insulation surface decreased with the decreases in the microporous defects. Hence, the results of the SEM figures and the TSC test corroborate each other.

The traps that had low trap levels are named "shallow traps". In Table 2, the shallow traps on the polymer surface were greater than those on the silicone rubber coatings. The photoemission of secondary electrons from the surfaces can be described as follows: the collisions of the high-energy photons detach the trap charges from the insulation surface and produce many high-energy secondary electrons, which then promote the development of the streamers [13,14]. Under the collision by the high-energy photons, a shallow trap emits high-energy secondary electrons more easily. The reason why the streamer propagation along the silicone rubber coating is more difficult than that along the polymer is that the photoemission of secondary electrons from the silicone rubber coating is weaker. The shallow traps on the Coating A were greater than those on Coating

B. Therefore, the stronger photoemission of secondary electrons from Coating A led to the streamer propagation more easily, which is consistent with the test results.

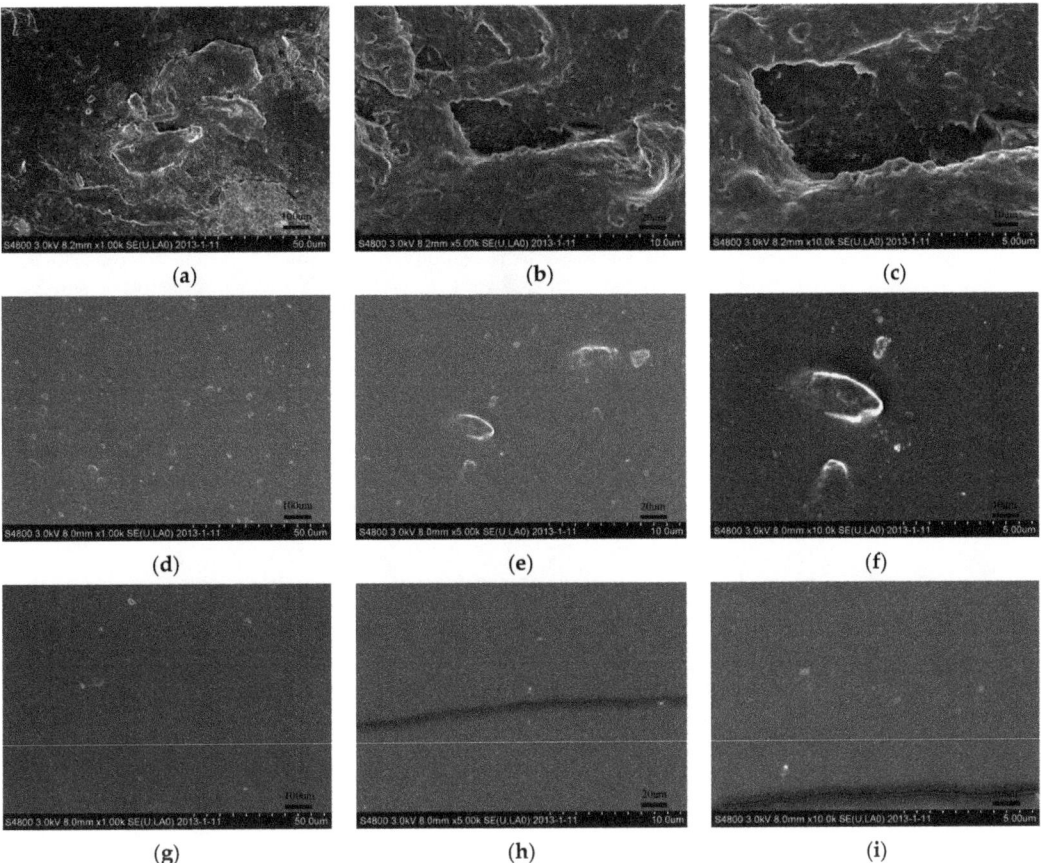

Figure 12. SEM figures of the insulation surfaces. (**a**–**c**) Polymer. (**d**–**f**) Coating A. (**g**–**i**) Coating B.

Table 2. Trap parameters of the dielectric materials measured by TSC test.

Parameter	Polymer	Coating A	Coating B
Current peak (PA)	1050	165	138
Trap charge (nC)	1879	246	225
Trap level (eV)	0.38	0.45	0.49

In a word, the polymer surface was more favorable for the streamer propagation than the silicone rubber coating surface from the perspectives of both the surface roughness and surface trap. The electric fields required for the streamer stable propagation along the silicone rubber coating were larger than that required for the streamer stable propagation along the polymer surface. The velocities of the streamer propagation along the silicone rubber coating were lower than that along the polymer surface under the same electric fields. From the perspective of surface properties, it was also a good explanation of the differences between the streamer stable propagation fields and velocities along the different silicone rubber coatings produced by the different manufacturers. There were differences in the characteristics of the streamer propagation along the different silicone rubber coatings, which indicates there are large differences in the insulation properties of the silicone rubber coatings belonging to different manufacturers. These differences can be found by measuring the characteristics of the streamer discharge. Therefore, tests of the streamer discharge can be used to evaluate the insulation properties of silicone rubber coatings.

The test results of the characteristics of streamer propagation along the different material surfaces have also taught us many things. Higher permittivity in a material is unfavorable for streamer propagation along it. Hence, materials with higher permittivity can be chosen to suppress the pre-discharge in some conditions of electrical power systems. It is difficult for streamer propagation to occur along materials with higher macroscopic surface roughness, so insulation surfaces can be made rougher to reduce pre-discharge in electrical power systems. The microporous defects on the insulation surface can affect the streamer propagation to a great extent. In the factory, more nanomaterials can be applied to insulation materials to fill the microporous defects on the insulation surface, thereby the pre-discharge will be prevented by the technology of reducing microporous defects. These results provide a theoretical basis for promoting the application of the nanomaterials.

5. Conclusions

The streamer 'stability' propagation fields for the polymer, Coating A and Coating B were 528 kV/m, 537 kV/m and 544 kV/m, respectively. The velocities of the 'surface' components of the streamer stable propagation along the polymer, Coating A and Coating B were 2.37×10^5 m/s, 2.50×10^5 m/s and 2.58×10^5 m/s, respectively. The velocities of the 'air' component of the streamer stable propagation along the polymer, Coating A and Coating B are 1.23×10^5 m/s, 1.34×10^5 m/s and 1.30×10^5 m/s, respectively. The streamer 'stability' propagation fields for the polymer were lower than those for the polymer with the silicone rubber coatings. Within the same electric fields, the velocities of streamer propagation along the polymer were higher than those along the polymer with the silicone rubber coatings.

Higher permittivity in a material was unfavorable for streamer propagation along it. The effects of permittivity on electric field distortion in front of needle tip, the effects of the surface charge accumulation before the development of a streamer on the distortion of the electric field in the gap and the effects of the charge attachment to the surface during the development of streamer were analyzed to determine the reason.

It is difficult for streamer propagation to occur along materials with higher macroscopic surface roughness. The effects of the surface charge accumulations before the development of streamers on the distortions of the electric field in the gap and the effects of the charge attachment to the surface during the development of streamers were also analyzed to determine the reason.

The streamer propagation along materials with more microporous defects on the insulation surface was easier. The reason is that the photoemission of secondary electrons from the surface increased with increased microporous defects, which would have promoted the development of the streamer.

There are large differences in the characteristics of surface streamers along the different silicone rubber coatings. Testing the streamer discharge can be used to evaluate the insulation properties of the silicone rubber coatings produced by the different manufacturers.

Author Contributions: Conceptualization, L.W.; methodology, L.W.; software, C.Z.; data curation, X.M.; writing—original draft preparation, X.M.; writing—review and editing, C.Z.; project administration, H.M.; funding acquisition, L.W. All authors have read and agreed to the published version of the manuscript.

Funding: This research was partially funded by the National Natural Science Foundation of China, grant number 51907178.

Institutional Review Board Statement: Not applicable.

Informed Consent Statement: Not applicable.

Data Availability Statement: Not applicable.

Conflicts of Interest: The authors declare no conflict of interest. The funders had no role in the design of the study; in the collection, analyses, or interpretation of the data; in the writing of the manuscript; nor in the decision to publish the results.

References

1. IEEE Standard 1523. *IEEE Guide for the Application, Maintenance, and Evaluation of Room Temperature Vulcanizing (RTV) Silicone Rubber Coatings for Outdoor Ceramic Insulators*; IEEE: Piscataway Township, NJ, USA, 2002.
2. Carberry, R.E.; Schneider, H.M. Evaluation of RTV coating for station insulators subjected to coastal contamination. *IEEE Trans. Power Deliv.* **1989**, *4*, 577–585. [CrossRef]
3. Gorur, R.S.; Orbeck, T. Surface dielectric behavior of polymeric insulation under HV outdoor conditions. *IEEE Trans. Electr. Insul.* **1991**, *26*, 1064–1072. [CrossRef]
4. Cherney, E.A.; Hackam, R.; Kim, S.H. Porcelain insulator maintenance with RTV silicone rubber coatings. *IEEE Trans. Power Deliv.* **1991**, *6*, 1177–1181. [CrossRef]
5. Deng, H.; Hackam, R. Low-molecular weight silicone fluid in RTV silicone rubber coatings. *IEEE Trans. Dielectr. Electr. Insul.* **1999**, *6*, 84–94. [CrossRef]
6. Cherney, E.A.; Gorur, R.S. RTV silicone rubber coatings for outdoor insulators. *IEEE Trans. Dielectr. Electr. Insul.* **1999**, *6*, 605–611. [CrossRef]
7. Homma, H.; Mirley, C.L.; Ronzello, J.; Boggs, S.A. Field and laboratory aging of RTV silicone insulator coatings. *IEEE Trans. Power Deliv.* **2000**, *15*, 1298–1303. [CrossRef]
8. Gao, H.; Jia, Z.; Guan, Z.; Wang, L.; Zhu, K. Investigation on field-aged RTV-coated insulators used in heavily contaminated areas. *IEEE Trans. Power Deliv.* **2007**, *22*, 1117–1124. [CrossRef]
9. He, S.J.; Wang, J.Q.; Hu, J.B.; Zhou, H.F.; Nguyen, H.; Luo, C.M.; Lin, J. Silicone rubber composites incorporating graphitic carbon nitride and modified by vinyl tri-methoxysilane. *Polym. Test* **2019**, *79*, 106005. [CrossRef]
10. Liao, Y.F.; Weng, Y.X.; Wang, J.Q.; Zhou, H.F.; Lin, J.; He, S.J. Silicone rubber composites with high breakdown strength and low dielectric loss based on polydopamine coated mica. *Polymers* **2019**, *11*, 2030. [CrossRef] [PubMed]
11. Mahajan, S.M.; Sudashan, T.S.; Dougal, R.A. Avalanches near a dielectric spacer in nitrogen gas. In *Gaseous Dielectrics V.*; Pergamon: Oxford, UK, 1987.
12. Gallimberti, I.; Marchesi, G.; Niemeyer, L. Streamer corona at an insulator surface. In Proceedings of the Seventh International Symposium on High Voltage Engineering, Dresden, Germany, 19–23 September 1991.
13. Allen, N.L.; Mikropoulos, P.N. Streamer propagation along insulating surfaces. *IEEE Trans. Dielectr. Electr. Insul.* **1999**, *6*, 357–362.
14. Akyuz, M.; Gao, L.; Cooray, V.; Gustavsson, T.G.; Gubanski, S.M.; Larsson, A. Positive streamer discharges along insulating surfaces. *IEEE Trans. Dielectr. Electr. Insul.* **2001**, *8*, 902–910.
15. Allen, N.L.; Hashem, A.; Rodrigo, H.; Tan, B.H. Streamer development on silicone-rubber insulator surfaces. *IEE Proc.-Sci. Meas. Technol.* **2004**, *151*, 31–38. [CrossRef]
16. Mikropoulos, P.N. Streamer propagation along room- temperature-vulcanised silicon-rubber-coated cylindrical insulators. *IET Sci. Meas. Technol.* **2008**, *2*, 187–195.
17. Meng, X.; Mei, H.; Chen, C.; Wang, L.; Guan, Z.; Zhou, J. Characteristics of Streamer Propagation along the Insulation Surface: Influence of Dielectric Material. *IEEE Trans. Dielectr. Electr. Insul.* **2015**, *22*, 1193–1203. [CrossRef]
18. Meng, X.; Mei, H.; Wang, L.; Guan, Z.; Zhou, J. Characteristics of Streamer Propagation along Insulation Surface: Quantitative Influence of Permittivity and Surface Properties. *IEEE Trans. Dielectr. Electr. Insul.* **2016**, *23*, 2867–2874. [CrossRef]
19. Srivastava, K.D.; Zhou, J.P. Surface charging and flashover of spacers in SF6 under impulse voltages. *IEEE Trans. Dielectr. Electr. Insul.* **1991**, *26*, 428–442. [CrossRef]
20. Chen, Z.H.; Yan, Y.G. Realization of fuzzy- filter control on a brushless AC generator. *Chin. J. Aeronaut.* **2000**, *13*, 172–176. (In Chinese)
21. Yamano, Y.; Kasuga, K.; Kobayashi, S.; Saito, Y. Surface flashover and charging characteristics on various kinds of alumina under nonuniform electric field in vacuum. In Proceedings of the 20th International Symposium on Discharges and Electrical Insulation in Vacuum, Tours, France, 1–5 July 2002.
22. Tumiran, M.; Maeyama, H.; Kobayashi, S.; Saito, Y. Flashover from surface charge distribution on alumina insulators in vacuum. *IEEE Trans. Dielectr. Electr. Insul.* **1997**, *4*, 400–406. [CrossRef]
23. Fujinami, H.; Takuma, T.; Yashima, M.; Kawamoto, T. Mechanism and effect of DC charge accumulation on SF6 gas insulated spacers. *IEEE Trans. Power Deliv.* **1989**, *4*, 1765–1772. [CrossRef]
24. Xie, J.; Chalmers, I.D. The influences of surface charge upon flash-over of particle-contaminated insulators in SF6 under impulsevoltage conditions. *J. Phys. D Appl. Phys.* **1997**, *30*, 1055–1063.

Article

Frequency and Temperature-Dependent Space Charge Characteristics of a Solid Polymer under Unipolar Electrical Stresses of Different Waveforms

Hanwen Ren [1], Qingmin Li [1,*], Yasuhiro Tanaka [2], Hiroaki Miyake [2], Haoyu Gao [1] and Zhongdong Wang [3]

[1] State Key Laboratory of Alternate Electrical Power System with Renewable Energy Sources, North China Electric Power University, Beijing 102206, China; rhwncepu@ncepu.edu.cn (H.R.); hygaoeee@ncepu.edu.cn (H.G.)
[2] Tokyo City University, 1-28-1, Tamazutsumi, Setagaya-ku, Tokyo 158-8557, Japan; ytanaka@tcu.ac.jp (Y.T.); hmiyake@tcu.ac.jp (H.M.)
[3] College of Engineering, Mathematics and Physical Sciences, University of Exeter, Exeter EX4 4QJ, UK; zhongdong.wang@exeter.ac.uk
* Correspondence: lqmeee@ncepu.edu.cn; Tel.: +86-10-61771413

Citation: Ren, H.; Li, Q.; Tanaka, Y.; Miyake, H.; Gao, H.; Wang, Z. Frequency and Temperature-Dependent Space Charge Characteristics of a Solid Polymer under Unipolar Electrical Stresses of Different Waveforms. *Polymers* **2021**, *13*, 3401. https://doi.org/10.3390/polym13193401

Academic Editor: Cesare Cametti

Received: 19 September 2021
Accepted: 30 September 2021
Published: 3 October 2021

Publisher's Note: MDPI stays neutral with regard to jurisdictional claims in published maps and institutional affiliations.

Copyright: © 2021 by the authors. Licensee MDPI, Basel, Switzerland. This article is an open access article distributed under the terms and conditions of the Creative Commons Attribution (CC BY) license (https://creativecommons.org/licenses/by/4.0/).

Abstract: In this paper, we studied the space charge phenomena of a solid polymer under thermal and electrical stresses with different frequencies and waveforms. By analyzing the parameter selection method of a protection capacitor and resistor, the newly built pulsed electro-acoustic (PEA) system can be used for special electrical stresses under 500 Hz, based on which the charge phenomena are studied in detail under positive and negative DC and half-wave sine and rectangular wave voltages. Experimental results show that the charge accumulated in the polyimide polymer under DC conditions mainly comes from the grounded electrode side, and the amount of charge accumulated with electric field distortion becomes larger in a high-temperature environment. At room temperature, positive charges tend to accumulate in low-frequency conditions under positive rectangular wave voltages, while they easily appear under high-frequency situations of negative ones. In contrast, the maximum electric field distortion and charge accumulation under both half-wave sine voltages occur at 10 Hz. When the measurement temperature increases, the accumulated positive charge decreases, with a more negative charge appearing under rectangular wave voltages, while a more positive charge accumulates at different frequencies of half-wave sine voltages. Therefore, our study of the charge characteristics under different voltage and temperature conditions can provide a reference for applications in the corresponding environments.

Keywords: space charge; polyimide polymer; unipolar electrical stress; temperature; frequency

1. Introduction

With the rapid development of the transmission and transformation grid, power electronic equipment has come to play a more important role in power operation. In contrast to traditional DC and AC power equipment, the insulation of electronic equipment usually has to deal with special electrical environments, including high-frequency rectangular wave and sine-like voltages [1,2]. Under a sine voltage at power frequency, traditional equipment can operate continuously for several years. In contrast, due to the coupling effect of high-frequency voltage and high temperature, the insulation of electronic equipment may deteriorate within 2 years [3–5]. Therefore, the effective evaluation of the insulation characteristics of electronic equipment has become a necessary to ensure its safe operation.

Under the coupling effect of electrical and thermal stresses, a great deal of space charge usually accumulates inside insulation, which can further lead to electrical tree development, electric field distortion, flashover, and breakdown phenomena [6]. The current research of the space charge inside insulation mainly relies on reliable measurement methods. Among the various measurement techniques used at present, the PEA method has been widely used in traditional DC and AC stresses at power frequency due to its simple structure and mature measurement technology. Many researchers have found that the voltage amplitude, waveform shape, and temperature conditions can directly affect the charge accumulation of insulating polymers [7–9]. Therefore, an evaluation of the charge characteristics under the voltage and temperature environments faced by the insulating polymer will be helpful for its future application and modification design for the corresponding environments.

For the special electrical stresses withstood by the insulation of electronic equipment, researchers have carried out some comparative studies on the charge characteristics of different polymers. From the results shown in [10], it is evident that the charge accumulation and electric field distortion inside cross-linked polyethylene polymers under unipolar half-wave sine and rectangular wave conditions are more serious than those under DC conditions. Meanwhile, the frequency can directly affect the charge characteristics. For the research of the polyamide-imide (PAI) polymer, it has also been found that the problem of accumulated charge is more severe under some unipolar voltage situations than that under DC stresses, and voltage polarity can also influence the charge accumulation [11,12]. A further comparison of the charge characteristics of polyester-imide, PAI, and polyimide (PI) polymers under special voltages was carried out in [13], with PI samples showing less space charge accumulation under rectangular wave stresses at 50 Hz. However, at present an analysis of the charge characteristics of this polymer under the effect of voltage frequency, temperature, and voltage waveform conditions is still lacking. Meanwhile, the existing research mainly focuses on the common conditions with low voltage frequencies and temperatures. PI material has been widely applied as the supporting insulation for high-frequency power electronic transformers, motors, and other electronic equipment. A comprehensive measurement of its charge characteristics under special electrical and high-temperature environments will aid in our understanding of the actual operating state of equipment insulation.

Based on the background described above, we carried out an investigation of the space charge phenomenon inside PI polymers under different temperature and voltage conditions. In contrast to the traditional PEA design for DC and AC conditions at 50 Hz, the key electrical component in the PEA system for high-frequency voltage conditions was analyzed and designed. After analyzing the component selection in the PEA system for high-frequency voltage conditions, the charge and electric field distributions under negative and positive DC and rectangular wave and half-wave sine voltages were studied. The effect of three temperature conditions was also compared. The research results of this paper firstly show the charge evolution under special unipolar electrical stresses in the space charge research field, which could be used to guide the design and modification of the insulating materials used for electrical performance improvement. Thus, this should provide an important reference for insulation system design in the manufacturing of power electronic equipment.

2. The PEA System Applied for High-Frequency Voltage Conditions

The basic PEA system can be seen in Figure 1a. The space charge from the two electrodes was injected into the sample under the effect of electrical stress. The pulse voltage from the pulse generator was simultaneously imposed on the sample. Then, the pressure wave was generated from space charge due to its vibration and finally measured by the oscilloscope. The whole measurement processes was carried out via a computer [14].

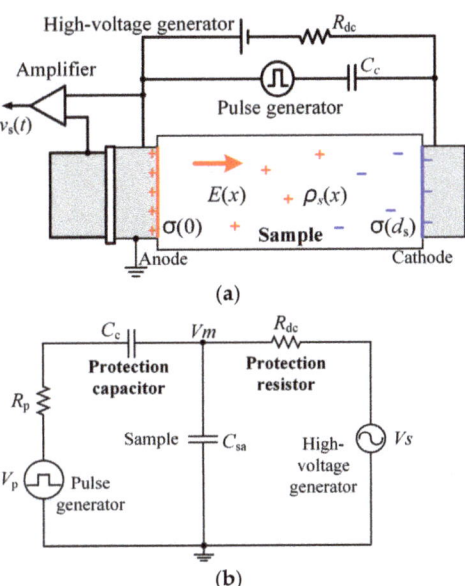

Figure 1. The PEA method for space charge measurement: (**a**) the measurement system, (**b**) the equivalent circuit.

Figure 1b shows the equivalent circuit of the PEA system based on the IEC Standard [15]. The values of protection resistor R_{dc} and capacitor C_c in Figure 1 can directly affect the voltage applied to the sample. According to the IEC standards, the protection capacitor should be smaller than 1 nF and the protection resistor should be larger than 10 kΩ. Based on this selection, the system has been widely used for DC or AC stresses at power frequency.

Since high-frequency voltages are studied in our research, the applicability of the traditional system was analyzed first. PI polymer was selected as the sample, and positive rectangular wave voltages with different frequencies were imposed on it. Figure 2 shows the voltage applied to the sample, as measured by a high-voltage probe. The amplitude of rectangular wave voltages was 500 V and the frequencies were 100, 250, and 500 Hz.

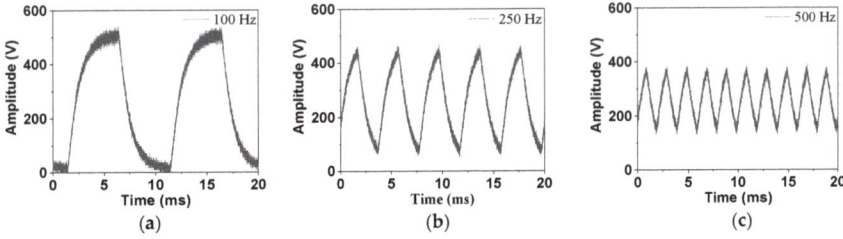

Figure 2. The voltages applied to the sample based on the traditional circuit design: (**a**) 100 Hz, (**b**) 250 Hz, (**c**) 500 Hz.

It can be seen from Figure 2 that the applied high-frequency voltage to the sample deforms obviously. The amplitude of the applied voltage decreases with frequency, which indicates that the system used for DC voltages cannot be applied under high-frequency voltages. Therefore, the system needs to be adjusted for use in this situation.

Based on the circuit of the PEA system shown in Figure 1b, the voltage V_m that was actually applied to the sample is represented by Equation (1).

$$\begin{cases} V_m = V_s - R_{dc}\frac{V_s}{R_{all}} \\ R_{all} = R_{dc} + \frac{1}{j\omega C_{sa}}(\frac{1}{j\omega C_c} + R_p) \\ \quad\quad = R_{dc} + \frac{j\omega R_p C_c + 1}{j\omega C_c + (j\omega R_p C_c + 1)j\omega C_{sa}} \end{cases} \quad (1)$$

where V_s (V) represents the output voltage from a high-voltage generator. R_{all} (Ω) is the resistance of the whole circuit. ω (Hz) represents the angular frequency. C_{sa} (Hz) is the capacitance of the sample. R_p (Ω) is the resistance of the pulse generator.

From this equation, the voltage applied to the sample can be calculated under the effects of the protection resistor and capacitor. Further, combined with the purchased components in the laboratory, we selected three protection capacitors of 560 pF in series and three protection resistors of 1 MΩ in parallel to establish a new system. The corresponding component values in the PEA system were 187 pF and 333 kΩ, respectively. This selection can decrease the probability of the flashover and breakdown of the components. Based on this design, the measurement of the voltage applied to the sample is shown in Figure 3.

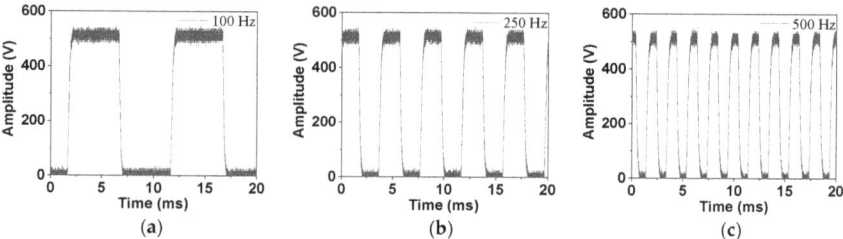

Figure 3. The voltages applied to the sample based on the improved circuit selection: (**a**) 100 Hz, (**b**) 250 Hz, (**c**) 500 Hz.

Using this component selection, we can see that the voltages actually applied to the sample with different frequencies are consistent with the expected input—i.e., the voltages still maintain a rectangular wave shape and their amplitude does not deform. Therefore, we used this system for the charge measurements under unipolar voltage conditions. In addition, the selected components were only used for electrical stresses under 500 Hz in the later research. Therefore, the voltage applied to the sample is analyzed within 500 Hz to judge the applicability of the components. The above three frequencies, including 100, 250, and 500 Hz, clearly represent the waveform distortion of applied voltages. The results obtained from this new system design are also compared with those of the traditional system for DC voltages, which verifies that the new design also maintains a good accuracy.

For the studied DC voltages, a high voltage was applied to the sample for 30 min with a later 30-min short-circuit state, during which the charge measurement was carried out every 5 seconds. For the other voltage waveforms, Figure 4 shows the measurement method used for positive situations. The pulse excitation voltage was applied to the maximum amplitude of the half-wave sine voltages and the center point of rectangular wave voltages. The half-wave sine voltage contained only half a cycle of the sine waveform, while the voltage in the other half was zero. To ameliorate the signal-to-noise ratio of the measurement, 200 continuously measured signals were used for the averaging process. In the following measurements, the PI material usually used for electronic equipment was determined as the sample, whose thickness was about 125 μm. The amplitude of all the following applied electric fields was set as 60 kV/mm. In addition, for the inevitable distortion of the measured signal by the equipment, the corresponding calibration method was also designed, which also shows a high precision compared to past methods [14]. The measurement below was repeated many times to ensure that the results and conclusions were accurate and reliable.

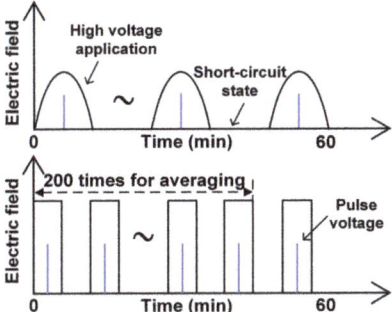

Figure 4. The measurement process for the two kinds of voltages.

3. Measurement Results under Different Voltage and Temperature Conditions
3.1. Results under Negative and Positive DC Stresses

The space charge characteristics under DC voltages with positive and negative polarities were firstly analyzed. Meanwhile, the effect of the three temperatures was also discussed, as shown in Figure 5. In the subfigures, the left part shows the voltage application state while the right part shows the short-circuit state when a high voltage was removed. The vertical axis represents the sample thickness direction from the grounded electrode to the upper electrode, while the horizontal one indicates the measurement time. The yellow and blue colors represent positive and negative charges, respectively.

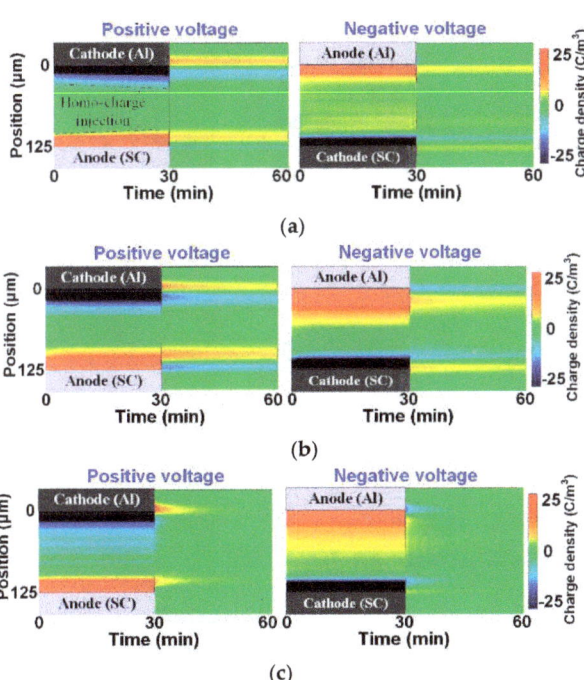

Figure 5. Space charge distributions under DC voltages: (**a**) room temperature, (**b**) 50 °C, (**c**) 80 °C.

From Figure 5, it can be seen that the polarity of the accumulated space charge inside the deep of the sample is the same as that of the grounded electrode. At room temperature, the charge injections from both of the electrodes could be found under the positive condition, while a small amount of positive charge accumulated in the polymer under the negative condition. When the temperature increases, more charge from the grounded electrode side migrates into the deep portion of the sample, indicating that the injection barrier at this side further decreases with temperature. From the results at the short-circuit stage, the extraction of space charge from the sample interfaces is slow at room temperature. However, the charge dissipates quickly from the electrodes after increasing the temperature, which means the extraction barriers of the electrode interfaces decrease and the migration of a large amount of space charge becomes very quick.

On the basis of the above charge distributions, the electric field results can be calculated using Poisson's equation, as shown in Equation (2):

$$E(z) = \frac{1}{\varepsilon_0 \varepsilon_r} \int_0^{d_s} \rho(z) dz \qquad (2)$$

where $E(z)$ (kV/mm) is the electric field distribution inside the sample. z (mm) represents the position. ε_0 (F/m) is the vacuum dielectric constant and ε_r is the relative permittivity of the sample. $\rho(z)$ (C/m^3) represents the space charge distribution.

Based on the above equation, the electric field distributions under DC voltages can be calculated, as shown in Figure 6. The results correspond to the last measurement of the voltage application state, which represents the field distribution along the direction of sample thickness. Under positive voltages, the electric field distributions found at room temperature and 50 °C are very close. The result obtained at 80 °C shows the maximum electric field distortion, whose value is about 67.09 kV/mm. Under negative voltages, the results at both 50 and 80 °C show high levels of electric field distortion. The maximum distortion also appears at 80 °C, the value of which is 68.64 kV/mm. Thus, the space charge accumulation and the electric field distortion are a little larger under the negative DC voltage application. This phenomenon also indicates that a higher temperature promotes space charge accumulation inside the PI sample. Meanwhile, combined with the results in Figure 5, charge accumulation and dissipation occur quicker under high-temperature conditions. This means that the electric field distortion could reach the states easily, as shown in Figure 6; thus, the material can be more easily damaged under high-temperature conditions.

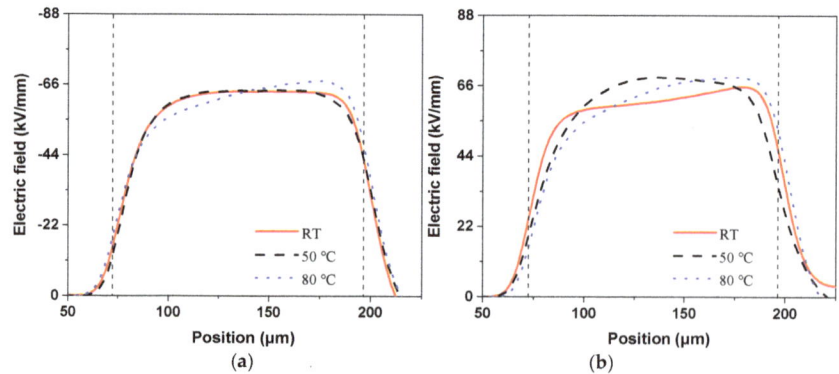

Figure 6. Electric field results under DC voltages: (**a**) positive voltages, (**b**) negative voltages.

3.2. Effect of Unipolar Rectangular Wave Voltages on Space Charge

The charge phenomena occurring under positive and negative rectangular wave voltages within a frequency ranging from 10 to 500 Hz were then measured. Figure 7 shows the results obtained at room temperature, 50 °C, and 80 °C; the measurement time is illustrated in Figure 4. The three results shown in the columns correspond to one frequency, and the four results in the rows correspond to one temperature.

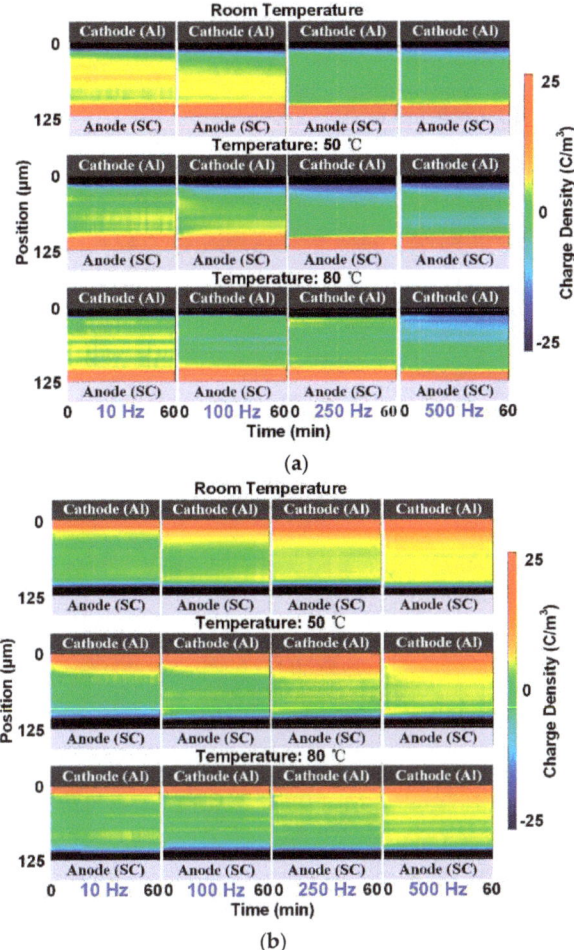

Figure 7. Charge results under rectangular wave voltages at different temperatures: (a) positive voltages, (b) negative voltages.

For the positive situations, positive charge accumulates inside the sample under the application of low-frequency voltage at room temperature from the upper electrode. At 250 and 500 Hz, barely any space charge appears in the deep of the material. At the higher temperature situation, it can be found that the amount of positive charge at the low-frequency voltage conditions decreases. In contrast, negative charge appears in the deep portion of the sample, especially at 500 Hz. This means that more negative charge accumulates under positive rectangular wave voltages when increasing the temperature, especially in high-frequency situations.

Similar to the results obtained under positive situations, positive charge appears in the sample under negative conditions at room temperature as well. However, the charge tends to accumulate under high-frequency conditions, especially at 500 Hz. When the temperature increases, the amount of positive charge decreases at each frequency condition.

This phenomenon may be due to the same reason as that in the positive situation—i.e., more negative charge accumulates in the sample. The PEA measurement only presents the net charge distribution. Therefore, under the rectangular wave stresses with two polarities, the increase in temperature can cause a greater accumulation of negative charge when the amount of positive charge is decreased.

From the above charge distributions, the electric field results can also be calculated at the last measurement of the voltage application, as shown in Figure 8. For simplicity, the most severely distorted electric field at each temperature condition is selected, as is also labeled in the figure.

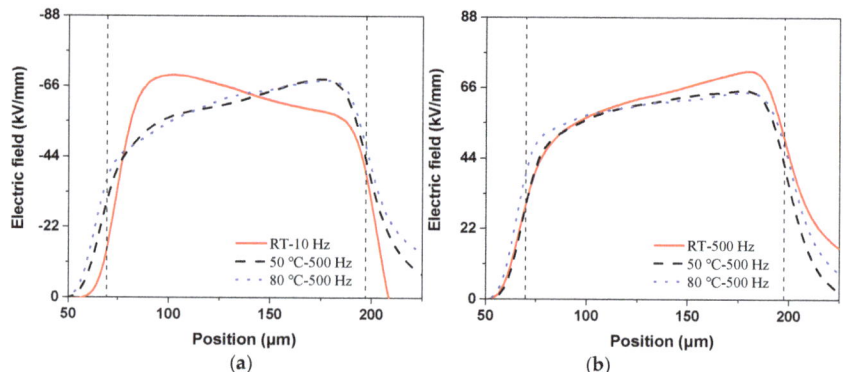

Figure 8. Electric field results under rectangular wave voltages at different temperatures: (**a**) positive voltages, (**b**) negative voltages.

Consistent with the charge results shown in Figure 7a, the maximum electric field distortion also appears at 10 Hz for the applied positive voltages, whose value is about 69.45 kV/mm. At 50 and 80 °C, the maximum field distortions are at the opposite site compared to the room temperature situation, which is due to the negative charge accumulation shown in Figure 7a. The two maximum distortions occur under the 500 Hz condition. Therefore, the increase in temperature reduces the electric field distortion at low-frequency situations and causes a more severe distortion in high-frequency conditions under positive rectangular wave voltages.

In contrast, the largest electric field distortions occur at the frequency of 500 Hz under negative voltages for each temperature condition. The maximum values of electric field at each temperature reach 71.32, 65.16, and 64.63 kV/mm, respectively. Therefore, the electric field distortion is alleviated a little with temperature due to the decrease in the positive charge, as shown in Figure 7b.

3.3. Results under Unipolar Half-Wave Sine Voltages

The results obtained under unipolar half-wave sine voltages were measured and discussed, as shown in Figure 9. The result at each condition corresponds to the maximum amplitude of the voltage waveform, as illustrated in Figure 4.

At room temperature, the space charge injected from the upper electrode layer accumulates in the deep portion of the sample under both the positive and negative voltages with a frequency of 10 Hz. When the voltage frequency increases, nearly no charge can be found in the deep portion, and only a small amount of charge from the two electrodes accumulates in their vicinity.

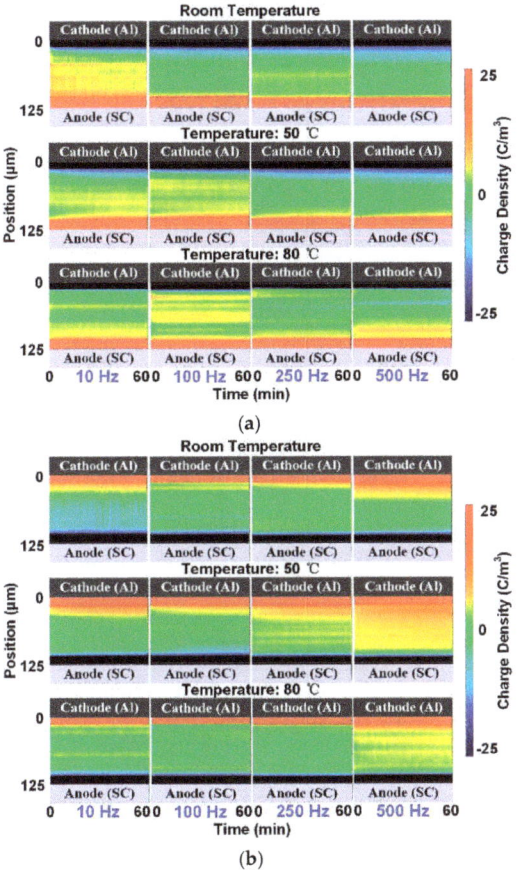

Figure 9. Charge results under half-wave sine voltages at different temperatures: (**a**) positive voltages, (**b**) negative voltages.

When the temperature increases, different charge phenomena are found. For positive situations, the amount of positive charge at the frequency of 10 Hz decreases, while the high-frequency situations including 100 and 500 Hz show some positive charge accumulation. In the negative voltage situations, barely any negative charge accumulates in the sample at 10 Hz compared to the room temperature condition. Meanwhile, a large accumulation of positive charge can be found at high-frequency conditions, especially at 500 Hz. Therefore, different from the results obtained under rectangular wave voltages, the increase in temperature causes a greater accumulation of positive charge under half-wave sine conditions with high frequencies.

Similarly, the electric field results calculated at the last measurement are shown in Figure 10.

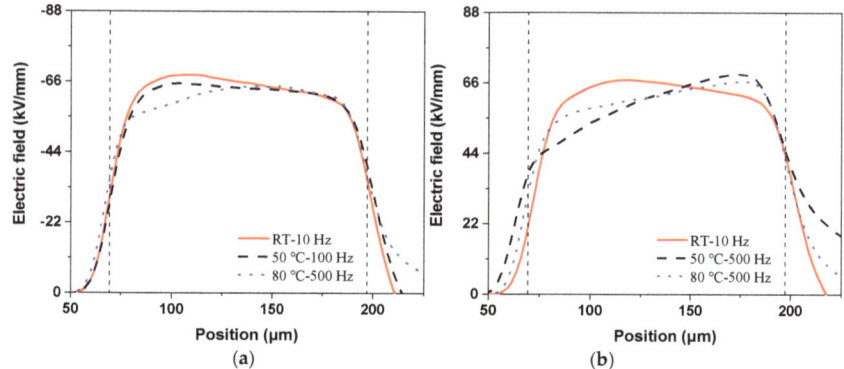

Figure 10. Electric field results under half-wave sine voltages at different temperatures: (**a**) positive voltages, (**b**) negative voltages.

For the positive conditions, the maximum electric field occurs under a condition of 10 Hz at room temperature, and the value is about 68.02 kV/mm. With temperature, the maximum electric field distortion appears at higher frequencies. In contrast, under negative voltage conditions, the maximum field distortion is found at the frequency of 500 Hz and the temperature of 80 °C, with a value of 68.82 kV/mm.

4. Discussion

From the above results, an interesting phenomenon can be seen; the charge accumulation is very quick under the conditions of special unipolar electrical stresses. The charge inside the deep portion of the polymer also dissipates quickly at the short-circuit stage of some DC conditions. This kind of rapid charge migration has also been found in previous charge measurements [16–18]. Under some DC and voltage transition conditions, space charge appears or dissipates immediately after the voltage state changes.

This phenomenon is caused by the rapid migration speed of space charge. In addition to the space charge with a mobility in the range from 10^{-17} to 10^{-13} m^2/(Vs), which has been observed by most measurement results, the charge in shallow traps can migrate quickly inside the sample [19]. According to the space charge phenomena observed by some researchers, it has been calculated that the mobility of rapid charge is nearly 10^{-9} m^2/(Vs) [20,21]. Meanwhile, some measured current results have also indicated that a part of space charge has a large migration speed [22]. From the analysis based on the results under the DC-superimposed pulsed voltage conditions [23], the charge inside the sample has been found to increase and decrease instantaneously after the pulsed voltage rises and falls. Therefore, this rapid charge migration exists under all kinds of voltage conditions. However, it is more obvious under special unipolar stresses due to the continuous and sharp changes in the applied voltage.

The charge accumulated inside the polymer withstands forces from many sources, which can generally be divided into two kinds—applied electric field force and the other force—as displayed in Figure 11 [23]. The former occurs due to the applied electrical stress. The latter mainly occurs due to the material itself, nearby charges, and other sources. Under stable voltage conditions, the forces withstood by the charge are also stable. Therefore, the charge can keep a relatively balanced state, can stay in traps for a long time, and have a low migration speed. In contrast, if the applied electric field continuously changes, the other force cannot always follow the applied electric field force due to the material deformation and relatively unchanged position between charges. Therefore, the charge obtains a faster migration speed more easily, due to the unbalanced force, and this phenomenon is more common under changing voltage conditions.

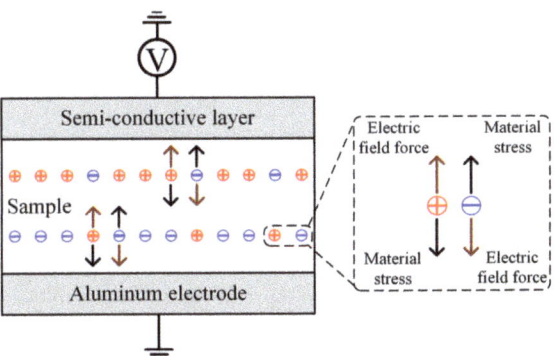

Figure 11. The forces withstood by the migrating space charge.

5. Conclusions

The space charge characteristics of PI polymer under different unipolar voltages and temperatures were studied. It was found that the accumulation process and evaluation law of charge are dependent on the voltage form, frequency, and temperature conditions. Some conclusions were obtained:

(1) The voltage that was actually applied to the sample was greatly affected by the protection capacitor and resistor in the PEA system. The circuit previously used for DC conditions is improper for high-frequency stresses, which can seriously deform the voltage applied to the sample when the frequency exceeds 100 Hz. After adjusting the protection capacitor and resistor to 187 pF and 333 kΩ, the PEA system could be used properly under 500 Hz.

(2) Experimental results achieved under DC voltages indicate that the space charge injected from the grounded electrode dominates in most cases. More charge accumulates inside the sample at a higher temperature, and the largest electric field distortions under the negative and positive stresses occur at 80 °C.

(3) In contrast to the charge phenomena under DC voltages, the results under special electrical stresses show frequency and temperature-dependent characteristics. For rectangular wave voltages at room temperature, positive charge accumulates easily under 100 Hz under positive conditions, while it accumulates more easily at higher frequencies under negative conditions. When the temperature increases, a negative net charge occurs under positive voltages, while the amount of positive charge is reduced under negative voltages, which indicates that an increase in temperature promotes the negative charge accumulation.

(4) For positive and negative half-wave sine voltages, the largest charge accumulations at room temperature occur at 10 Hz. When the temperature increases, a more positive charge is found inside the sample under different frequency conditions, which is completely opposite to the charge phenomenon that occurs under rectangular wave voltages. Meanwhile, the results under unipolar electrical stresses show the phenomenon of rapid charge accumulation. The special mechanisms of space charge accumulation and evolution under different voltages still need to be further researched.

Author Contributions: Conceptualization, H.R. and Q.L.; methodology, Y.T. and H.M.; software, H.R. and H.G.; validation, H.G.; formal analysis, H.R.; investigation, H.R.; data curation, Y.T. and H.M.; writing—original draft preparation, H.R.; writing—review and editing, Z.W.; supervision, Q.L.; project administration, Q.L.; funding acquisition, Q.L. All authors have read and agreed to the published version of the manuscript.

Funding: The work was funded by Beijing Natural Science Foundation (3202031) and National Natural Science Foundation of China (No. 51737005, 51929701, 52081330507).

Institutional Review Board Statement: Not applicable.

Informed Consent Statement: Not applicable.

Data Availability Statement: The data presented in this study are available on request from the corresponding author.

Conflicts of Interest: The authors declare no conflict of interest.

References

1. Chen, H.; Prasai, A.; Moghe, R.; Chintakrinda, K.; Divan, D. A 50 kVA three-phase solid-state transformer based on the minimal topology: Dyna-C. *IEEE Trans. Power Electron.* **2016**, *31*, 8126–8137. [CrossRef]
2. Liu, T.; Li, Q.; Huang, X.; Lu, Y.; Asif, M.; Wang, Z. Partial discharge behavior and ground insulation life expectancy under different voltage frequencies. *IEEE Trans. Dielectr. Electr. Insul.* **2018**, *25*, 603–613. [CrossRef]
3. Nguyen, M.; Malec, D.; Mary, D.; Werynski, P.; Gornicka, B.; Therese, L.; Guillot, P. Investigations on dielectric properties of enameled wires with nanofilled varnish for rotating machines fed by inverters. In Proceedings of the 2009 IEEE Electrical Insulation Conference, Montreal, QC, Canada, 31 May–3 June 2009; pp. 377–381.
4. Bonnett, A. Analysis of the impact of pulse-width modulated inverter voltage waveforms on AC induction motors. *IEEE Trans. Ind. Applicat.* **1996**, *32*, 386–392. [CrossRef]
5. Oliver, A.; Stone, G. Implications for the application of adjustable speed drive electronics to motor stator winding insulation. *IEEE Electr. Insul. Mag.* **1995**, *11*, 32–36. [CrossRef]
6. Li, J.; Du, B.; Su, J.; Liang, H.; Liu, Y. Surface layer fluorination-modulated space charge behaviors in HVDC cable accessory. *Polymers* **2018**, *10*, 500. [CrossRef] [PubMed]
7. Mima, M.; Iwata, T.; Miayke, H.; Tanaka, Y. Evaluation of electrical insulation characteristics of polyamide-imide film using measurement results of space charge distributions with an applied square wave voltage. In Proceedings of the 2017 3rd International Conference on Condition Assessment Techniques in Electrical Systems (CATCON), Rupnagar, India, 16–18 November 2017; pp. 69–74.
8. Wang, S.; Zhou, Q.; Liao, R.; Xing, L.; Wu, N.; Jiang, Q. The impact of cross-linking effect on the space charge characteristics of cross-linked polyethylene with different degrees of cross-linking under strong direct current electric field. *Polymers* **2019**, *11*, 1149. [CrossRef] [PubMed]
9. Zhou, Y.; Dai, C.; Huang, M. Space charge characteristics of oil-paper insulation in the electro-thermal aging process. *CSEE J. Power Energy Syst.* **2016**, *2*, 40–46. [CrossRef]
10. Wang, X.; Shu, Z.; Duan, S.; Wang, H.; Liu, S. Study of space charge accumulation property in polyethylene under applied voltage of square wave. In Proceedings of the 2019 2nd International Conference on Electrical Materials and Power Equipment (ICEMPE), Guangzhou, China, 7–10 April 2019; pp. 178–181.
11. Takizawa, K.; Suetsugu, T.; Miyake, H.; Tanaka, Y. Space charge behavior in covering insulating material for motor windings under applied voltage of square wave. In Proceedings of the International Symposium on Electrical Insulating Materials, Niigata, Japan, 1–5 June 2014; pp. 409–412.
12. Mima, M.; Miyake, H.; Tanaka, Y. Measurement of space charge distribution in polyamide-imide film under square wave voltage practical environment for inverter-fed motor. In Proceedings of the 2018 IEEE 2nd International Conference on Dielectrics (ICD), Budapest, Hungary, 1–5 July 2018; pp. 1–4.
13. Mima, M.; Miyake, H.; Tanaka, Y. Dependence of duty ratio of square wave voltage on space charge accumulation property in motor winding coating insulating material. In Proceedings of the International Conference the Properties and Applications of Dielectric Materials, Xi'an, China, 20–24 May 2018; pp. 102–106.
14. Ren, H.; Li, Q.; Wang, Z. An improved calibration method for the measurement of space charge inside insulating materials. *IEEE Trans. Instrum. Meas.* **2020**, *69*, 1652–1663. [CrossRef]
15. IEC/TS Standard 62758. *Calibration of Space Charge Measuring Equipment Based on the Pulsed Electro-Acoustic (PEA) Measurement Principle*; IEC Commission: Geneva, Switerzerland, 2012.
16. Zhang, Y.; Liu, P.; Feng, H.; Zhang, S.; Li, D.; Peng, Z. Space charge dynamics in epoxy resin impregnated crepe paper multilayer under voltage polarity reversal. *IEEE Trans. Dielectr. Electr. Insul.* **2019**, *26*, 253–260. [CrossRef]
17. Vu, T.; Tsyssedre, G.; Roy, S.; Laurent, C. Space charge criteria in the assessment of insulation materials for HVDC. *IEEE Trans. Dielectr. Electr. Insul.* **2017**, *24*, 1405–1415. [CrossRef]
18. Tanaka, Y.; Fujitomi, T.; Kato, T.; Miyake, H.; Mori, H.; Kikuchi, S.; Yagi, Y. Packet-like charge formation in cable insulating materials at polarity reversal. *IEEE Trans. Dielectr. Electr. Insul.* **2017**, *24*, 1372–1379. [CrossRef]
19. Ren, H.; Takada, T.; Uehara, H.; Iwata, S.; Li, Q. Research on charge accumulation characteristics by PEA method and Q(t) method. *IEEE Trans. Instrum. Meas.* **2021**, *70*, 6004209. [CrossRef]
20. Xu, M.; Montanari, G.; Fabiani, D.; Dissada, L.; Krivda, A. Supporting the electromechanical nature of ultra-fast charge pulses in insulating polymer conduction. In Proceedings of the 2011 International Symposium on Electrical Insulating Materials, Kyoto, Japan, 6–10 September 2011; pp. 1–4.
21. Montanari, G.; Fabiani, D.; Dissado, L. Fast charge pulses: The evidence and its interpretation. In Proceedings of the 2013 IEEE International Conference on Solid Dielectrics (ICSD), Bologna, Italy, 30 June–4 July 2013; pp. 10–14.

22. Zheng, F.; Lin, C.; Dong, J.; An, Z.; Zhang, Y. Study of ultra-fast space charge in thin dielectric films. In Proceedings of the 2013 Annual Report Conference on Electrical Insulation and Dielectric Phenomena, Shenzhen, China, 20–23 October 2013; pp. 287–290.
23. Zhang, T.; He, D.; Meng, F.; Li, Q. Space charge behavior in cable insulation under a DC-superimposed pulsed electric field. In Proceedings of the 2020 IEEE International Conference on High Voltage Engineering and Application (ICHVE 2020), Beijing, China, 6–10 September 2020; pp. 1–4.

Article

Research on the Compound Optimization Method of the Electrical and Thermal Properties of SiC/EP Composite Insulating Material

Xupeng Song [1], Xiaofeng Xue [1], Wen Qi [1], Jin Zhang [2], Yang Zhou [2], Wei Yang [3], Yiran Zhang [4], Boyang Shen [5], Jun Lin [1] and Xingming Bian [1,*]

[1] State Key Laboratory of Alternate Electrical Power System with Renewable Energy Sources, North China Electric Power University, Beijing 102206, China; songxupeng96@163.com (X.S.); xxf7002@163.com (X.X.); qiwen0602@163.com (W.Q.); Jun.lin@ncepu.edu.cn (J.L.)
[2] State Grid Corporation of China, Co., Ltd., Beijing 100031, China; jin-zhang@sgcc.com.cn (J.Z.); yang-zhou@sgcc.com.cn (Y.Z.)
[3] State Key Laboratory of Advanced Power Transmission Technology, Global Energy Interconnection Research Institute, Co., Ltd., Beijing 102211, China; 19630100@163.com
[4] State Grid Hubei Maintenance Company Yichang Operation Maintenance Branch, Yichang 430050, China; yr495459435@163.com
[5] Electrical Engineering Division, University of Cambridge, Cambridge CB3 0FA, UK; bs506@cam.ac.uk
* Correspondence: bianxingming@ncepu.edu.cn

Abstract: In this paper, in order to improve the electrical and thermal properties of SiC/EP composites, the methods of compounding different crystalline SiC and micro-nano SiC particles are used to optimize them. Under different compound ratios, the thermal conductivity and breakdown voltage parameters of the composite material were investigated. It was found that for the SiC/EP composite materials of different crystal types of SiC, when the ratio of α and β silicon carbide is 1:1, the electrical performance of the composite material is the best, and the breakdown strength can be increased by more than 10% compared with the composite material filled with single crystal particles. For micro-nano compound SiC/EP composites, different total filling amounts of SiC correspond to different optimal ratios of micro/nano particles. At the optimal ratio, the introduction of nanoparticles can increase the breakdown strength of the composite material by more than 10%. Compared with the compound of different crystalline SiC, the advantage is that the introduction of a small amount of nanoparticles can play a strong role in enhancing the break-down field strength. For the filled composite materials, the thermal conductivity mainly depends on whether an effective heat conduction channel can be constructed. Through experiments and finite element simulation calculations, it is found that the filler shape and particle size have a greater impact on the thermal conductivity of the composite material, when the filler shape is rounder, the composite material can more effectively construct the heat conduction channel.

Keywords: SiC crystal form; micro-nano compound; thermal conductivity; breakdown field strength

1. Introduction

The filled high thermal conductivity composite material is to dope inorganic fillers or metal fillers with high thermal conductivity into the polymer matrix to build a thermally conductive network. Filled thermally conductive polymers are simple to operate, low in cost, suitable for industrial production and have become the mainstream development direction of high thermally conductive insulating materials. [1–7].

The type, distribution and filling amount of the fillers incorporated have a great impact on the electrical and thermal properties of the composite material [8–11]. Many studies modified the epoxy matrix by introducing two or more types of powder fillers into the resin [12–15]. For example, Zhou et al. [16] added 6 wt% multi-walled carbon nanotubes (MWCNT) or 71.7 wt% micron-sized silicon carbide (SiC) to the epoxy resin to maximize the thermal conductivity of the composites, and then partially replaced the micron-sized fillers with nano-sized fillers. This paper took the advantage of the large aspect ratio of one-dimensional structure of MWCNTs, making MWCNTs effectively act as thermal conductivity channels between micron SiC particles and form a more effective three-dimensional infiltration network for heat flow. Zhao et al. [17] used two-dimensional boron nitride nanosheets (BNNS) and zero-dimensional boron nitride microspheres (BNMS) to build a three-dimensional thermal conductivity network, which can effectively improve the thermal conductivity of epoxy composites. Tu et al. [18] studied the surface charge transfer characteristics and the control mechanism of the coating to illustrate the influence of the surface coating on the surface charge accumulation characteristics of epoxy resin. The surface charge of epoxy resin can be effectively suppressed by coating 1–3 wt% micron SiO_2 particles or 3 wt% nano SiO_2 particles. Takahiro et al. [19] prepared micro-nano epoxy composites by dispersing nano-layered silicate filler and micron silica (SiO_2) filler in the epoxy matrix. Compared with the conventional SiO_2 filling, the internal structure of the composite obtained by micro-nano composite was tighter, and the breakdown strength of the composite was 7% higher than that of the conventional filled epoxy composite. Huang et al. [20] prepared alumina (Al_2O_3)/silicon carbide (SiC)/epoxy composites by sol-gel method. The results showed that the introduction of dense and uniform Al_2O_3 on the SiC surface improved the interface adhesion between the epoxy matrix and SiC particles, and the thermal resistance of the filler-matrix interface was reduced. Wu J [21] et al. magnetized the SiC particles so that the paramagnetic SiC particles were aligned along the magnetic field lines under the action of an external magnetic field. After thermosetting, an ordered epoxy/silicon carbide composite material was obtained, and the relationship between the distribution state of silicon carbide and the dielectric properties and thermal conductivity of the composite material was explored. Seenaa Hussein [22] et al. studied the modification method of poly(vinyl alcohol) (PVA)/poly(vinyl pyrrolidone) (PVP) polymer film. By using a solution blending method to incorporate nano-graphene fillers into PVP/PVA, the thermal conductivity and mechanical properties of the composite material are greatly improved. At the same time, as the graphene content increases, the thermal weight loss rate of the composite material decreases, which proves that there is a strong interface interaction between the nano-graphene and the matrix. Alaa M. Abd-Elnaiem [23] et al. used sol-gel polymerization method to coat multi-walled carbon nanotubes (MWCNTs) with tetraethyl orthosilicate (TEOS) to form a covalent bond with the epoxy resin matrix when used as a filler to enhance the interface effect. Epoxy-TEOS/MWCNTs composites filled with different proportions were prepared. When the filling ratio was 4 wt%, the mechanical, optical and thermal properties of the epoxy composites reached the best. Nadia A [24] et al. prepared polymethyl methacrylate (PMMA)/MWCNTs composites using a solution blending method. The covalent bond and hydrogen bond between PMMA and MWCNTs were analyzed by infrared spectroscopy, which proved the strong interaction between the filler and the matrix. By adding an appropriate proportion of MWCNTs filler, the mechanical properties and electrical conductivity of the composite material have been greatly improved. Seenaa I. Hussein [25] et al. used graphene, carbon nanotubes and carbon fibers as fillers, respectively, to fabricate epoxy resin-based composite materials. Under the action of the filler, the wear rate of the composite material is significantly reduced, while the thermal conductivity is greatly improved. Among them, graphene and carbon nanotubes have the most significant increase in the thermal conductivity of composite materials.

The SiC-filled composite dielectric has nonlinear conductivity characteristics [26,27]. When the local electric field in the insulating medium is distorted, it can homogenize the electric field [28,29]. At the same time, filling with SiC can significantly improve the thermal conductivity of the composite material, so SiC/EP has broad application prospects in the packaging of high-capacity power devices and the insulation of high-voltage equipment. However, the introduction of SiC will decrease the breakdown strength of composite materials [30]. In the current research on SiC/EP composite materials, attention has been paid to the electrical and thermal properties of the composite materials filled with various crystal forms of SiC, but there is still a lack of relevant research on the combined effects of different crystal forms. In view of the above problems, this paper studies the effects of different crystalline SiC compound and micro-nano particle compound on the breakdown field strength and thermal conductivity of SiC/EP composites. Suggestions for improving the electrical and thermal properties of SiC/EP composites are provided from the perspective of particle compounding.

2. Materials and Methods

2.1. Materials

Bisphenol A epoxy resin was used as the matrix. This kind of epoxy resin can fuse well with many additives, with excellent bonding strength and strong corrosion resistance. Anhydride curing agent has a good mixing property with epoxy, the viscosity of the mixed liquid is low, which is convenient for the mixing of filler particles, and the electrical and mechanical properties of the cured epoxy composites are improved to a certain extent. Silicon carbide (SiC) with different crystal shape and particle size was selected as filler. In order to improve the binding degree of nano-SiC particles and epoxy resin and enhance the dispersion of filler particles, the surface modification of nano-SiC particles was carried out by using silane coupling agent (KH-560). The raw materials used in the experiment are listed in Table 1.

Table 1. Information of the raw materials.

Name	Parameter	Manufacturer
E51 type epoxy resin	The epoxy value 0.48~0.54 eq/100 g Viscosity 10~16 Pa·s (25 °C)	Shanghai Xiongrun Resin Co., LTD, Shanghai, China
Methyl tetrahydrophthalic anhydride (curing agent)	Purity > 80% (GC)	TCI Chemical Industrial Development Co., LTD, Toyko, Japan
2,4,6-tris (dimethylaminomethyl) phenol (accelerator)	Purity > 80% (GC)	TCI Chemical Industrial Development Co., LTD, Toyko, Japan
Micron silicon carbide (α,β-SiC)	Particle size: 1.5, 10 μm	Qinhuangdao One nuo Company, Qinhuangdao, China
Nano silicon carbide (n-SiC)	Particle size: 50 nm	Qinhuangdao One nuo Company, Qinhuangdao, China

2.2. Sample Preparation

In order to study the effect of different crystalline SiC composite and micro/nano SiC composite on the electrical and thermal characteristics of the material. Under the premise that the overall filling ratio remains unchanged, we changed the proportions of different crystalline SiC and micro/nano SiC particles in the filler. Since the physical and chemical properties of silicon carbide and epoxy resin are different, the particles in epoxy are easy to agglomerate. However, the silicon carbide should evenly disperse in the matrix as far as possible. The flow chart is shown in Figure 1, and the specific experimental steps are as follows:

(1) The required mass of epoxy resin was weighed and poured into a three-port flask. The oil bath was heated to 60 °C to improve the fluidity of the matrix, and the water vapor adsorbed by the epoxy was mechanically stirred for 30 min to discharge the water vapor absorbed by the epoxy during storage.

(2) With the mass ratio of epoxy: curing agent of 100:85, the curing agent was weighed into the epoxy, and the silicon carbide filler particles were weighed according to the corresponding ratio. After full grinding, the epoxy curing agent mixed solution was added, and the mixture was mixed in the oil bath at 60 °C at a constant speed of 360 r/min for 60 min to ensure uniform distribution of filler particles.

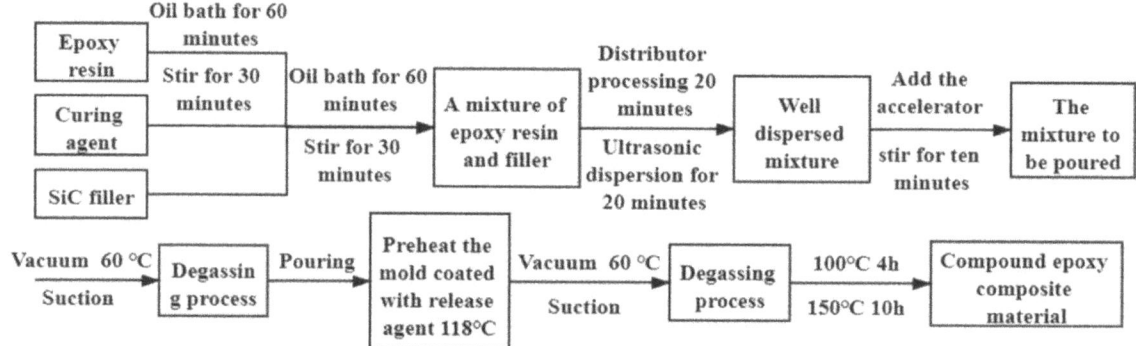

Figure 1. Epoxy composite preparation process.

(3) After the matrix and the filler were fully mixed, the matrix was dispersed by ultrasonic for 20 min, and the high-speed disperser was treated for 20 min. The mass ratio of epoxy:accelerator = 100:1 drop and the accelerator were added, and the mixture was mixed at 260 r/min at a constant speed for 10min.

(4) Pour out the mixed solution and put it in a vacuum drying box, keep it at 60 °C for vacuum operation, and extract it for many times until no obvious bubbles overflow.

(5) Put the mold sprayed with release agent into the blast dryer for preheating, and the temperature was 100 °C for 30 min. Pour the vacuum pumped mixed solution into the mold and put it in the vacuum oven for degassing again. The temperature was 60 °C for 30 min. The silicon carbide epoxy composite was obtained.

When the filling ratio of filler particles reaches 20 vol%, the thermal conductivity of the composite material is significantly improved, but when the filling ratio exceeds 30 vol%, the liquid mixture is very viscous, and it is difficult to vacuum out the internal bubbles and cast molding. Therefore, we chose the most representative 20 vol% and 30 vol% as the total filling amount, which have better comprehensive performance. The filling ratios of different filler are shown in Tables 2 and 3:

Table 2. SiC/EP composites filled with different crystal forms of SiC particles.

Sample	α-SiC	β-SiC	SiC Volume Fraction
α-SiC/EP$_{20vol\%}$	20 vol%	0	
3α-SiC/1β-SiC/EP$_{20vol\%}$	15 vol%	5 vol%	
1α-SiC/1β-SiC/EP$_{20vol\%}$	10 vol%	10 vol%	20 vol%
1α-SiC/3β-SiC/EP$_{20vol\%}$	5 vol%	15 vol%	
β-SiC/EP$_{20vol\%}$	0	20 vol%	
α-SiC/EP$_{30vol\%}$	30 vol%	0	
3α-SiC/1β-SiC/EP$_{30vol\%}$	22.5 vol%	7.5 vol%	
1α-SiC/1β-SiC/EP$_{30vol\%}$	15 vol%	15 vol%	30 vol%
1α-SiC/3β-SiC/EP$_{30vol\%}$	7.5 vol%	22.5 vol%	
β-SiC/EP$_{30vol\%}$	0	30 vol%	

In order to improve the bonding degree between the surface of SiC particles and epoxy resin matrix, the silane coupling agent KH-560 was used to organically modify the surface of SiC particles [7]. The modification method was as follows:

(1) According to the volume ratio of ethanol and ultra-pure water of 19:1, 400 mL solution was mixed and poured into the flask. 50 g SiC particles were weighed and dried at 80 °C for 6 h and added into the flask. The SiC particles were stirred at 60 °C for 10 min to fully disperse in the solvent;

Table 3. SiC/EP composites filled with micro/nano particles.

Sample	β-SiC Mass Fraction	n-SiC Mass Fraction	Overall Mass Fraction
19.5M/0.5N/EP	19.5 wt%	0.5 wt%	
19M/1N/EP	19 wt%	1 wt%	
18.5M/1.5N/EP	18.5 wt%	1.5 wt%	20 wt%
18M/2N/EP	18 wt%	2 wt%	
17.5M/2.5N/EP	17.5 wt%	2.5 wt%	
29.5M/0.5N/EP	29.5 wt%	0.5 wt%	
29M/1N/EP	29 wt%	1 wt%	
28.5M/1.5N/EP	28.5 wt%	1.5 wt%	30 wt%
28M/2N/EP	28 wt%	2 wt%	
27.5M/2.5N/EP	27.5 wt%	2.5 wt%	

(2) 2.5 g silane coupling agent KH-560 was dropped into SiC/ethanol/aqueous solution and stirred at 60 °C for 6 h to fully react.

(3) Pour the solution into the plastic reagent bottle, use centrifuge to separate SiC particles and solvent, take out after drying at 50 °C for 8 h, and grind with agate mortar for reserve.

2.3. Material Characterization and Testing

Scanning electron microscope (SEM) Quanta FEG 250 was used to observe α-SiC and β-SiC packing particles with different crystal types. α-SiC has hexagonal crystal structure, and its surface morphology is more distinct, β-SiC is cubic crystal, and its surface morphology is rounder. The SEM of SiC particles are shown in Figures 2 and 3.

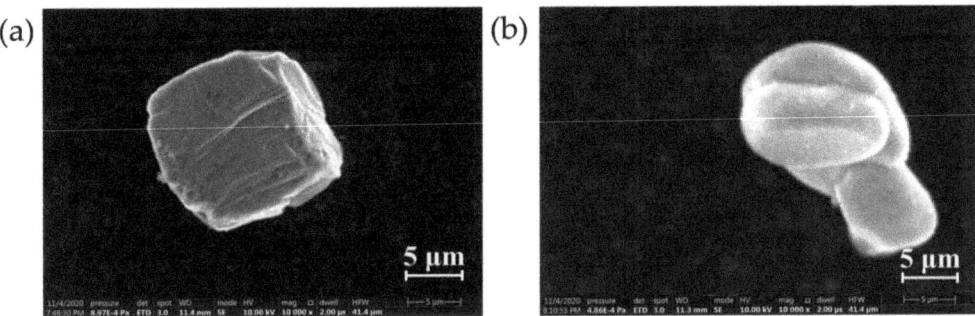

Figure 2. Electron microscope of SiC particles with different crystal types. (a) 10 μm α-SiC; (b) 10 μm β-SiC.

Figure 3. Electron microscopy of micron and nanometer SiC particles. (a) 1.5 μm β-SiC; (b) 50 nm β-SiC.

Transient plane heat source method was used to measure the thermal conductivity of the material using the thermal conductivity meter TPS 2500S. The sample used for the test was a wafer with the radius of 35 mm and thickness of 3 mm. In the room temperature, the breakdown strength tester HCDJC-50kV was used for the sample breakdown test. The voltage was increased at a rate of 2 kV/s until electric breakdown occurred. The samples used to measure the breakdown voltage were discs with a thickness of 1 mm and a diameter of 50 mm. 10 samples were tested for each group of materials.

3. Electrical and Thermal Properties of Epoxy Resin Composite with Different Crystal Forms of SiC

Different crystal types of SiC have different atomic packings, so their macroscopic electrical and thermal properties are different. By preparing epoxy resin composite materials with different crystal forms of SiC, and testing their electrical and thermal properties, the influence of the interaction between the filler and matrix on the properties of the composite materials was explored.

3.1. Breakdown Strength Analysis of Composite Materials

Two-parameter Weibull distribution was used to analyze the breakdown data of SiC/EP composites. The equation of two-parameter Weibull distribution is as follows:

$$F(x) = 1 - \exp\left(-\left(\frac{x}{\alpha}\right)^\beta\right) \tag{1}$$

where x is a variable, representing the breakdown data of the sample obtained by the test; β is the shape parameter, the dispersion of the data decreases with the increase of the value; α is the scale parameter, representing the breakdown strength when the breakdown probability is 63.2%, which is also called the average breakdown strength.

The least square method and Ross failure probability distribution function were used to solve the correlation coefficient, and the critical value criterion of the correlation coefficient was used to judge whether the breakdown data complied with the two-parameter Weibull distribution.

Ross failure probability distribution function is:

$$F(i, n) = \frac{i - 0.44}{n + 0.25} \times 100\% \tag{2}$$

The experimental breakdown voltage data were arranged according to their size and processed by formula. Least square method for linear fitting was used to solve the correlation coefficient.

$$X_i = \ln(E_i) \tag{3}$$

$$Y_i = \ln\left\{-\ln\left(1 - \frac{F(i,n)}{100}\right)\right\} \tag{4}$$

As shown in Figure 4, the data processing results showed that the breakdown strength data of SiC/EP composites conformed to Weibull distribution.

By comparing the two groups of composites with different filling quantities in Figure 4, the breakdown strength showed a consistent trend with the change of the ratio of the two crystal forms. The breakdown strength of β-SiC composite is higher than α-SiC. When the ratio of α-SiC to β-SiC is 3:1 or 1:3, the breakdown strength is significantly lower than the composite filled with single crystal SiC. When the ratio of α-SiC to β-SiC is 1:1, the breakdown strength is the highest. It can be seen in Figure 5 that when the total filling ratio is 20 vol%, the breakdown field strength of the composite material with a ratio of α-SiC:β-SiC of 1:1 is increased by 41.7% compared with single α-SiC crystal type filling, and 32.9% compared with single β-SiC crystal type filling. When the total filling ratio is 30 vol%, compared with the single α-SiC crystal type filling, the breakdown field strength of the composite material with a ratio of α-SiC:β-SiC of 1:1 is increased by 34.0%, and compared with the single β-SiC crystal type filling, it is increased by 10.4%.

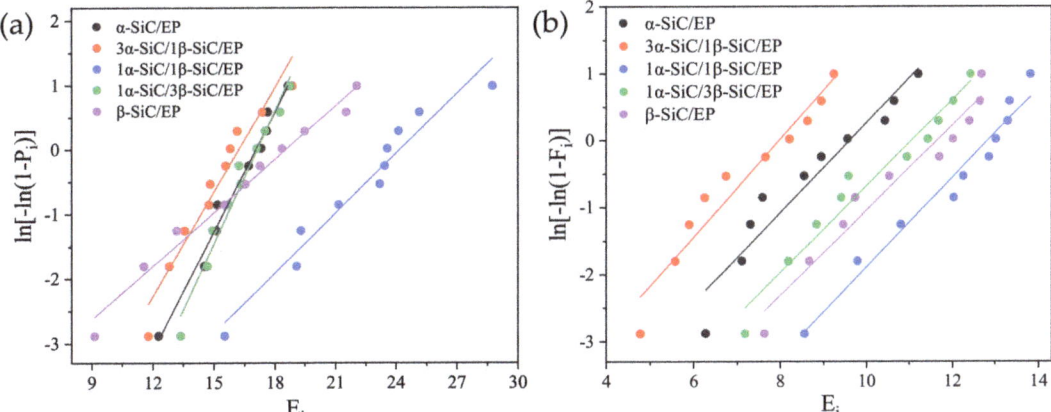

Figure 4. Weibull distribution of composite material breakdown strength. (**a**) 20 vol% SiC/EP; (**b**) 30 vol% SiC/EP.

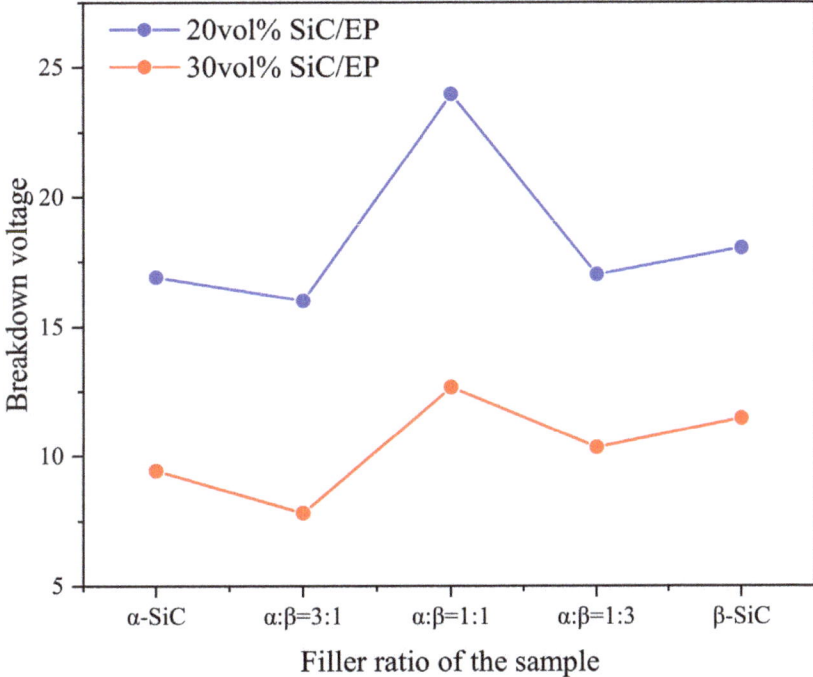

Figure 5. Average breakdown strength of composite material.

It can be seen from the SEM images of two different crystal forms of SiC that the morphologies of the two fillers are not completely regular, the shape of α-SiC is relatively sharp, and the surface of β-SiC is comparatively round. When the composite material uses two different crystal forms of SiC as the filler of the composite, it will lead to internal defects of the material, which results in a significant decrease in the breakdown strength of the composite material. When the compounding ratio is 1:1, the internal defects of the material are obviously reduced, and the bonding degree of the interface between the filler and the matrix is obviously improved. The breakdown path cannot form along the air gap defect and the development path extend, which improves the breakdown strength of the material to a certain extent. This conjecture was verified by analyzing the infrared spectrum.

In order to compare the relative size of each absorption peak more intuitively and quantitatively, the infrared absorption spectrum was converted into a projection spectrum, and the method of baseline correction was adopted to make the spectral values at both ends of each absorption peak lying on a straight line. Enlarge the characteristic peak area, as shown in Figures 6 and 7.

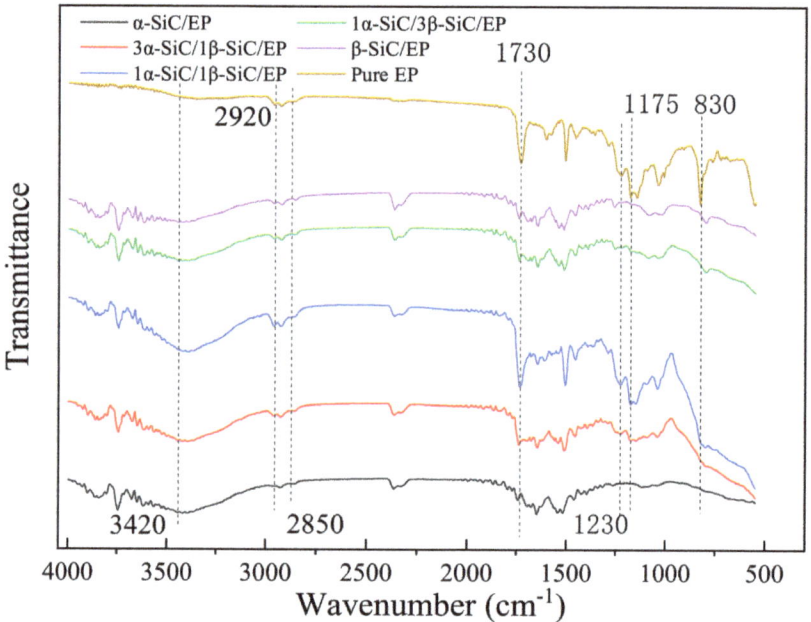

Figure 6. Infrared spectrum of SiC/EP composite material.

The wave numbers of 1230 cm^{-1} and 1170 cm^{-1} are the absorption bands of the ester [31–33]. The wave number of 2850 cm^{-1} represents the symmetrical stretching vibration of C-H in CH3 [24]. The wave number of 2920 cm^{-1} is attributable to the vibration of aromatic protons, and the broad peak at 3420 cm^{-1} is hydroxyl and hydrogen bonds [23]. By comparing the infrared spectra of the five formula systems, it can be found that the crosslinking degree of epoxy resin composite material has decreased to varying degrees compared with the pure epoxy resin, which also reflects the reduction of the breakdown strength of the composite material. A higher degree of cross-linking indicates that the epoxy resin molecules are more closely linked, and the defects between the insulating material and the filler in the system are reduced, the breakdown path is difficult to develop along the interface [34,35], which increases the breakdown strength of the composite material. For the five composite material systems, when the volume ratio of α-SiC to β-SiC was 1:1, the crosslinking degree of the composite material was the highest, so it also had a higher breakdown strength. This may be due to the different crystal forms of α-SiC and β-SiC and the better overlap between the particles, which reduced the agglomeration of the same crystal particles and played a certain mutual coordination during the condensation reaction, so as to improve the crosslinking degree of epoxy

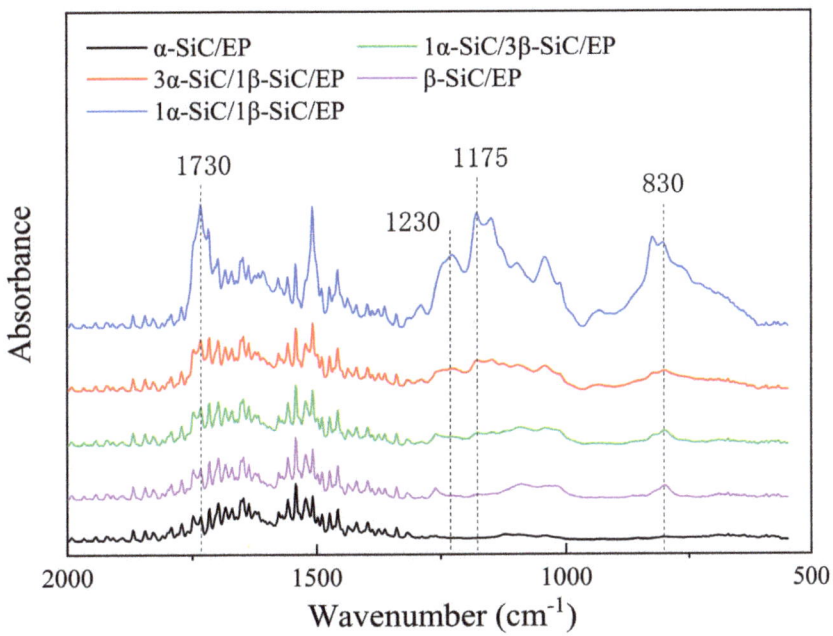

Figure 7. Partial enlarged view of baseline correction of different crystal SiC/EP infrared spectra.

Compared with 20 vol% composite materials, the overall breakdown strength of composite materials filled with 30 vol% was reduced. As a crystal with high electrical conductivity, SiC has high breakdown strength, therefore, it is the epoxy resin matrix that determines the breakdown strength of the composite material. When the SiC filling amount increased, due to the micro-particles and the matrix interface were not tightly bonded, the internal air gap defects greatly increased, the development path of the electrical tree in the epoxy was shortened, and the breakdown strength was significantly reduced [36,37]. When other inorganic fillers such as SiO_2 and BN are used [38,39], the breakdown field strength of the composite material exhibits consistent characteristics at high filling ratios, and this trend is not conducive to the application of thermally conductive insulating materials.

In order to investigate the effect of crystal configuration on the crosslinking degree in terms of microstructure, the cross sections of the composites with different crystal configuration were observed by scanning electron microscopy. The focus was on the state of the interface between the filler particles and the matrix and the dispersibility of the filler particles. The SEM image is shown in Figure 8:

It can be seen from the microscopic morphology that the filler particles are uniformly distributed in the matrix without delamination or particle aggregation. In the filling system, α-SiC particles have more edges and corners, while the shape of β-SiC is more rounded, and there are obvious air gap defects at the interface between the filler particles and the matrix. When the ratio of α-SiC to β-SiC was 1:1, the bonding degree between the filler and the matrix was better and the defects at the interface were significantly reduced. There are no protruding particles or obvious pits at the cross-section, which appears to be compact structure. This shows that the two crystal particles overlap and cooperate with each other in the mixing and solidification process under the appropriate ratio, which can improve the uniformity of the mixed system. The degree of cross-linking of epoxy resin between particles is increased, which significantly reduces defects at the interface of the composite material [40].

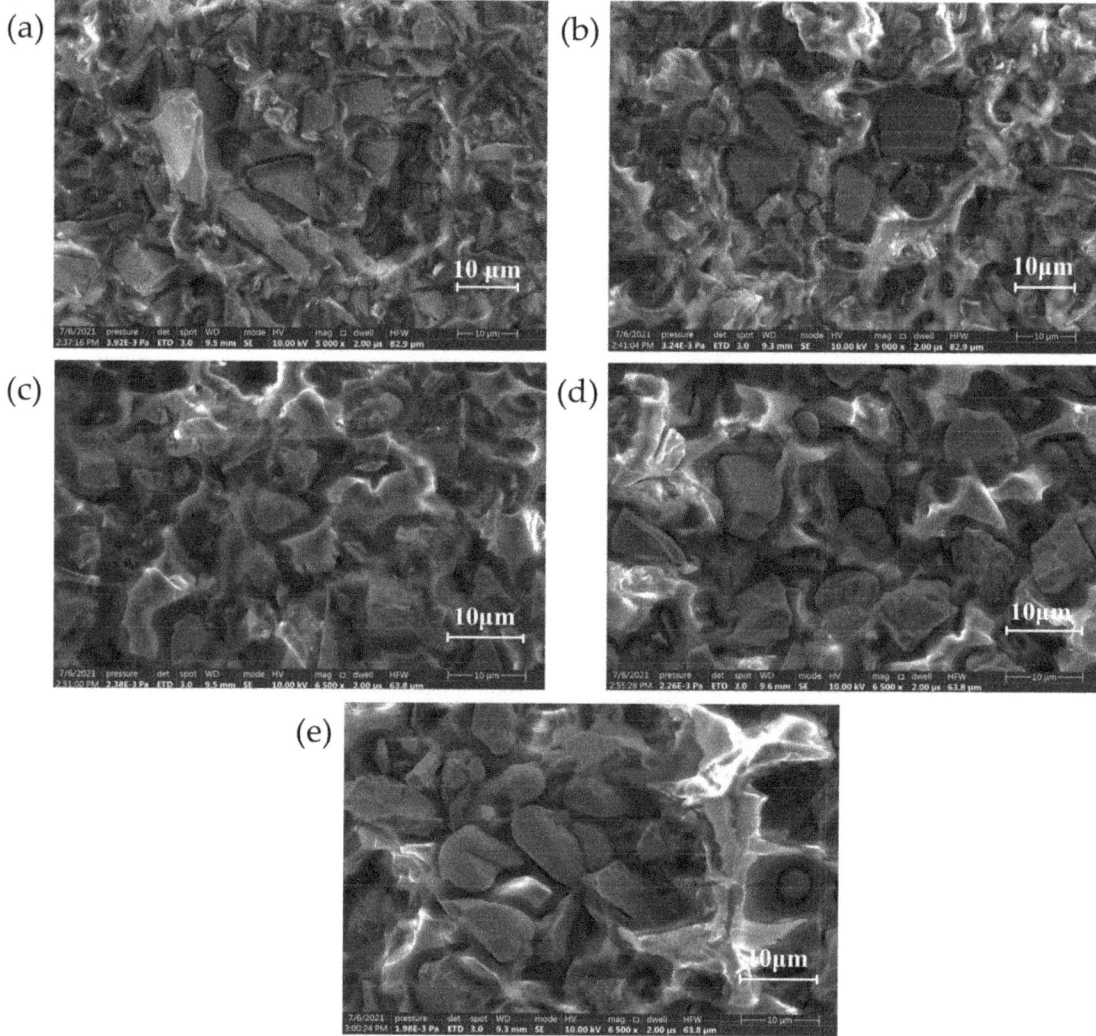

Figure 8. Sectional electron micrograph of SiC/EP composite material. (**a**) α-SiC/EP; (**b**) α-SiC:β-SiC = 3:1; (**c**) α-SiC:β-SiC = 1:1; (**d**) α-SiC:β-SiC = 1:3; (**e**) β-SiC/EP.

3.2. Thermal Conductivity of SiC/EP Composites

The Figure 9 shows the variation of thermal conductivity of two crystal SiC composites in five proportions. Since the epoxy resin has low cross-linking crystallinity and disorder, only the thermal motion of macromolecular chains and groups cannot provide a way for the rapid movement of phonons, so the thermal conductivity of pure epoxy resin material is only 0.196 W/(m·K). The thermal conductivity of β-SiC composite was significantly higher than α-SiC composite when the volume fraction of β-SiC was fixed and the ratio of two different crystal forms was changed. The thermal conductivity of β-SiC/epoxy composites was 0.7433 W/(m·K) and 0.9019 W/(m·K), respectively, which was 61.13% and 48.58% higher than α-SiC/epoxy composites with the same amount of filling, and 279.23% and 360.15% higher than pure epoxy resin.

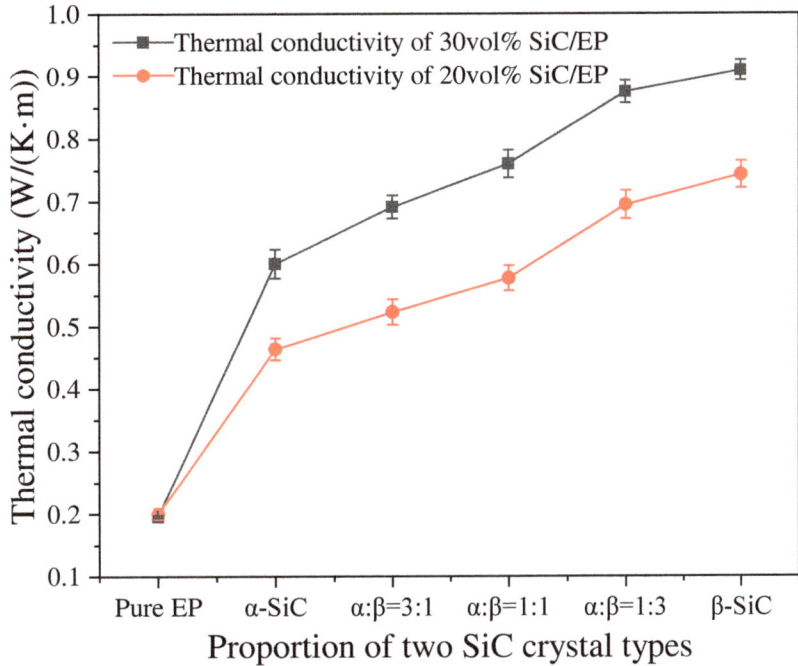

Figure 9. Thermal conductivity of SiC composite filled with different crystal forms.

The thermal conductivity of the composite material of the two crystal forms increased with the increasing of the proportion of β-SiC. At present, the thermal path theory is the most common theory used to explain the thermal conduction mechanism of filled thermally conductive composites. This theory describes that the high thermal conductivity filler particles in the matrix overlap each other to form a thermal conduction path, and the heat flux is transferred from the high temperature direction to the low temperature along the network built by the thermal conductive filler [41,42]. The results showed that the crosslinking degree of epoxy matrix had little effect on the thermal conductivity of filled composites, but the shape of filler had much effect on the thermal conductivity of filled composites. Compared with α-SiC, β-SiC particles have more rounded shapes, better bonding degree with epoxy resin matrix, and lower interface thermal resistance. Furthermore, β-SiC particles have more effective contact area between particles than those with sharp edges and angles, which is more conducive to the formation of thermal conductivity channels.

3.3. Simulation Study on the Influence of Different Shapes of SiC Particles on the Thermal Conductivity of Composites

In order to study the influence of the shape of filler particles on the construction of the internal heat conduction channel of the composite material, the finite element method [5,43,44] was used to construct a solid heat transfer model for the internal representative volume elements (RVE) of the filled composite material. In the steady state, according to Fourier's law of heat conduction, the differential equation for solving the internal heat conduction of the composite material was obtained, as in the following equation:

$$\frac{\partial^2 T}{\partial^2 x} + \frac{\partial^2 T}{\partial^2 y} + \frac{\partial^2 T}{\partial^2 z} = 0 \tag{5}$$

We set the side surface of the composite material as thermal insulation, which was also the second type of boundary condition, as in the following formula:

$$\left.\frac{\partial T}{\partial n}\right|_\Gamma = 0 \tag{6}$$

In the formula, the subscript Γ denoted the side surface of the composite material. Set the initial temperature on the upper surface (S_1) to 60 °C, and set the temperature on the lower surface (S_2) to 20 °C, which was also the first type of boundary condition, as in the following formula:

$$T|_{s_1} = 60\ °C;\ T|_{s_2} = 20\ °C \tag{7}$$

The setting of boundary conditions is shown in Figure 10.

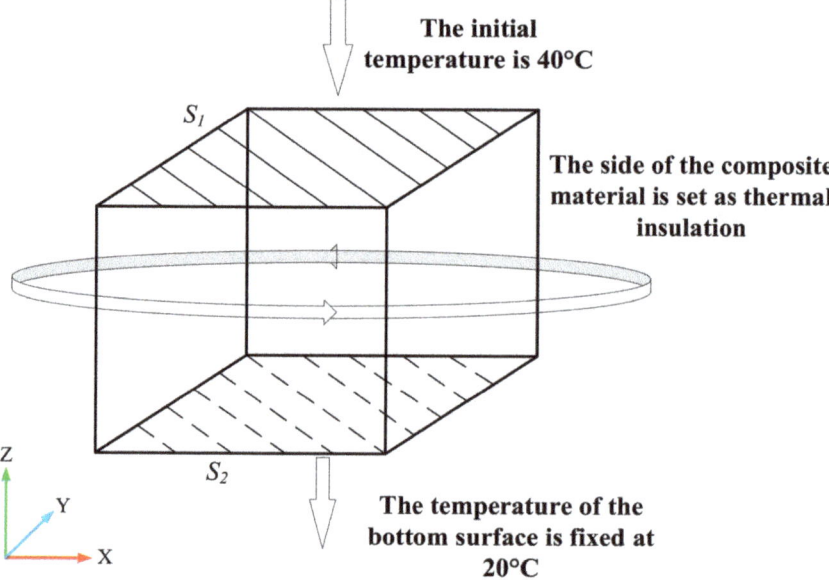

Figure 10. Model boundary settings.

Using the finite element method, by solving the heat conduction equation (1), the conduction heat flux inside the composite material was obtained. According to the Fourier law of heat conduction, the thermal conductivity of the composite material was calculated by the following formula [45]:

$$K = \frac{QL}{\Delta T} \tag{8}$$

In the formula, K is the equivalent thermal conductivity; Q is the average conductive heat flux in the z-direction inside the composite material; L is the thickness of the model in the z-axis direction; ΔT is the temperature difference between the upper surface and the bottom surface in the z-axis direction.

In order to study the influence of particle shape on the heat flux of the surrounding matrix, a single particle was placed in a cube RVE with a side length of 30 μm. Spherical and cubic particles of the same size were placed, and the cubes were placed in parallel and rotated 45° to make the calculation results representative. As shown in Figure 11.

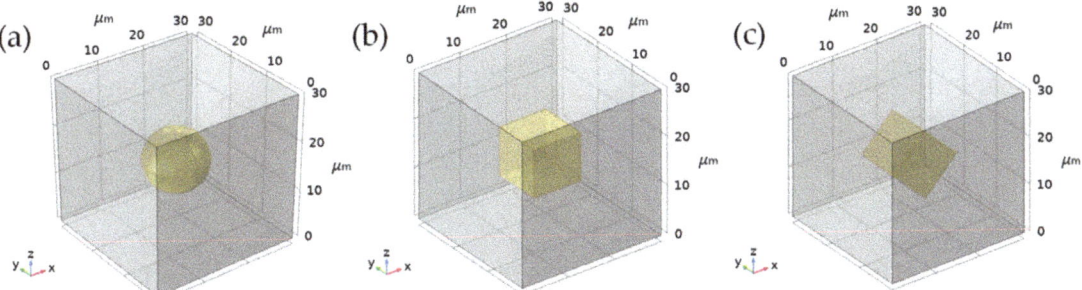

Figure 11. The geometric model of a composite material filled with a single particle. (**a**) A single spherical particle; (**b**) A single cubic particle placed horizontally; (**c**) A single cubic particle rotated 45°.

Taking the cross section parallel to the *xz* plane and *y* = 15 μm to study the conductive heat flux of the filler and its surrounding matrix, as shown in Figure 12:

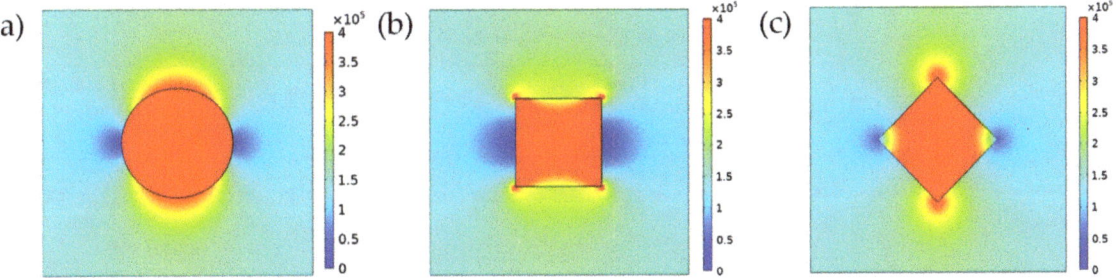

Figure 12. Cross-section conductive heat flux cloud diagram of composite materials. (**a**) Filler particle is small sphere; (**b**) Filler particle is horizontally placed cube; (**c**) Filler particle cube is rotated 45°.

Considering that the filler particles construct a heat conduction channel through effective contact, it is believed that when the conductive heat flux of the matrix between the fillers reaches 80% of the filler [5], the matrix between the fillers can participate in the construction of an effective heat conduction channel. We used the ImageJ software to process the conduction heat flux cloud image. Taking the conduction heat flux at the center of the filler particles as the reference, the conduction heat flux at the center of the particle was red in the chromatogram, and the corresponding color gamut value was 0; the point where the conduction heat flux was 0 is blue in the chromatogram, and the corresponding color gamut value was 158. When the conductive heat flux in the matrix was 80–100% of the heat flux in the center of the particle, the color gamut value range of the matrix in the three cases can be obtained by calculation, and the area within this range was integrated in ImageJ, as shown in Figure 13:

The effective contact areas obtained in the three cases were: (a) 9.941, (b) 1.34 and (c) 2.405. It can be seen that when the filler is spherical particles, because the surface was more rounded, according to the principle of heat transfer, the effective heat transfer area of the surrounding matrix was larger, while the effective heat transfer area of the matrix surrounding the cubic particles was much smaller.

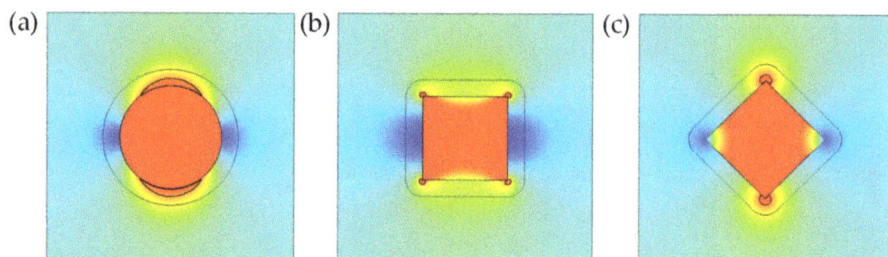

Figure 13. Cloud iamge of conductive heat flux after ImageJ processing. (**a**) Filler particle is small sphere; (**b**) Filler particle is horizontally placed cube; (**c**) Filler particle cube is rotated 45°.

In order to study the influence of multiple particles on the thermal conductivity of composite materials, a three-dimensional random particle distribution model was established. We took the composite material RVE as a cube with a side length of 50 μm, and filled it with 20 vol% particles, corresponding to the composite material of different crystal types, and the numbers of particles were set as shown in Table 4.

Table 4. The number of filler particles in the RVE geometric model.

Corresponding Crystalline Compound Composite Material	Number of Cube Fillers	Number of Spherical Fillers
α/EP	0	25
$3\alpha/1\beta$/EP	7	18
$1\alpha/1\beta$/EP	13	12
$1\alpha/3\beta$/EP	19	6
β/EP	25	0

The established RVE model is shown in Figure 14:

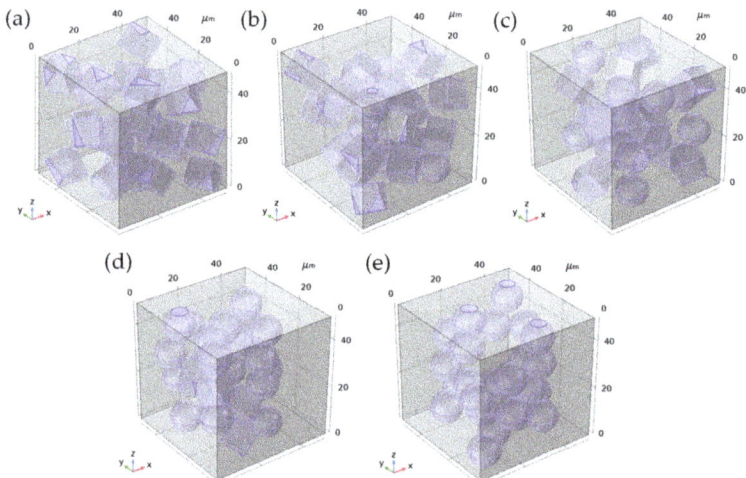

Figure 14. RVE geometric model of randomly filled composite materials with particles. (**a**) Model corresponding to β/EP; (**b**) Model corresponding to $1\alpha/3\beta$/EP; (**c**) Model corresponding to $1\alpha/1\beta$/EP; (**d**) Model corresponding to $3\alpha/1\beta$/EP; (**e**) Model corresponding to α/EP.

Comparing the thermal conductivity of the volume representative element (RVE) obtained by the simulation with the experimental results is shown in the Figure 15. The results show that the finite element simulation can be consistent with the experimental results. Therefore, the simulation results can be used to explain the influence of the shape of the filler particles on the heat of the composite material. As the proportion of spherical particles increases, the thermal conductivity of the composite material increases. This is due to the effective heat conduction area of the matrix around the spherical filler particles is larger, which is beneficial to participate in the construction of an effective heat conduction channel.

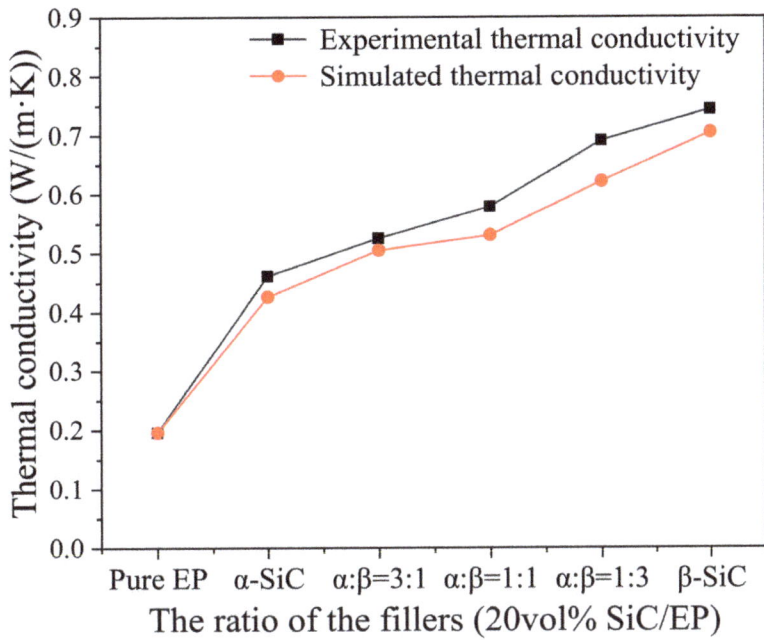

Figure 15. Comparison of experimental and simulated composite thermal conductivity results.

4. Electrical and Thermal Properties of Epoxy Resin Composites with Micro-Nano Compound

Compared with micro-particles, nanoparticles have a larger specific surface area, and their particle sizes are small. There is a scale effect at the interface, which can significantly improve the degree of bonding between the particles and the matrix, thereby affecting the electrical and thermal performance of the composite material [46]. Nano particles are small in size, large in surface energy and packed loosely, making it difficult to measure their actual density. Therefore, the proportion of the mass fraction of micro-nano particles is used as a variable in the study of micro-nano composite.

4.1. Changes and Analysis of the Breakdown Strength of Composite Materials

By comparing the breakdown voltage of SiC/EP composites with a total filling mass fraction of 20 wt% and 30 wt%, it can be found that under different total filling ratios, the optimal relative ratio of the micro-nano compound has also changed. It can be seen from the current experimental results that for the composite with the total filling mass fraction of 20 wt%, when the ratio of micro and nano was 18.5 wt% and 1.5 wt%, the breakdown strength of the composite material was the highest, and the breakdown strength was increased by 14.9% compared with the case when no nanoparticles were added. For the composite with the total filling mass fraction of 30 wt%, when the ratio of micro and nano was 29.5 wt% and 0.5 wt%, the breakdown strength of the composite material was the highest, and the breakdown strength was increased by 32.4% compared with the case when no nanoparticles were added. Weibull distribution of micro-nano composite SiC/EP breakdown strength is shown in Figure 16.

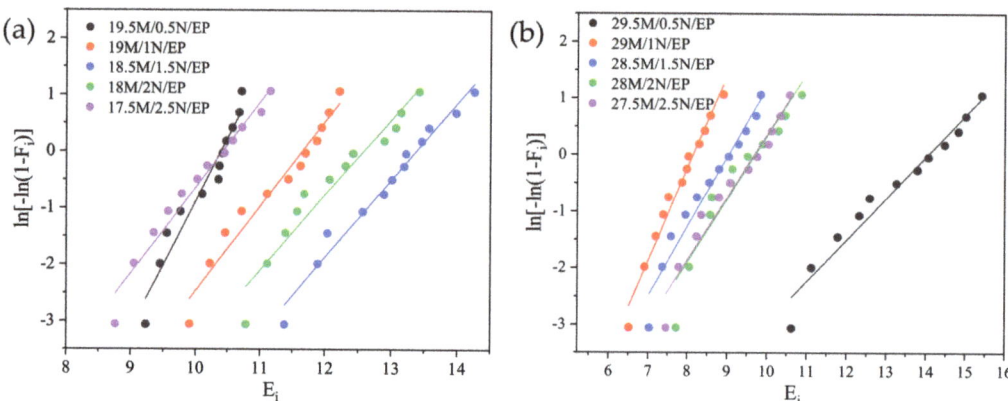

Figure 16. Weibull distribution of micro-nano composite SiC/EP breakdown strength. (**a**) 20 wt% SiC/EP; (**b**) 30 wt% SiC/EP.

In micron SiC/EP composites, there will be more air gap defects at the interface between the micron particles and the epoxy resin matrix, which leads to the development of partial discharge channels along the two-phase interface. Thus, the electrical breakdown path of the composite material is formed, and the breakdown strength of the composite material is reduced. After the introduction of nanoparticles, due to the small radius of curvature of the nanoparticles, the groups on the surface of the particles are more likely to react with the epoxy resin instead of bonding with each other [47]. The interface bonding effect between the nanoparticles and the matrix is stronger, and the breakdown path is blocked, so it is unable to develop along the interface [19,48].

It can be seen from the data in Figure 17 that after adding nano-SiC particles to SiC/EP composites, the breakdown strength of the composites changed with the amount of nano-SiC particles, and there was a certain correspondence with the infrared spectral characteristics of the composites.

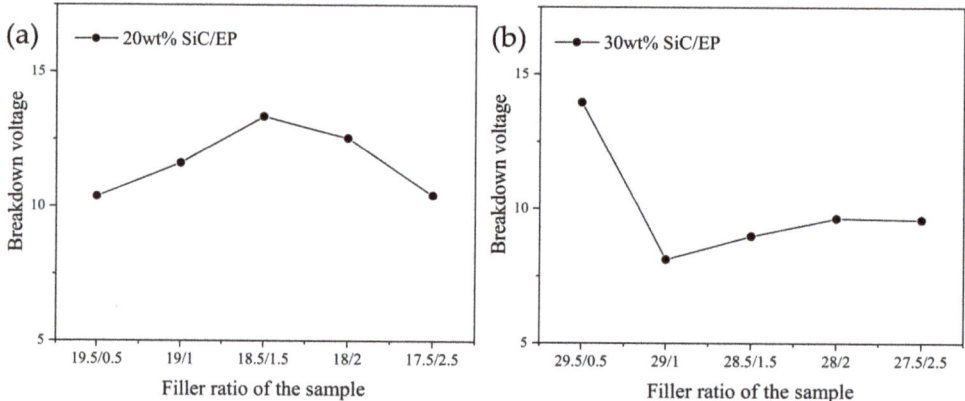

Figure 17. Average breakdown strength of micro-nano composite SiC/EP. (**a**) 20 wt% SiC/EP; (**b**) 30 wt% SiC/EP.

The same processing method in Figure 7 was adopted. We converted infrared absorption spectrum into projection spectrum and performed baseline correction and local amplification to visually compare the relative size of each absorption peak, as shown in Figures 18 and 19.

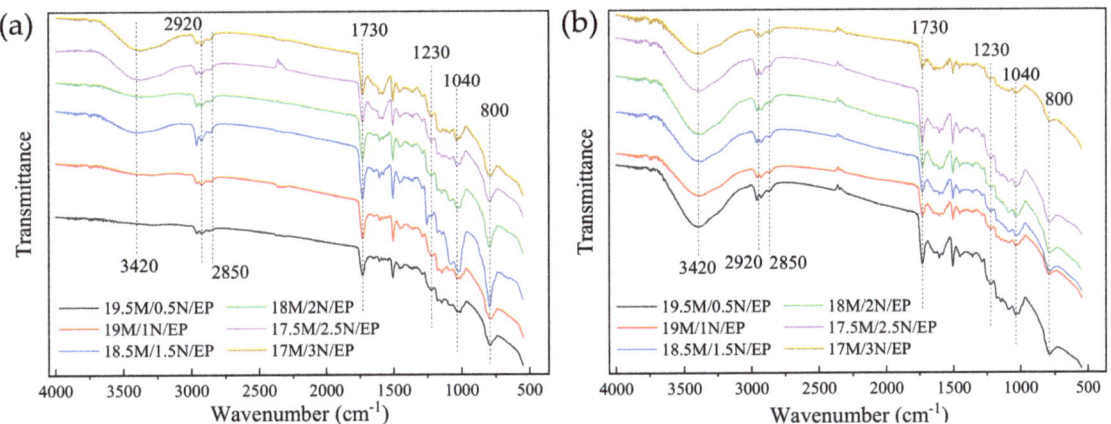

Figure 18. Infrared spectrum of SiC/EP composite material. (**a**) 20 wt% SiC; (**b**) 30 wt% SiC.

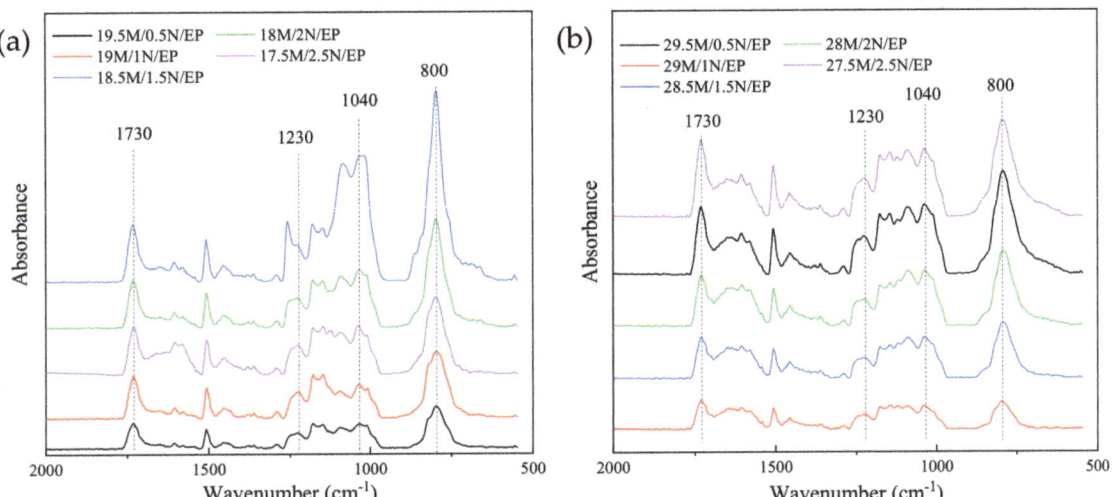

Figure 19. Partial enlarged view of baseline correction of the micro/nano SiC/EP infrared spectra. (**a**) 20 wt% SiC/EP; (**b**) 30 wt% SiC/EP.

Wavenumber 800 cm^{-1} corresponds to C-N single bond, and the stronger absorption peak represents the greater amide concentration, which indicates that the reaction between tertiary amine promoter and anhydride curing agent was more completed. The reaction rate of tertiary amine accelerator and anhydride curing agent changed when the filling amount of nanoparticles was changed, which also proves that the filling of nanoparticles has a significant effect on the curing reaction of epoxy resin matrix. The wavenumbers of 1230 cm^{-1} and 1170 cm^{-1} are the absorption bands of esters, and the wavenumbers of 1040 cm^{-1} are the absorption bands of primary alcohols. The peak with a wavenumber of 2850 cm^{-1} is caused by the symmetric stretching of C-H in CH3 [24]. The vibrational band at 2920 cm^{-1} is attributed to the vibration of aromatic protons, and the broad peak at 3420 cm^{-1} is hydroxyl and hydrogen bonds [23]. After curing, the higher the crosslinking degree is, the stronger the absorption peaks of corresponding bands are [31–33].

When the amount of addition was appropriate, the SiC nanoparticles were uniformly dispersed, whose surface was bonded with the epoxy matrix [19]. The crosslinking degree of the epoxy resin was also improved. However, when the filling amount of nano-SiC particles was too large, the potential energy of the nanoparticle interface will be larger, which will cause agglomeration [49,50]. The equivalent radius of the agglomerate increased, leading to a large number of void defects formed inside, which caused the epoxy crosslinking degree to decrease, and the decrease of breakdown strength of the composite material.

Figure 20 showed that the filler particles are relatively uniformly distributed in the matrix, but when the filling ratio of the nanoparticles exceeded the optimal ratio, the nanoparticles will agglomerate more seriously. The nanoparticles will agglomerate together, and more air gap defects will be formed inside the agglomerated particles. [31] The degree of bonding between the particles and the matrix decreased, resulting in more defects at the cross section. At the same time, in a nano-composite system, when the density of micron SiC particles became larger, the nanoparticles used as reinforcing fillers were more likely to cause agglomeration problems. This was due to the micro- and nanoparticles collide with each other during the mixing process. Since the size of micro-particles and nanoparticles were very different, the movement range of nanoparticles was limited to the gap between the micro-particles, and there was a greater chance for nanoparticles to agglomerate together. Once agglomerating, it was less likely to be separated by the impact force of the stirring process.

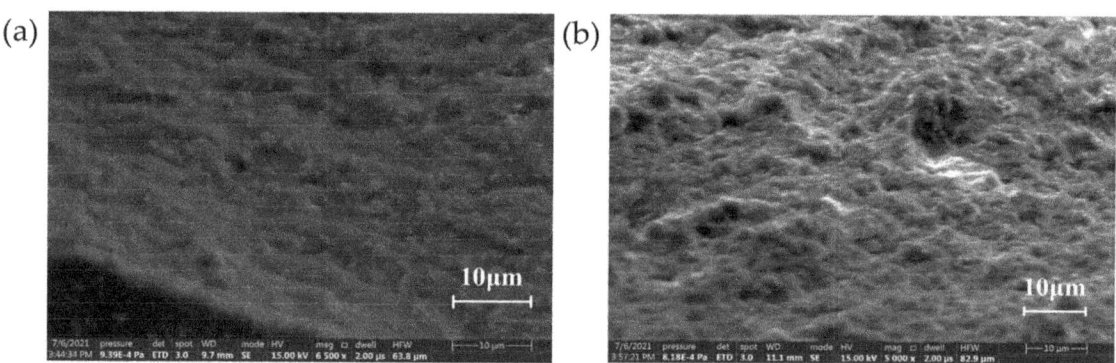

Figure 20. The cross-sectional microscopic electron microscope image of the micro-nano composite system. (**a**) 29.5% m-SiC/0.5%n-SiC/EP; (**b**) 27.5% m-SiC/2.5%n-SiC/EP.

4.2. Effect of Nanoparticles on Thermal Conductivity of Composites

When nanoparticles were added to the SiC/EP composite material, the thermal conductivity of composite material changed as shown in the Figure 21:

With the increase of the doping amount of nanoparticles, the thermal conductivity of the composite material decreased. Some previous studies also showed similar results: in the micro-nano composite system, the contribution of nanoparticles to the thermal conductivity is not as good as that of micro-particles [12,39]. This was because nanoparticles were too small in size compared to micro particles. Under the same mass, although the number of nanoparticles was larger, their dispersibility was stronger and they cannot form effective contact similar to larger micron particles, thereby constructing a heat conduction channel. Therefore, as more nanoparticles replaced micron particles, the thermal conductivity of composites showed a downward trend.

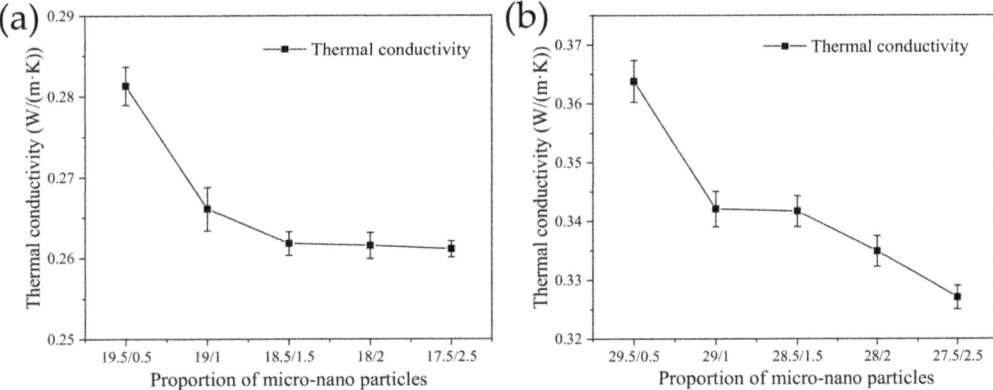

Figure 21. Thermal conductivity of micro-nano compound SiC/EP composite. (**a**) 20 wt% SiC/EP; (**b**) 30 wt% SiC/EP.

5. Conclusions

In this paper, in order to improve the electrical and thermal properties of SiC/EP composites, the effects of SiC crystal form and micro nano scale of SiC particles on the breakdown strength and thermal conductivity properties were studied. The main conclusions are as follows:

(1) The compounding of different crystal particles and the compounding of micro-nano particles will significantly affect the bonding degree between the filler particles and the matrix at the interface. For composite materials with different crystal type particles, when the filled particle ratio of α-SiC to β-SiC is 1:1, the breakdown strength of the composite material is more than 10% higher than that of the composite material filled with a single crystal type.

(2) For micro/nano composites, different SiC content corresponds to different optimum ratio of micro to nano. When the amount of SiC increases, the motion space of nanoparticles will be squeezed, which leads to more serious agglomeration and the decrease of corresponding optimal nanometer ratio. Under the optimal ratio, the breakdown strength of the composite material is improved. Compared with the compounding of different crystal types, the introduction of a small amount of nanoparticle can significantly increase the breakdown strength of the micro-nano composite material.

(3) Whether an effective thermal conductivity channel can be formed is the key to determine the thermal conductivity of composites. The experimental results and the finite element simulation analysis show that the shape and particle size of the filler have a greater impact on the thermal conductivity of the composite. When the shape of the particles is more rounded, the heat conduction channel can be constructed more effectively.

(4) In the filled composite material, the method of compounding different crystalline SiC particles and micro-nano particles can increase the breakdown field strength of the composite, and at the same time make the composite material have a relatively high thermal conductivity, showing certain advantages.

Author Contributions: Conceptualization, X.S.; methodology, X.S.; validation, X.X., W.Q. and W.Y.; formal analysis, X.S., J.Z., Y.Z. (Yiran Zhang) and Y.Z. (Yang Zhou); investigation, X.S., Y.Z. (Yiran Zhang) and Y.Z. (Yang Zhou); resources, J.L. and X.B.; data curation, X.X. and W.Q.; writing—original draft preparation, X.S.; writing—review and editing, B.S., W.Y., J.L. and X.B.; supervision, X.B.; project administration, X.B.; funding acquisition, X.B. All authors have read and agreed to the published version of the manuscript.

Funding: This work was supported by the National Key Research and Development Program of China (No. 2017YFB0903800), the State Key Laboratory of Reliability and Intelligence of Electrical Equipment (No. EERIKF2019002) and the Fundamental Research Funds for the Central Universities (Grant No. 2019MS011).

Conflicts of Interest: The authors declare no conflict of interest.

References

1. Yao, H.; Fan, Z.; Cheng, H.; Guan, X.; Wang, C.; Sun, K.; Ouyang, J. Recent Development of Thermoelectric Polymers and Composites. *Macromol. Rapid Commun.* **2018**, *39*, e1700727. [CrossRef] [PubMed]
2. Li, A.; Zhang, C.; Zhang, Y.F. Thermal conductivity of graphene-polymer composites: Mechanisms, properties, and applications. *Polymers* **2017**, *9*, 437. [CrossRef] [PubMed]
3. Bian, X.; Tuo, R.; Yang, W.; Zhang, Y.; Xie, Q.; Zha, J.; Lin, J.; He, S. Mechanical, thermal, and electrical properties of BN–epoxy composites modified with carboxyl-terminated butadiene nitrile liquid rubber. *Polymers* **2019**, *11*, 1548. [CrossRef] [PubMed]
4. Huang, X.; Iizuka, T.; Jiang, P.; Ohki, Y.; Tanaka, T. Role of Interface on the Thermal Conductivity of Highly Filled Dielectric Epoxy/AlN Composites. *J. Phys. Chem. C* **2012**, *116*, 13629–13639. [CrossRef]
5. Wu, J.; Song, X.; Gong, Y.; Yang, W.; Chen, L.; He, S.; Lin, J.; Bian, X. Analysis of the heat conduction mechanism for Al_2O_3/Silicone rubber composite material with FEM based on experiment observations. *Compos. Sci. Technol.* **2021**, *210*, 108809. [CrossRef]
6. Dang, J.; Wang, R.; Yang, L.; Gao, L.; Zhang, Z.; Zha, M. Preparation of β-SiCw/BDM/DBA composites with excellent comprehensive properties. *Polym. Compos.* **2014**, *35*, 1875–1878. [CrossRef]
7. Tang, D.; Su, J.; Kong, M.; Zhao, Z.; Yang, Q.; Huang, Y.; Liao, X.; Niu, Y. Preparation and properties of epoxy/BN highly thermal conductive composites reinforced with SiC whisker. *Polym. Compos.* **2016**, *37*, 2611–2621. [CrossRef]
8. Hwang, Y.; Kim, M.; Kim, J. Effect of Al_2O_3 coverage on SiC particles for electrically insulated polymer composites with high thermal conductivity. *RSC Adv.* **2014**, *4*, 17015. [CrossRef]
9. Lin, C.; Xu, H.-F.; He, S.-J.; Du, Y.-F.; Yu, N.-J.; Du, X.-Z.; Lin, J.; Nazarenko, S. Thermal conductivity performance of polypropylene composites filled with polydopamine-functionalized hexagonal boron nitride. *PLoS ONE* **2017**, *12*, e0170523.
10. Lee, Y.-S.; Lee, S.-Y.; Kim, K.S.; Noda, S.; Shim, S.E.; Yang, C.-M. Effective heat transfer pathways of thermally conductive networks formed by one-dimensional carbon materials with different sizes. *Polymers* **2019**, *11*, 1661. [CrossRef]
11. Lin, C.Y.; Kuo, D.H.; Liou, G.S.; Cheng, J.Y.; Jhou, Y.C. Thermal conductive performance of organosoluble polyimide/BN and polyimide/(BN+ ALN) composite films fabricated by a solution-cast method. *Polym. Compos.* **2013**, *34*, 252–258.
12. Bian, W.; Yao, T.; Chen, M.; Zhang, C.; Shao, T.; Yang, Y. The synergistic effects of the micro-BN and nano-Al2O3 in micro-nano composites on enhancing the thermal conductivity for insulating epoxy resin. *Compos. Sci. Technol.* **2018**, *168*, 420–428. [CrossRef]
13. Fabiani, D.; Montanari, G.; Testa, L. Effect of aspect ratio and water contamination on the electric properties of nanostructured insulating materials. *IEEE Trans. Dielectr. Electr. Insul.* **2010**, *17*, 221–230. [CrossRef]
14. Mai, V.-D.; Lee, D.-I.; Park, J.-H.; Lee, D.-S. Rheological properties and thermal conductivity of epoxy resins filled with a mixture of alumina and boron nitride. *Polymers* **2019**, *11*, 597. [CrossRef] [PubMed]
15. Kim, Y.; Kim, J. 3D Interconnected Boron Nitride Networks in Epoxy Composites via Coalescence Behavior of SAC305 Solder Alloy as a Bridging Material for Enhanced Thermal Conductivity. *Polymers* **2020**, *12*, 1954. [CrossRef]
16. Zhou, T.; Wang, X.; Liu, X.; Xiong, D. Improved thermal conductivity of epoxy composites using a hybrid multi-walled carbon nanotube/micro-SiC filler. *Carbon* **2010**, *48*, 1171–1176. [CrossRef]
17. Zhao, L.; Yan, L.; Wei, C.; Li, Q.; Huang, X.; Wang, Z.; Fu, M.; Ren, J. Synergistic Enhanced Thermal Conductivity of Epoxy Composites with Boron Nitride Nanosheets and Microspheres. *J. Phys. Chem. C* **2020**, *124*, 12723–12733. [CrossRef]
18. Tu, Y.; Zhou, F.; Jiang, H.; Bai, F.; Wang, C.; Lin, J.; Cheng, Y. Effect of nano-TiO_2/EP composite coating on dynamic characteristics of surface charge in epoxy resin. *IEEE Trans. Dielectr. Electr. Insul.* **2018**, *25*, 1308–1317. [CrossRef]
19. Imai, T.; Sawa, F.; Nakano, T.; Ozaki, T.; Shimizu, T.; Kozako, M.; Tanaka, T. Effects of nano- and micro-filler mixture on electrical insulation properties of epoxy based composites. *IEEE Trans. Dielectr. Electr. Insul.* **2006**, *13*, 319–326. [CrossRef]
20. Huang, X. Influence of nanoparticle surface treatment on the electrical properties of cycloaliphatic epoxy nanocomposites. *IEEE Trans. Dielectr. Electr. Insul.* **2010**, *17*, 635. [CrossRef]
21. Wu, J.; Zhang, Y.; Gong, Y.; Wang, K.; Chen, Y.; Song, X.; Lin, J.; Shen, B.; He, S.; Bian, X. Analysis of the Electrical and Thermal Properties for Magnetic Fe_3O_4-Coated SiC-Filled Epoxy Composites. *Polymers* **2021**, *13*, 3028. [CrossRef]
22. Hussein, S.I.; Abd-Elnaiem, A.M.; Nadia, A.; Abdelazim, M. Enhanced Thermo-Mechanical Properties of Poly (vinyl alcohol)/Poly (vinyl pyrrolidone) Polymer Blended with Nanographene. *Curr. Nanosci.* **2020**, *16*, 994–1001. [CrossRef]
23. Abd-Elnaiem, A.M.; Hussein, S.I.; Assaedi, H.; Mebed, A.M. Fabrication and evaluation of structural, thermal, mechanical and optical behavior of epoxy–TEOS/MWCNTs composites for solar cell covering. *Polym. Bull.* **2021**, *78*, 3995–4017. [CrossRef]
24. Ali, N.A.; Hussein, S.I.; Asafa, T.B.; Abd-Elnaiem, A.M. Mechanical Properties and Electrical Conductivity of Poly(methyl methacrylate)/Multi-walled Carbon Nanotubes Composites. *Iran. J. Sci. Technol. Trans. A Sci.* **2020**, *44*, 1567–1576. [CrossRef]
25. Hussein, S.; Abd-Elnaiem, A.M.; Asafa, T.B.; Jaafar, H.I. Effect of incorporation of conductive fillers on mechanical properties and thermal conductivity of epoxy resin composite. *Appl. Phys. A* **2018**, *124*, 475. [CrossRef]
26. Donzel, L.; Greuter, F.; Christen, T. Nonlinear resistive electric field grading Part 2: Materials and applications. *IEEE Electr. Insul. Mag.* **2011**, *27*, 18–29. [CrossRef]

27. Hu, H.; Zhang, X.; Zhang, D.; Gao, J.; Hu, C.; Wang, Y. Study on the Nonlinear Conductivity of SiC/ZnO/Epoxy Resin Micro- and Nanocomposite Materials. *Materials* **2019**, *12*, 761. [CrossRef] [PubMed]
28. Liang, H.; Du, B.; Li, J.; Li, Z.; Li, A. Effects of non-linear conductivity on charge trapping and de-trapping behaviours in epoxy/SiC composites under DC stress. *IET Sci. Meas. Technol.* **2018**, *12*, 83–89. [CrossRef]
29. Han, Y.; Li, S.; Frechette, M.; Min, D. Nonlinear Conductivity of Polymer Nanocomposites: A Study on Epoxy Resin\/Silicon Carbide Materials. *IEEE Nanotechnol. Mag.* **2018**, *12*, 23–32. [CrossRef]
30. Du, B.X.; Han, C.; Li, Z.L. Effect of Mechanical Stretching on Nonlinear Conductivity and Dielectrics Breakdown Strength of SiR/SiC Composites. *IEEE Trans. Dielectr. Electr. Insul.* **2021**, *28*, 996–1004. [CrossRef]
31. Poisson, N.; Lachenal, G.; Sautereau, H. Near- and mid-infrared spectroscopy studies of an epoxy reactive system. *Vib. Spectrosc.* **1996**, *12*, 237–247. [CrossRef]
32. Yamasaki, H.; Morita, S. Identification of the epoxy curing mechanism under isothermal conditions by thermal analysis and infrared spectroscopy. *J. Mol. Struct.* **2014**, *1069*, 164–170. [CrossRef]
33. Fu, J.H.; Schlup, J.R. Mid- and near-infrared spectroscopic investigations of reactions between phenyl glycidyl ether (PGE) and aromatic amines. *J. Appl. Polym. Sci.* **1993**, *49*, 219–227. [CrossRef]
34. Iyer, G.; Gorur, R.S.; Richert, R.; Krivda, A.; Schmidt, L.E. Dielectric properties of epoxy based nanocomposites for high voltage insulation. *IEEE Trans. Dielectr. Electr. Insul.* **2011**, *18*, 659–666. [CrossRef]
35. Fang, L.; Wu, C.; Qian, R.; Xie, L.; Yang, K.; Jiang, P. Nano–micro structure of functionalized boron nitride and aluminum oxide for epoxy composites with enhanced thermal conductivity and breakdown strength. *RSC Adv.* **2014**, *4*, 21010–21017. [CrossRef]
36. Zhang, J. Small-scale effect on the piezoelectric potential of gallium nitride nanowires. *Appl. Phys. Lett.* **2014**, *104*, 889. [CrossRef]
37. Ali, F.; Ugurlu, B.; Kawamura, A. Polymer nanocomposite dielectrics-the role of the interface. *IEEE Trans. Dielectr. Electr. Insul.* **2005**, *12*, 629–643.
38. Chi, Q.G.; Cui, S.; Zhang, T.D.; Yang, M.; Chen, Q.G. SiC/SiO$_2$ filler reinforced EP composite with excellent nonlinear conductivity and high breakdown strength. *IEEE Trans. Dielectr. Electr. Insul.* **2020**, *27*, 535–541. [CrossRef]
39. Zhang, C.; Xiang, J.; Wang, S.; Yan, Z.; Cheng, Z.; Fu, H.; Li, J. Simultaneously Enhanced Thermal Conductivity and Breakdown Performance of Micro/Nano-BN Co-Doped Epoxy Composites. *Materials* **2021**, *14*, 3521. [CrossRef]
40. Cao, H.; Liu, B.; Ye, Y.; Liu, Y.; Li, P. Study on the relationships between microscopic cross-linked network structure and properties of cyanate ester self-reinforced composites. *Polymers* **2019**, *11*, 950. [CrossRef] [PubMed]
41. Gao, B.; Xu, J.; Peng, J.; Kang, F.; Du, H.; Li, J.; Chiang, S.; Xu, C.; Hu, N.; Ning, X. Experimental and theoretical studies of effective thermal conductivity of composites made of silicone rubber and Al2O3 particles. *Thermochim. Acta* **2015**, *614*, 1–8. [CrossRef]
42. Amende, T.; Friedrich, M.; Endres, M.; Pihale, S.; Schmidt, R. Thermal conductivity of Al$_2$O$_3$ substrates and precise 3D layer reconstruction—Key parameters for matching FEM simulations with thermal measurements. In Proceedings of the CPIS 2016, 9th International Conference on Integrated Power Electronics Systems, Nuremberg, Germany, 8–10 March 2016; pp. 1–6.
43. Tsekmes, I.A.; Kochetov, R.; Morshuis, P.H.F.; Smit, J.J. Modeling the thermal conductivity of polymeric composites based on experimental observations. *IEEE Trans. Dielectr. Electr. Insul.* **2014**, *21*, 412–423. [CrossRef]
44. Zhai, S.; Zhang, P.; Xian, Y.; Zeng, J.; Shi, B. Effective thermal conductivity of polymer composites: Theoretical models and simulation models. *Int. J. Heat Mass Transf.* **2018**, *117*, 358–374. [CrossRef]
45. Min, H.; Demei, Y.; Jianbo, W. Thermal conductivity determination of small polymer samples by differential scanning calorimetry. *Polym. Test.* **2007**, *26*, 333–337.
46. Coetzee, D.; Venkataraman, M.; Militky, J.; Petru, M.; Venkataraman, M. Influence of nanoparticles on thermal and electrical conductivity of composites. *Polymers* **2020**, *12*, 742. [CrossRef]
47. Tanaka, T.; Kozako, M.; Fuse, N.; Ohki, Y. Proposal of a multi-core model for polymer nanocomposite dielectrics. *IEEE Trans. Dielectr. Electr. Insul.* **2005**, *12*, 669–681. [CrossRef]
48. Fuse, N.; Ohki, Y.; Kozako, M.; Tanaka, T. Possible mechanisms of superior resistance of polyamide nanocomposites to partial discharges and plasmas. *IEEE Trans. Dielectr. Electr. Insul.* **2008**, *15*, 161–169. [CrossRef]
49. Park, J.J.; Lee, J.Y. Effect of epoxy-modified silicone-treated micro-/nano-silicas on the electrical breakdown strength of epoxy/silica composites. *IEEE Trans. Dielectr. Electr. Insul.* **2017**, *24*, 3794–3800. [CrossRef]
50. Lei, Z.; Men, R.; Wang, F.; Li, Y.; Song, J.; Shahsavarian, T.; Li, C.; Fabiani, D. Surface modified nano-SiO$_2$ enhances dielectric properties of stator coil insulation for HV motors. *IEEE Trans. Dielectr. Electr. Insul.* **2020**, *27*, 1029–1037. [CrossRef]

Article

Bioinspired Dielectric Film with Superior Mechanical Properties and Ultrahigh Electric Breakdown Strength Made from Aramid Nanofibers and Alumina Nanoplates

Qiu-Wanyu Qing [1], Cheng-Mei Wei [1], Qi-Han Li [2], Rui Liu [3], Zong-Xi Zhang [3] and Jun-Wen Ren [1,*]

[1] College of Electrical Engineering, Sichuan University, Chengdu 610065, China; 2018141441129@stu.scu.edu.cn (Q.-W.Q.); weichengmei@stu.scu.edu.cn (C.-M.W.)
[2] College of Aviation Engineering, Civil Aviation Flight University of China, Guanghan 618307, China; cunzhangfangyang@163.com
[3] State Grid Sichuan Electric Power Research Institute, Chengdu 610072, China; mbchaoren@163.com (R.L.); 2019223035137@stu.scu.edu.cn (Z.-X.Z.)
* Correspondence: myboyryl@scu.edu.cn; Tel.: +86-177-8071-2606

Citation: Qing, Q.-W.; Wei, C.-M.; Li, Q.-H.; Liu, R.; Zhang, Z.-X.; Ren, J.-W. Bioinspired Dielectric Film with Superior Mechanical Properties and Ultrahigh Electric Breakdown Strength Made from Aramid Nanofibers and Alumina Nanoplates. *Polymers* 2021, 13, 3093. https://doi.org/10.3390/polym13183093

Academic Editor: Shaojian He

Received: 14 August 2021
Accepted: 31 August 2021
Published: 14 September 2021

Publisher's Note: MDPI stays neutral with regard to jurisdictional claims in published maps and institutional affiliations.

Copyright: © 2021 by the authors. Licensee MDPI, Basel, Switzerland. This article is an open access article distributed under the terms and conditions of the Creative Commons Attribution (CC BY) license (https://creativecommons.org/licenses/by/4.0/).

Abstract: Materials with excellent thermal stability, mechanical, and insulating properties are highly desirable for electrical equipment with high voltage and high power. However, simultaneously integrating these performance portfolios into a single material remains a great challenge. Here, we describe a new strategy to prepare composite film by combining one-dimensional (1D) rigid aramid nanofiber (ANF) with 2D alumina (Al_2O_3) nanoplates using the carboxylated chitosan acting as hydrogen bonding donors as well as soft interlocking agent. A biomimetic nacreous 'brick-and-mortar' structure with a 3D hydrogen bonding network is constructed in the obtained ANF/chitosan/Al_2O_3 composite films, which provides the composite films with exceptional mechanical and dielectric properties. The ANF/chitosan/Al_2O_3 composite film exhibits an ultrahigh electric breakdown strength of 320.1 kV/mm at 15 wt % Al_2O_3 loading, which is 50.6% higher than that of the neat ANF film. Meanwhile, a large elongation at break of 17.22% is achieved for the composite film, integrated with high tensile strength (~233 MPa), low dielectric loss (<0.02), and remarkable thermal stability. These findings shed new light on the fabrication of multifunctional insulating materials and broaden their practical applications in the field of advanced electrics and electrical devices.

Keywords: aramid nanofiber; hydrogen bonds; electric breakdown strength; mechanical strength; alumina nanoplates

1. Introduction

Polymers-based dielectrics are widely utilized in advanced electronics and electric power systems by virtue of their irreplaceable advantages, such as easy processing, light weight, and excellent mechanical properties [1–8]. The rapid development of those modern devices with high power density, high integration, and high voltage has caused escalating hot-spot temperatures, causing a great challenge to the heat resistance of polymer dielectrics present in applications including high-frequency motors, high-voltage transformers, electric vehicles, 5G equipment, and pulsed power apparatuses, etc. [9–13]. However, most of traditional polymer dielectrics are limited to unsatisfactory temperature stability, which usually causes a remarkable deterioration in performance at a high temperature. Therefore, excellent thermal stability, mechanical, and insulating properties become the inevitable requirements for the next generation dielectric materials. Unfortunately, simultaneously integrating these properties portfolios into a single material remains a great challenge.

As one of the high performance fibers, aramid fiber, constructed by highly aligned molecular chains of poly (paraphenylene terephthalamide) (PPTA), is well known for its outstanding mechanical properties, high heat resistance, and excellent electrical insulation properties [14]. At present, aramid fibers and/or aramid pulp fibers are widely used to make insulating papers, but their mechanical properties and dielectric strength are still inadequate due to the poor interfacial interactions between microscale aramid fibers [15,16]. It has been found that aramid fibers can be completely split into uniform high aspect ratio aramid nanofibers (ANFs) by controlled deprotonation [17]. The obtained ANFs inherit the excellent properties of aramid fiber and has emerged as a promising nanoscale building block to fabricate advanced materials owing to its high thermal stability and excellent electrical insulation [17–22]. For example, Hu et al. reported a composite film with supreme electromagnetic interference shielding efficiency and exceptional Joule heating performance by combining the ANFs with carbon nanotube and hydrophobic fluorocarbon [23]. Wu et al. and Wang et al. fabricated highly thermoconductive and thermostable polymer nanocomposite films by engineering ANFs with boron nitride nanosheets [1,22]. Zhang and coworkers found that ANFs-based composite films had a potential application as high-performance nanofluidic osmotic power generators [24]. Therefore, incorporating functional fillers into ANFs matrix composite films is an effective strategy to improve its performance. To access the extraordinary properties of ANFs, elegant design the architecture of composites is necessary.

Over the past decade, the aligned "brick-and-mortar" layered structure of nature nacre have demonstrated an effective architecture to achieve remarkable properties. Inspired by the hierarchical microstructures of nature nacre, Zeng et al. successfully fabricated a highly thermally conductive nacre-like papers based on noncovalent functionalized boron nitride nanosheets and poly (vinyl alcohol) via a vacuum-assisted self-assembly technique [25]. Wang et al. claimed that composite films with exceptional insulating properties could be prepared by constructing three-dimensional "brick-and-mortar" layered structures using ANFs and mica nanoplates. As a result, a high dielectric breakdown strength of 164 kV/mm was achieved for the composite film [15]. In addition, the precise design of the inorganic–organic interface is another important factor to fulfill the composite's properties. Yu et al. proposed a multiscale soft-rigid polymer dual-network interfacial design strategy to reinforce the nanoscale building blocks, which endows the resultant nacreous nanocomposite with superior mechanical enhancement and improved stability under high humidity and temperature conditions [26].

In this study, by learning from the hierarchical microstructure of natural nacre, we fabricated mechanically strong and electrical insulating films by combining ANFs with alumina (Al_2O_3) nanoplates using vacuum-assisted filtration, followed by a hot-pressing technique. The underlying rationale for using Al_2O_3 nanoplatelets is that Al_2O_3 platelets have an excellent dielectric properties with wide band-gap. In addition, the two-dimensional structure of Al_2O_3 nanoplates is beneficial for forming a highly ordered arrangement in ANFs framework. Meanwhile, chitosan, a natural cationic polymer obtained by deacetylation of chitin extracted from the shells of shrimp and crabs, has been utilized to enhance the interfacial interaction between ANFs and Al_2O_3 nanoplates by constructing hierarchical hydrogen bonds. The ANF/chitosan/Al_2O_3 composite film with unique "brick-and-mortar" structure and three-dimensional hydrogen bonds was successfully prepared. The obtained composite film exhibits an ultrahigh electric breakdown strength of 320.1kV/mm and a large elongation at break of 17.22% at 15 wt % filler loading, which was 50.6% and 89.9% higher than those of neat ANFs film, respectively. Moreover, high tensile strength, low dielectric loss, high thermal decomposition temperature are achieved for the composite film simultaneously. It is believed that the biomimetic approach is of great importance for the fabrication and practical application of multifunctional dielectric materials in electrical equipment.

2. Materials and Methods

2.1. Materials

Al_2O_3 nanoplates were purchased from Jicang Nano Technology Co., Ltd., Nanjing, China. Potassium hydroxide (KOH) and carboxylated chitosan were purchased from Aladdin Biochemical Technology Co., Ltd., Shanghai, China. Kevlar® 29 fibers were purchased from DuPont (Wilmington, DE, USA). Dimethyl sulfoxide (DMSO), ethylalcohol, and deionized water (DI H_2O) were obtained from Chengdu Kelong Chemical Reagent Co., Ltd., Chengdu, China, and were used as received.

2.2. Preparation of ANFs, ANF/Chitosan, and ANF/Chitosan/Al_2O_3 Composite Films

ANFs were fabricated by treating chopped Kevlar® 29 fibers with a DMSO/KOH solution according to the typical method explored by Kotov et al. [17]. First, 1.6 g of chopped Kevlar® yarn and 2.4 g of KOH were added into the 320 mL of DMSO. Then, the mixture was magnetically stirred at 30 °C at 800 rpm for 1 week, yielding a clear dark red ANF/DMSO dispersion. Then, 100 mL of the obtained ANF/DMSO dispersion was injected into 500 mL of H_2O to form the colloidal ANF. The filtrate was filtered out with a Buchner funnel, and then the ANF was repeatedly washed with DI H_2O until the filtrate was neutral, and the purified colloidal ANF was obtained. A stable ANF slurry was obtained by adding 400 mL H_2O and stirring it at 8000 rpm for 10 min. The pure ANF film was prepared by simple vacuum-assisted filtration with a 0.2 μm pore PTFE membrane. Then, the obtained ANF film was hot-pressed at 150 °C for 5 min and vacuum-dried at 45 °C for 48 h.

The ANF/chitosan composite films were fabricated using the same procedure as for ANFs, with the addition of a certain amount of carboxylated chitosan. Typically, 3 g of carboxylated chitosan was dispersed in DI H_2O and magnetically stirred for 15 min to obtain a chitosan/H_2O solution with a concentration of 3 mg/mL, after which other concentrations required could be obtained by dilution with DI H_2O. The required content of carboxylated chitosan/H_2O solution was uniformly dispersed in ANF/DMSO solution by sonicating for 3 h, and 500 mL H_2O was added to obtain ANF/chitosan suspension. The obtained suspension was repeatedly washed with DI H_2O to make the filtrate neutral. The filtrate was then treated in a high-speed homogenizer at 10,000 rpm for 10 min to obtain homogeneous ANF/chitosan slurry. Then, with the aid of vacuum, ANF/chitosan film was formed on a 0.2 μm pore PTFE membrane. Finally, ANF/chitosan film was further hot-pressed at 150 °C for 5 min and dried at 45 °C for 48 h.

The ANF/chitosan/Al_2O_3 composite films were fabricated by a simple vacuum-assisted filtration of a uniformly distributed suspension containing ANFs, chitosan, and Al_2O_3 nanoplates. First, The Al_2O_3 powder was dispersed in DI H_2O and then added into the ANF/chitosan suspension. The ANF/chitosan/Al_2O_3 slurry was then treated in a high-speed homogenizer at 10,000 rpm for 10 min to obtain homogeneous ANF/chitosan/Al_2O_3 slurry. The ANF/chitosan/Al_2O_3 films were prepared by direct filtration of the ANF/chitosan/Al_2O_3 slurry with the same procedure as ANF/chitosan. The obtained ANF/chitosan/Al_2O_3 films were further hot-pressed at 150 °C for 5 min and dried at 45 °C for 48 h. The preparation processes for ANFs and their composite films are illustrated in Figure 1a–h.

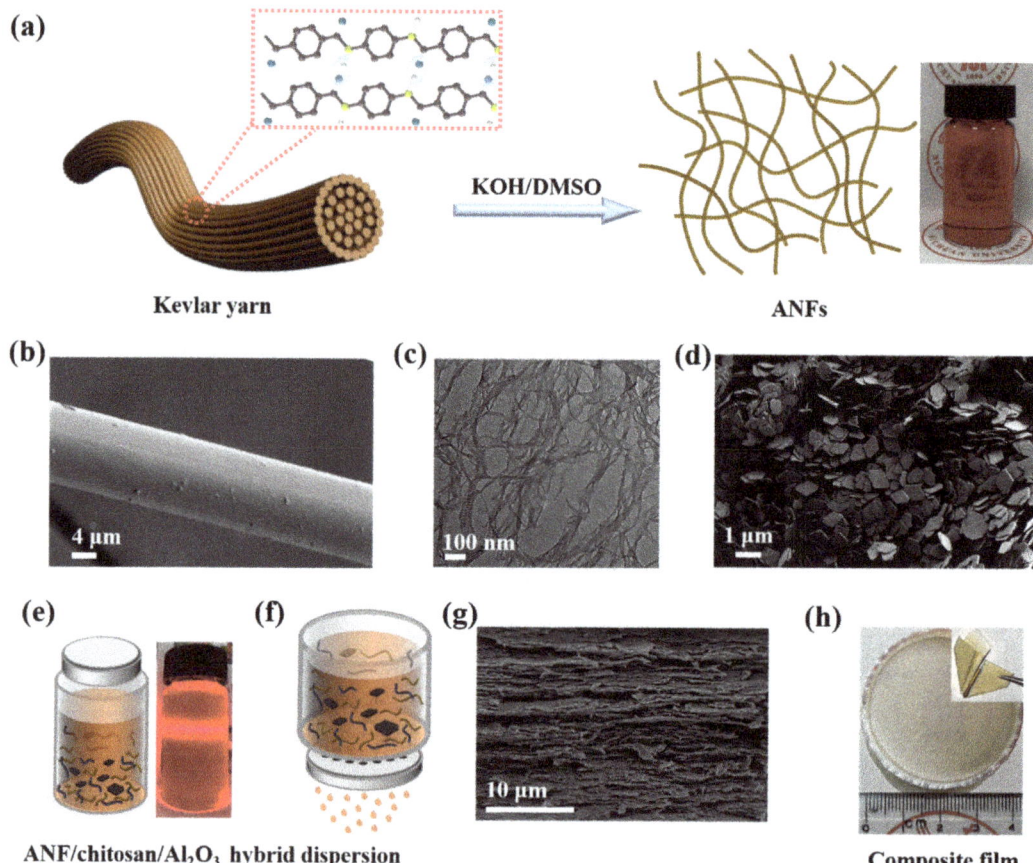

Figure 1. Fabrication of ANF/chitosan/Al$_2$O$_3$ composite film. (**a**) Schematic diagram for the preparation of ANF using KOH/DMSO dissociating method, and the photograph of the obtained ANF/DMSO dispersion. (**b**) SEM image of Kevlar fiber. (**c**) TEM image of the ANFs. (**d**) SEM image of Al$_2$O$_3$ nanoplates. (**e**) ANF/chitosan/Al$_2$O$_3$ hybrid dispersion with strong Tyndall effect. (**f**) Fabrication of composite films by vacuum-assisted filtration. (**g**) SEM image of the cross-section morphology of the ANF/chitosan/Al$_2$O$_3$ composite film. (**h**) Photographs of ANF/chitosan/Al$_2$O$_3$ composite films and its mechanical flexibility.

2.3. Characterization

The microstructure and morphology of ANF, Al$_2$O$_3$ nanoplates, and ANF/chitosan/Al$_2$O$_3$ composite films were characterized by transmission electron microscopy (TEM, JEM2100F, JEOL, Beijing, China) and scanning electron microscopy (SEM, Quanta 250 FEG, FEI, Shanghai, China). Thermal gravimetric analysis (TGA) was performed on the composite films with TG 2950 (NETZSCH, Selb, Germany) at a heating rate of 10 °C/min and N$_2$ flow rate of 20 mL/min. The mechanical properties of the composite films were tested at room temperature by the universal testing machine (Instron 5967, Norwood, MA, USA). DDJ-50 kV electric breakdown tester (Kelang Measuring Instrument Co., Ltd., Beijing, China) was used to test the electrical breakdown performance of the composite films at DC high voltage, and the voltage boost rate was 500 V/s. The dielectric response of the composite films was analyzed using the Concept 80 broadband dielectric impedance relaxation spectrometer (Novocontrol GmbH, Montabaur, Germany) in the frequency range of $10^2 \sim 10^6$ Hz. The Nicolet-5700 Fourier transform infrared spectrometer (Thermo Nicolet Corporation, Madison, SD, USA) was used to collect the Fourier transform infrared (FT-IR) spectra of the composite films.

3. Results and Discussion

The Kevlar fiber was spilt into ANFs by consistent stirring in a KOH/DMSO system for a week, resulting in a dark red colloidal dispersion as schematically shown in Figure 1a. During the dissociating process, the intermolecular hydrogen-bonding interactions between PPTA molecular backbones were weakened due to the deprotonated effect (Figure 1a, inset) [17,18,27]. Consequently, the original Kevlar fiber with a diameter of ~15 μm (Figure 1b) was dissociating into curly nanofibers with length in micrometer scale and diameter in the range of 20–30 nm, as revealed in the TEM image in Figure 1c. Kotov et al. found that the ANFs not only inherit the exceptional properties of Kevlar fiber, but also possess a large number of functional groups on their surface [17]. The nanoscale, high aspect ratio, surface activity, and good dispersibility of ANFs render them promising nanoscale building blocks to prepare advanced materials [1,22,23,28]. The Al_2O_3 nanoplates, with a lateral size of approximately 1 μm and mean thickness of 100 nm (Figure 1d), were utilized to enhance the performance of ANFs based films, owing to their excellent dielectric properties [29,30]. As can be seen from Figure 1e, the hybrid suspension of ANFs, chitosan, and Al_2O_3 nanoplates show strong Tyndall effect, which indicates the homogeneous suspension and good interaction between ANFs, chitosan, and Al_2O_3 nanoplates. The ANFs/chitosan/Al_2O_3 can be made ready by the vacuum-assisted filtration method, as schematically shown in Figure 1f, to form a yellow film with typical lamellar microstructure (Figure 1g), which exhibit excellent flexibility and fold-ability (Figure 1h).

Constructing hydrogen-bonding is a feasible and effective approach to improving the properties of composites [31]. Here, carboxylated chitosan was chosen as molecular modifier to improve the mechanical properties of ANFs, because strong hydrogen bonding can be generated between ANFs and the abundant functional groups (carboxyl, hydroxyl and amino groups) of carboxylated chitosan (Figure 2a). In addition, the carboxylated chitosan not only acts as a hydrogen bonding donor, but also as an interlocking agent to connect the ANFs. Figure 2b,e presents the difference in the mechanical properties of ANF/chitosan composite films with various chitosan contents. It is noted that the mechanical properties of ANF films can be improved remarkably by employing the carboxylated chitosan (Figure 2b). The tensile strength and the elongation at break of the ANF films reach a high value of 360 MPa and 13.34%, which are 78.2% and 47.1% higher than that of neat ANF film (202 MPa and 9.07%), respectively. The intermolecular hydrogen bonding between ANFs and chitosan can be invoked as being responsible for the excellent mechanical properties of ANF/chitosan films. These hydrogen bonds greatly enhanced the intermolecular forces. In addition, due to the existence of hydrogen bonds, the compatibility between ANF and chitosan was brilliant. Chitosan can be uniformly dispersed in ANF, making the internal structure of the composite films compact. The defects and pores are significantly decreased in the compact structure, which is beneficial for enhancing the mechanical properties. The tensile strength and elongation at break of ANF/chitosan composite films decreased slightly when the content of chitosan was higher than 5 wt %, but all of them were higher than that of the neat ANF film. When the content of chitosan was 10 wt %, the tensile modulus of ANF/chitosan composite films decreased by 5.4% (from 7608 MPa to 7201 MPa). The mechanical properties of ANF/chitosan composite films are mainly determined by hydrogen bonding and physical entanglement between soft chitosan molecular chains and relatively rigid ANFs framework. The deterioration of mechanical properties at high chitosan content can be attributed to the saturation of hydrogen bonds and the soft nature of chitosan. Similar phenomena were found in the ANFs/polyvinyl alcohol (PVA) and resol/PVA systems, as reported in the works of E et al. [32] and Chen et al. [26]. Based on the above analysis, when the content of chitosan was 5 wt %, ANF/chitosan composite films showed the best mechanical properties. Therefore, in the present work, we fixed the mass fraction of chitosan as 5 wt % and added more Al_2O_3 nanoplates to prepare ANF/chitosan/Al_2O_3 composite films.

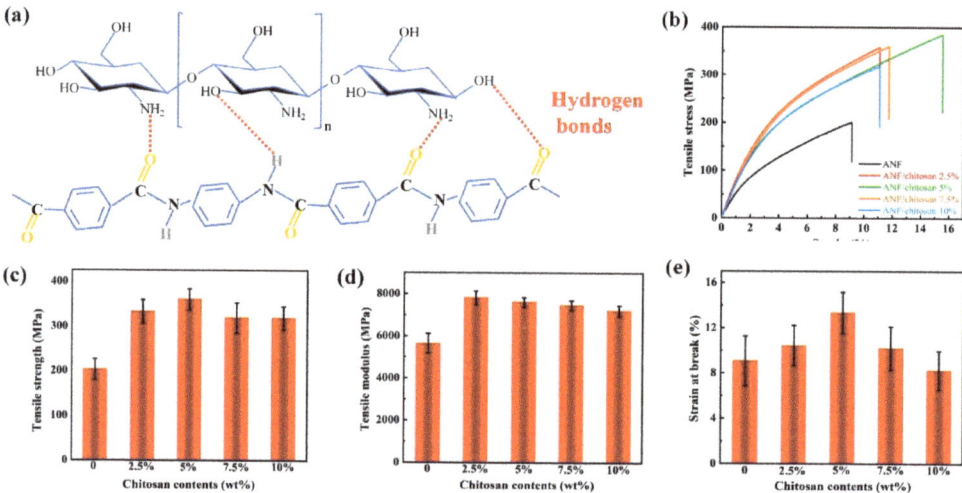

Figure 2. Mechanical properties of ANF/chitosan composite films. (**a**) Schematic representation of the formation of intermolecular hydrogen bonding between chitosan and ANF. (**b**) Typical stress–strain curves of ANF/chitosan composite films at different chitosan contents. (**c**–**e**) Tensile strength, tensile modulus, and break elongation of ANF/chitosan composite at different chitosan contents.

In the ANF/chitosan/Al_2O_3 composite films, the Al_2O_3 nanoplates orderly embedded into the framework of the ANFs, and generated a natural nacre-like "brick-and-mortar" structure (Figure 3a). In this special structure, the embedded Al_2O_3 nanoplates were glued with ANFs framework together by soft chitosan molecules, and three-dimensional hydrogen bonds were established between ANFs, chitosan, and Al_2O_3 nanoplates (Figure 3b). This can be confirmed by means of FT-IR spectra (Figure 4), since the vibrational peaks of functional groups are closely related to the intermolecular interactions. A comparison of the FT-IR spectra of ANF and ANF/chitosan/Al_2O_3 shows a remarkable red shift that is representative of the deformation of N–H (from 3316 cm^{-1} of ANF to 3313 cm^{-1} of ANF/chitosan/Al_2O_3). This shift is indicative of the interaction between the Al_2O_3 and amino (N–H) groups of ANF, resulting in the formation of hydrogen bonds between the ANF and Al_2O_3 nanoplates. The nacre-inspired "brick-and-mortar" structure endowed ANF/chitosan/Al_2O_3 composites with excellent ductile deformation behavior. As shown in Figure 3c, a large elongation at break of 17.22% was achieved for the ANF/chitosan/Al_2O_3 composite film at the filler contents of 15 wt %, which is 89.9% higher than that of neat ANF film (9.07%). The elongation at break was far superior to that of the conventional commercial Nomex insulating paper [15]. As demonstrated in Figure 1g, the composite film could be arbitrarily folded without breakage. The large ductility indicates that the composite film has wonderful manipulation reliability, which is essential to the dielectric materials.

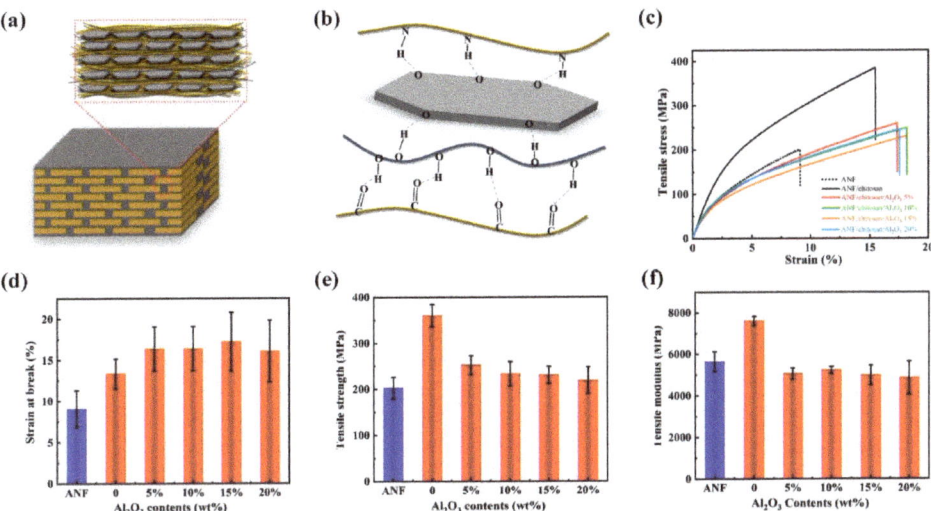

Figure 3. Mechanical properties of ANF/chitosan/Al$_2$O$_3$ composite films. (**a**) Typical stress–strain curves of ANF/chitosan/Al$_2$O$_3$ composite films at different Al$_2$O$_3$ contents. (**b–d**) Tensile strength, tensile modulus, and break elongation of ANF/chitosan composite at different Al$_2$O$_3$ contents. (**e**) Schematic diagrams of the structure of ANF/chitosan/Al$_2$O$_3$ films. (**f**) Schematic representation of the formation of intermolecular hydrogen bonding between chitosan, ANF, and Al$_2$O$_3$ nanoplates.

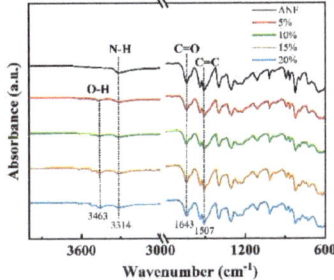

Figure 4. FT-IR curves of ANF/chitosan/Al$_2$O$_3$ composite films with different contents of Al$_2$O$_3$ nanoplates.

It is noted that the dramatic improvement of the elongation at break of the composite films with the addition of Al$_2$O$_3$ content is accompanied by a significant decrease in the tensile strength and modulus. Typical stress–strain curves, tensile strength, tensile modulus, and elongation at break of ANF/chitosan/Al$_2$O$_3$ composite films were shown in Figure 3c–f. The addition of Al$_2$O$_3$ nanoplates caused a decrease in the tensile strength and modulus of the composite films. The deterioration of the tensile strength and modulus of ANF/chitosan/Al$_2$O$_3$ can be attributed to the decrease in intermolecular interaction in the composite films. On the one hand, part of the hydrogen bonds between ANF and chitosan were replaced by the hydrogen bonds between ANF/Al$_2$O$_3$ and chitosan/Al$_2$O$_3$. Although the new three-dimensional hydrogen bonds were formed, their strength was far lower than that of the hydrogen bonds between ANFs and chitosan. As a result, the intermolecular force was greatly reduced, leading to a decrease in tensile strength of composite films. On the other hand, the free volume of molecular chains is inversely proportional to the compact of the films. The incorporation of Al$_2$O$_3$ nanoplates will increase the free volume of the composite film, resulting in a decrease in tensile strength. Moreover, the agglomeration will occur with the increase in Al_2O_3 nanoplates, which will act as stress concentration point and lead to the degeneration of the mechanical properties. Although the addition of Al_2O_3 nanoplates led to the deterioration of mechanical strength, the tensile strength of ANF/chitosan/Al_2O_3 composite film at 15 wt % filler contents (232 MPa) was still 14.9% higher than that of neat ANFs film (202 MPa).

Next, we turned to investigate the dielectric properties of the ANF/chitosan/Al_2O_3 composite films. The frequency-dependent dielectric constants, dielectric losses, and AC conductivity of the composite films are shown in Figure 5a–c. It is observed that the addition of Al_2O_3 nanoplates caused a slight increase in dielectric constants of composite film at low filler contents, which can be attributed to the increase in interfacial polarization in ANF/chitosan/Al_2O_3 composite films [33–35]. In addition, the differential dielectric constants of ANFs and Al_2O_3 nanoplates might generate a lot of mini-capacitors, contributing to an increase in the dielectric constant [36]. However, the values of tanδ of the ANF/chitosan/Al_2O_3 composite films were below 0.02 (100 Hz, Figure 5b). The AC conductivity of ANF/chitosan/Al_2O_3 composites increase linearly with the increase in frequency and no DC plateau is situated in the low frequency (Figure 5c). These results indicate that ANF/chitosan/Al_2O_3 composite films possess high insulation capability and low charge carriers mobility [37].

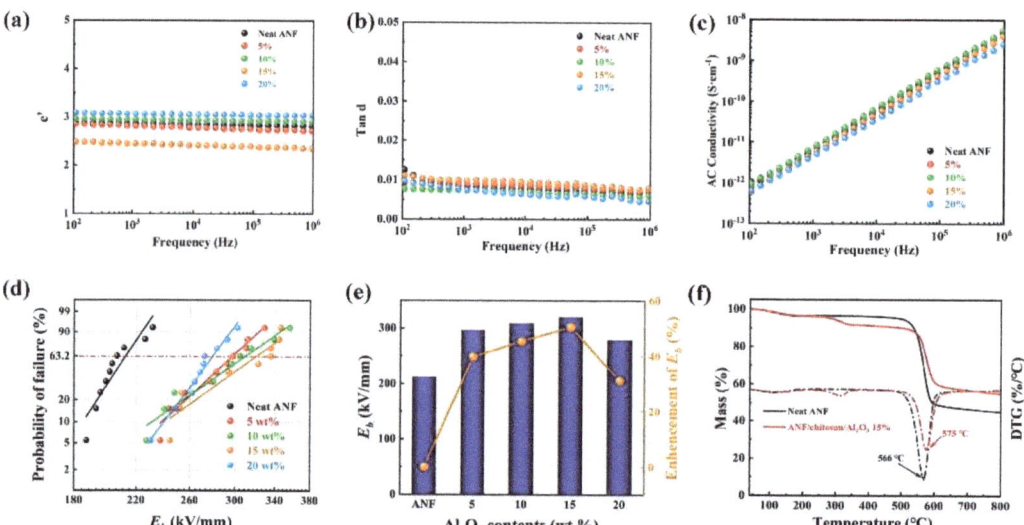

Figure 5. Dielectric properties of ANF/chitosan/Al_2O_3 composite films. (**a**) The dielectric constant (ε') and (**b**) loss tangent of ANF/chitosan/Al_2O_3 composite films as a function of frequency. (**c**) AC conductivity of ANF/chitosan/Al_2O_3 composite films as a function of frequency. (**d**) Breakdown strength of ANF/chitosan/Al_2O_3 composite with Weibull distributions. (**e**) The enhancement of the breakdown strength of ANF/chitosan/Al_2O_3 composites compared with that of the neat ANF film. (**f**) TGA curves of ANF and ANF/chitosan/Al_2O_3 composite films at 15 wt % filler loading.

The two-parameter Weibull cumulative probability function was utilized to analyze the dielectric breakdown strength of ANF/chitosan/Al_2O_3 composite films according to Equation (1):

$$P_f = 1 - \exp[-(E/E_0)\beta] \quad (1)$$

where P_f represents the cumulative breakdown probability of the electrical system; E is the experimental breakdown strength; β is the shape parameter, reflecting the breakdown voltage dispersion degree; and E_0 is the characteristic breakdown intensity, reflecting the size of the breakdown field intensity when the cumulative breakdown probability is 63.2%. The dielectric breakdown strength of neat ANF film and ANF/chitosan/Al_2O_3 composite films were shown in Table 1 and Figure 5d. It can be observed that the breakdown strength composite films increase remarkably with the addition of Al_2O_3 nanoplates. The ANF/chitosan/Al_2O_3 composite film with 15 wt % filler contents exhibits the highest dielectric strength of 320.1 kV/mm, which is 50.6% higher than that of neat ANF film (Figure 5e). This indicates that Al_2O_3 nanoplates, as a typical dielectrics, can significantly increase the dielectric strength of the ANF film. Similar results are reported in previous research; Zeng et al. have showed that the nacre-mimetics ANF/mica films possess much better dielectric performance than the neat ANF [15]. However, the dielectric strength of ANF/mica is significantly lower than that of ANF/chitosan/Al_2O_3 film. This can be attributed to a denser structure of ANF/chitosan/Al_2O_3, owing to the hot press under high temperature and high pressure. The prominent improvement of dielectric breakdown strength of composite films was mainly due to the formation of deep traps caused by the addition of Al_2O_3 nanoplates. The existence of deep traps can inhibit charge injection and hot electron formation, which will be beneficial for enhancing the dielectric breakdown strength. With the increase in the loading of Al_2O_3 nanoplates, the density of the trap increased, leading to the increase in the dielectric breakdown strength. Additionally, the electrons are easier to attract with the wide band gap Al_2O_3 nanoplates, leading to more internal charge consumption and less accumulation of space charge in the composite films. As a result, the electric branches migrated to the direction of nanoparticles, which was conducive to the improvement of the breakdown characteristics of the composite films. In addition, the special biomimetic nacreous "brick-and-mortar" structure of ANF/chitosan/Al_2O_3 can distribute the electrical stress homogeneously and avoiding the concentration of electrical field. In such a well-arranged architecture, the Al_2O_3 nanoplates were orderly embedded into the ANFs framework, which efficiently impedes the growth of electric tree in composite films. Therefore, the breakdown strength of ANF/chitosan/Al_2O_3 composite films is significantly improved. However, a decrease in electric breakdown strength is observed when the contents of Al_2O_3 are higher than 20 wt %. This can be ascribed to the agglomeration of fillers at high concentration, which will act as a weak point under the high electric field and contribute to the deterioration of the breakdown performance of the composite films. Although there is a slight decrease in electric breakdown strength at high filler loading, the value is still much higher than that of the neat ANF film. Therefore, the ANF/chitosan/Al_2O_3 composite films have a great promising as insulating materials application in high-voltage electric power systems. More importantly, ANF/chitosan/Al_2O_3 composite film exhibited outstanding thermal stabilities. owing to the high thermal durability of the ANF and Al_2O_3 (Figure 5f). Compared to the decomposition temperature (T_d) of neat ANF film, the T_d of ANF/chitosan/Al_2O_3 composite film increases by 9 °C from 566 °C of ANF to 575 °C of ANF/chitosan/Al_2O_3, which can be attributed to the "tortuous path effect" caused by the "brick-and-mortar" structure. In addition, in the composite film, the Al_2O_3 nanoplates preferably absorbed heat due to its high intrinsic heat capacity, which effectively retarded the volatilization of the PPTA chains [22].

Table 1. Mechanical properties of ANF and ANF/chitosan/Al_2O_3 composite films.

Content (wt %)	β	E_b (kV/mm)
0	16.3	212.6
5	10.2	297.1
10	7.7	309.1
15	7.9	320.1
20	14.7	279.2

4. Conclusions

In summary, a series of ductile composite films consisting of ANFs, chitosan, and Al_2O_3 nanoplates was successfully fabricated by vacuum-assisted filtration followed by hot-pressing. A special biomimetic nacreous "brick-and-mortar" structure was constructed in the ANF/chitosan/Al_2O_3 composite films, which effectively restrained the accumulation of space charge and prorogation paths of electric branches in the films. This contributed to a prominent improvement of dielectric breakdown strength of ANF/chitosan/Al_2O_3 composite films. An ultrahigh electric breakdown strength of 320.1 kV/mm was achieved for ANF/chitosan/Al_2O_3 composite film with 15 wt % Al_2O_3 loading, which is 50.6% higher than that of the neat ANF film. In addition, favorable three-dimensional hydrogen bonds have formed between the ANFs, chitosan, and Al_2O_3 nanoplates, which imparts an excellent flexibility of composite film, and a large elongation at break of 17.22% was achieved. Furthermore, low dielectric constant, low dielectric loss (<0.02), high tensile strength (~230 MPa), and remarkable thermal stability (Td ~575 °C) were simultaneously achieved for the ANF/chitosan/Al_2O_3 composite film. Those admirable features confirmed that the ANF/chitosan/Al_2O_3 film, as a typical dielectric material, shows great potential for application in high power apparatuses operating at high temperatures.

Author Contributions: Conceptualization, Q.-W.Q. and J.-W.R.; methodology, Q.-W.Q. and J.-W.R.; formal analysis, C.-M.W.; investigation, C.-M.W., Q.-H.L. and R.L.; data curation, Z.-X.Z.; writing—original draft preparation, Q.-W.Q.; writing—review and editing, J.-W.R.; supervision, J.-W.R.; All authors have read and agreed to the published version of the manuscript.

Funding: This research received no external funding.

Institutional Review Board Statement: Not applicable.

Informed Consent Statement: Not applicable.

Data Availability Statement: The data presented in this study are available upon request from the corresponding author.

Acknowledgments: The authors are grateful for the financial support from the Postdoctoral Science Foundation of China (2018M643475), the Postdoctoral Interdisciplinary Innovation Foundation of Sichuan University (0030304153008), the key research and development program of Sichuan province (2021YFG0284), and the Fundamental Research Funds for the Central Universities (YJ201655).

Conflicts of Interest: No potential conflict of interest was reported by the authors.

References

1. Wu, K.; Wang, J.; Liu, D.; Lei, C.; Liu, D.; Lei, W.; Fu, Q. Highly Thermoconductive, Thermostable, and Super-Flexible Film by Engineering 1D Rigid Rod-Like Aramid Nanofiber/2D Boron Nitride Nanosheets. *Adv. Mater.* **2020**, *32*. [CrossRef]
2. Li, Q.; Chen, L.; Gadinski, M.R.; Zhang, S.; Zhang, G.; Li, U.; Iagodkine, E.; Haque, A.; Chen, L.Q.; Jackson, N.; et al. Flexible high-temperature dielectric materials from polymer nanocomposites. *Nature* **2015**, *523*, 576–579. [CrossRef] [PubMed]
3. Feng, C.P.; Chen, L.B.; Tian, G.L.; Wan, S.S.; Bai, L.; Bao, R.Y.; Liu, Z.Y.; Yang, M.B.; Yang, W. Multifunctional Thermal Management Materials with Excellent Heat Dissipation and Generation Capability for Future Electronics. *ACS Appl. Mater. Interfaces* **2019**, *11*, 18739–18745. [CrossRef]
4. Liu, Z.-J.; Yin, C.-G.; Cecen, V.; Fan, J.-C.; Shi, P.-H.; Xu, Q.-J.; Min, Y.-L. Polybenzimidazole thermal management composites containing functionalized boron nitride nanosheets and 2D transition metal carbide MXenes. *Polymer* **2019**, *179*, 121613. [CrossRef]
5. Krause, B.; Rzeczkowski, P.; Potschke, P. Thermal Conductivity and Electrical Resistivity of Melt-Mixed Polypropylene Composites Containing Mixtures of Carbon-Based Fillers. *Polymers* **2019**, *11*, 1073. [CrossRef]
6. Wang, X.; Yu, Z.; Bian, H.; Wu, W.; Xiao, H.; Dai, H. Thermally Conductive and Electrical Insulation BNNS/CNF Aerogel Nano-Paper. *Polymers* **2019**, *11*, 660. [CrossRef]
7. Bian, X.; Tuo, R.; Yang, W.; Zhang, Y.; Xie, Q.; Zha, J.; Lin, J.; He, S. Mechanical, Thermal, and Electrical Properties of BN-Epoxy Composites Modified with Carboxyl-Terminated Butadiene Nitrile Liquid Rubber. *Polymers* **2019**, *11*, 1548. [CrossRef] [PubMed]
8. He, S.; Hu, J.; Zhang, C.; Wang, J.; Chen, L.; Bian, X.; Lin, J.; Du, X. Performance improvement in nano-alumina filled silicone rubber composites by using vinyl tri-methoxysilane. *Polym. Test.* **2018**, *67*, 295–301. [CrossRef]
9. Gupta, R.; Smith, L.; Njuguna, J.; Deighton, A.; Pancholi, K. Insulating MgO-Al2O3-LDPE nanocomposites for offshore medium-voltage DC cables. *ACS Appl. Electron. Mater.* **2020**, *2*, 1880–1891. [CrossRef]
10. Huang, X.; Ding, Z.A.; Wang, J.; Wang, J.; Li, Q. The impacts of chemical modification on the initial surface creepage discharge behaviors of polyimide insulating film in power electronics. *ACS Appl. Electron. Mater.* **2020**, *2*, 3418–3425. [CrossRef]
11. Zhou, L.; Yang, Z.; Luo, W.; Han, X.; Jang, S.H.; Dai, J.; Yang, B.; Hu, L. Thermally conductive, electrical insulating, optically transparent bi-layer nanopaper. *ACS Appl. Mater. Interfaces* **2016**, *8*, 28838–28843. [CrossRef]
12. He, X.; Huang, Y.; Wan, C.; Zheng, X.; Kormakov, S.; Gao, X.; Sun, J.; Zheng, X.; Wu, D. Enhancing thermal conductivity of polydimethylsiloxane composites through spatially confined network of hybrid fillers. *Compos. Sci. Technol.* **2019**, *172*, 163–171. [CrossRef]

13. Li, L.; Ma, Z.; Xu, P.; Zhou, B.; Li, Q.; Ma, J.; He, C.; Feng, Y.; Liu, C. Flexible and alternant-layered cellulose nanofiber/graphene film with superior thermal conductivity and efficient electromagnetic interference shielding. *Compos. Part A Appl. Sci. Manuf.* **2020**, *139*, 106134. [CrossRef]
14. Xie, C.; He, L.; Shi, Y.; Guo, Z.X.; Qiu, T.; Tuo, X. From Monomers to a Lasagna-like Aerogel Monolith: An Assembling Strategy for Aramid Nanofibers. *ACS Nano* **2019**, *13*, 7811–7824. [CrossRef]
15. Zeng, F.; Chen, X.; Xiao, G.; Li, H.; Xia, S.; Wang, J. A Bioinspired Ultratough Multifunctional Mica-Based Nanopaper with 3D Aramid Nanofiber Framework as an Electrical Insulating Material. *ACS Nano* **2020**, *14*, 611–619. [CrossRef] [PubMed]
16. Li, M.; Zhu, Y.; Teng, C. Facial fabrication of aramid composite insulating paper with high strength and good thermal conductivity. *Compos. Commun.* **2020**, *21*, 100370. [CrossRef]
17. Yang, M.; Cao, K.; Sui, L.; Qi, Y.; Zhu, J.A.; Waas, A.; Arruda, E.M.; Kieffer, J.; Thouless, M.D.; Kotov, N.A. Dispersions of Aramid Nanofibers: A New Nanoscale Building Block. *ACS Nano* **2011**, *5*, 6945–6954. [CrossRef] [PubMed]
18. Kwon, S.R.; Harris, J.; Zhou, T.; Loufakis, D.; Boyd, J.G.; Lutkenhaus, J.L. Mechanically Strong Graphene/Aramid Nanofiber Composite Electrodes for Structural Energy and Power. *ACS Nano* **2017**, *11*, 6682–6690. [CrossRef]
19. Gong, Y.J.; Heo, J.W.; Lee, H.; Kim, H.; Cho, J.; Pyo, S.; Yun, H.; Kim, H.; Park, S.Y.; Yoo, J.; et al. Nonwoven rGO Fiber-Aramid Separator for High-Speed Charging and Discharging of Li Metal Anode. *Adv. Energy Mater.* **2020**, *10*. [CrossRef]
20. Yang, B.; Wang, L.; Zhang, M.; Luo, J.; Lu, Z.; Ding, X. Fabrication, Applications, and Prospects of Aramid Nanofiber. *Adv. Funct. Mater.* **2020**, *30*, 2000186. [CrossRef]
21. Zhao, L.; Liao, C.; Liu, Y.; Huang, X.; Ning, W.; Wang, Z.; Jia, L.; Ren, J. A combination of aramid nanofiber and silver nanoparticle decorated boron nitride for the preparation of a composite film with superior thermally conductive performance. *Compos. Interfaces* **2021**, 1–17. [CrossRef]
22. Wang, T.; Wei, C.; Yan, L.; Liao, Y.; Wang, G.; Zhao, L.; Fu, M.; Ren, J. Thermally conductive, mechanically strong dielectric film made from aramid nanofiber and edge-hydroxylated boron nitride nanosheet for thermal management applications. *Compos. Interfaces* **2020**, 1–14. [CrossRef]
23. Hu, P.; Lyu, J.; Fu, C.; Gong, W.B.; Liao, J.; Lu, W.; Chen, Y.; Zhang, X. Multifunctional Aramid Nanofiber/Carbon Nanotube Hybrid Aerogel Films. *ACS Nano* **2020**, *14*, 688–697. [CrossRef]
24. Zhang, Z.; Yang, S.; Zhang, P.; Zhang, J.; Chen, G.; Feng, X. Mechanically strong MXene/Kevlar nanofiber composite membranes as high-performance nanofluidic osmotic power generators. *Nat. Commun.* **2019**, *10*, 1–9. [CrossRef]
25. Zeng, X.; Ye, L.; Yu, S.; Li, H.; Sun, R.; Xu, J.; Wong, C.P. Artificial nacre-like papers based on noncovalent functionalized boron nitride nanosheets with excellent mechanical and thermally conductive properties. *Nanoscale* **2015**, *7*, 6774–6781. [CrossRef]
26. Chen, S.-M.; Gao, H.-L.; Sun, X.-H.; Ma, Z.-Y.; Ma, T.; Xia, J.; Zhu, Y.-B.; Zhao, R.; Yao, H.-B.; Wu, H.-A.; et al. Superior Biomimetic Nacreous Bulk Nanocomposites by a Multiscale Soft-Rigid Dual-Network Interfacial Design Strategy. *Matter* **2019**, *1*, 412–427. [CrossRef]
27. Ma, T.; Zhao, Y.; Ruan, K.; Liu, X.; Zhang, J.; Guo, Y.; Yang, X.; Kong, J.; Gu, J. Highly Thermal Conductivities, Excellent Mechanical Robustness and Flexibility, and Outstanding Thermal Stabilities of Aramid Nanofiber Composite Papers with Nacre-Mimetic Layered Structures. *ACS Appl Mater. Interfaces* **2020**, *12*, 1677–1686. [CrossRef] [PubMed]
28. Zhao, L.H.; Liao, Y.; Jia, L.C.; Wang, Z.; Huang, X.L.; Ning, W.J.; Zhang, Z.X.; Ren, J.W. Ultra-Robust Thermoconductive Films Made from Aramid Nanofiber and Boron Nitride Nanosheet for Thermal Management Application. *Polymers* **2021**, *13*, 2028. [CrossRef]
29. Agrawal, A.; Satapathy, A. Thermal, mechanical, and dielectric properties of aluminium oxide and solid glass microsphere-reinforced epoxy composite for electronic packaging application. *Polym. Compos.* **2019**, *40*, 2573–2581. [CrossRef]
30. Li, W.; Hillborg, H.; Gedde, U.W. Influence of process conditions and particle dispersion on the ac breakdown strength of polyethylene-aluminium oxide nanocomposites. *IEEE Trans. Dielectr. Electr. Insul.* **2015**, *22*, 3536–3542. [CrossRef]
31. Ren, J.; Yu, D. Effects of enhanced hydrogen bonding on the mechanical properties of poly (vinyl alcohol)/carbon nanotubes nanocomposites. *Compos. Interfaces* **2017**, *25*, 205–219. [CrossRef]
32. E, S.; Ma, Q.; Huang, J.; Jin, Z.; Lu, Z. Enhancing mechanical strength and toughness of aramid nanofibers by synergetic interactions of covalent and hydrogen bonding. *Compos. Part A Appl. Sci. Manuf.* **2020**, *137*, 106031. [CrossRef]
33. Zhou, W.; Li, T.; Yuan, M.; Li, B.; Zhong, S.; Li, Z.; Liu, X.; Zhou, J.; Wang, Y.; Cai, H.; et al. Decoupling of inter-particle polarization and intra-particle polarization in core-shell structured nanocomposites towards improved dielectric performance. *Energy Storage Mater.* **2021**, *42*, 1–11. [CrossRef]
34. On, S.Y.; Kim, M.S.; Kim, S.S. Effects of post-treatment of meta-aramid nanofiber mats on the adhesion strength of epoxy adhesive joints. *Compos. Struct.* **2017**, *159*, 636–645. [CrossRef]
35. Wang, F.; Wu, Y.; Huang, Y. High strength, thermostable and fast-drying hybrid transparent membranes with POSS nanoparticles aligned on aramid nanofibers. *Compos. Part A Appl. Sci. Manuf.* **2018**, *110*, 154–161. [CrossRef]
36. Ren, J.; Li, Q.; Yan, L.; Jia, L.; Huang, X.; Zhao, X.; Ran, Q.; Fu, M. Enhanced thermal conductivity of epoxy composites by introducing graphene@boron nitride nanosheets hybrid nanoparticles. *Mater. Des.* **2020**, *191*, 108663. [CrossRef]
37. Ren, J.; Yu, D.; Feng, L.; Wang, G.; Lv, G. Nanocable-structured polymer/carbon nanotube composite with low dielectric loss and high impedance. *Compos. Part A Appl. Sci. Manuf.* **2017**, *98*, 66–75. [CrossRef]

MDPI
St. Alban-Anlage 66
4052 Basel
Switzerland
Tel. +41 61 683 77 34
Fax +41 61 302 89 18
www.mdpi.com

Polymers Editorial Office
E-mail: polymers@mdpi.com
www.mdpi.com/journal/polymers

www.ingramcontent.com/pod-product-compliance
Lightning Source LLC
LaVergne TN
LVHW070718100526
838202LV00013B/1118